Annals of Mathematics Studies
Number 192

Contents

Non-archimedean tame topology and stably dominated types

Chapter One

Introduction

Model theory rarely deals directly with topology; the great exception is the theory of o-minimal structures, where the topology arises naturally from an ordered structure, especially in the setting of ordered fields. See [11] for a basic introduction. Our goal in this work is to create a framework of this kind for valued fields.

A fundamental tool, imported from stability theory, will be the notion of a definable type; it will play a number of roles, starting from the definition of a point of the fundamental spaces that will concern us. A definable type on a definable set V is a uniform decision, for each definable subset U (possibly defined with parameters from larger base sets), of whether $x \in U$; here x should be viewed as a kind of ideal element of V. A good example is given by any semi-algebraic function f from $\mathbb{R}$ to a real variety V. Such a function has a unique limiting behavior at ∞: for any semi-algebraic subset U of V, either $f(t) \in U$ for all large enough t, or $f(t) \notin U$ for all large enough t. In this way f determines a definable type.

One of the roles of definable types will be to be a substitute for the classical notion of a sequence, especially in situations where one is willing to refine to a subsequence. The classical notion of the limit of a sequence makes little sense in a saturated setting. In o-minimal situations it can often be replaced by the limit of a definable curve; notions such as definable compactness are defined using continuous definable maps from the field R into a variety V. Now to discuss the limiting behavior of f at ∞ (and thus to define notions such as compactness), we really require only the answer to this dichotomy—is $f(t) \in U$ for large t?—uniformly, for all U; i.e. knowledge of the definable type associated with f. For the spaces we consider, curves will not always be sufficiently plentiful to define compactness, but definable types will be, and our main notions will all be defined in these terms. In particular the limit of a definable type on a space with a definable topology is a point whose every neighborhood is large in the sense of the definable type.

A different example of a definable type is the generic type of the valuation ring $\mathcal{O}$, or of a closed ball B of K, for K a non-archimedean valued field, or of $V(\mathcal{O})$ where V is a smooth scheme over $\mathcal{O}$. Here again, for any definable subset U of $\mathbb{A}^1$, we have $v \in U$ for all sufficiently generic $v \in V$, or else $v \notin U$ for all sufficiently generic $v \in V$; where "sufficiently generic" means "having residue outside Z_U" for a certain proper Zariski closed subset Z_U of $V(\mathbf{k})$, depending only on U. Here $\mathbf{k}$ is the residue field. Note that the generic type of $\mathcal{O}$ is invariant under multiplication by $\mathcal{O}^*$ and addition by $\mathcal{O}$, and hence induces a definable type on any closed ball. Such definable types are *stably dominated*, being determined by a function into objects over the residue field, in this case the residue map into $V(\mathbf{k})$. They can also be characterized as *generically stable*. Their basic properties were developed in [20]; some results are now seen more easily using the general theory of NIP, [27].

Let V be an algebraic variety over a field K. A valuation or ordering on K induces a topology on K, hence on K^n, and finally on $V(K)$. We view this topology as an object of the definable world; for any model M, we obtain a topological space whose set of points is $V(M)$. In this sense, the topology is on V.

In the valuative case however, it has been recognized since the early days of the theory that this topology is inadequate for geometry. The valuation topology is totally disconnected, and does not afford a useful globalization of local questions. Various remedies have been proposed, by Krasner, Tate, Raynaud, Berkovich and Huber. Our approach can be viewed as a lifting of Berkovich's to the definable category. We will mention below a number of applications to classical Berkovich spaces, that indeed motivated the direction of our work.

The fundamental topological spaces we will consider will not live on algebraic varieties. Consider instead the set of semi-lattices in K^n. These are $\mathcal{O}^n$-submodules of K^n isomorphic to $\mathcal{O}^k \oplus K^{n-k}$ for some k. Intuitively, a sequence Λ_n of semi-lattices approaches a semi-lattice Λ if for any a, if $a \in \Lambda_n$ for infinitely many n then $a \in \Lambda$; and if $a \notin \mathcal{M}\Lambda_n$ for infinitely many n, then $a \notin \mathcal{M}\Lambda$. The actual definition is the same, but using definable types. A definable set of semi-lattices is *closed* if it is closed under limits of definable types. The set of closed balls in the affine line $\mathbb{A}^1$ can be viewed as a closed subset of the set of semi-lattices in K^2. In this case the limit of a decreasing sequence of balls is the intersection of these balls; the limit of the generic type of the valuation ring $\mathcal{O}$ (or of small closed balls around generic points of $\mathcal{O}$) is the closed ball $\mathcal{O}$. We also consider subspaces of these spaces of semi-lattices. They tend to be definably connected and compact, as tested by definable types. For instance the set of all semi-lattices in K^n cannot be split into two disjoint closed nonempty definable subsets.

To each algebraic variety V over a valued field K we will associate in a canonical way a projective limit $\widehat{V}$ of spaces of the type described above. A point of $\widehat{V}$ does not correspond to a point of V, but rather to a stably dominated definable type on V. We call $\widehat{V}$ the *stable completion* of V. For instance when $V = \mathbb{A}^1$, $\widehat{V}$ is the set of closed balls of V; the stably dominated type associated to a closed ball is just the generic type of that ball (which may be a point, or larger). In this case, and in general for curves, $\widehat{V}$ is definable (more precisely, a definable set of some imaginary sort), and no projective limit is needed.

While V admits no definable functions of interest from the value group Γ, there do exist definable functions from Γ to $\widehat{\mathbb{A}^1}$: for any point a of $\mathbb{A}^1$, one can consider the closed ball $B(a;\alpha) = \{x : \mathrm{val}(a-x) \geq \alpha\}$ as a definable function of $\alpha \in \Gamma$. These functions will serve to connect the space $\widehat{\mathbb{A}^1}$. In [19] the imaginary sorts were classified, and moreover the definable functions from Γ into them were classified; in the case of $\widehat{\mathbb{A}^1}$, essentially the only definable functions are the ones mentioned above. It is this kind of fact that is the basis of the geometry of imaginary sorts that we study here.

At present we remain in a purely algebraic setting. The applications to Berkovich spaces are thus only to Berkovich spaces of algebraic varieties. This limitation has the merit of showing that Berkovich spaces can be developed purely algebraically; historically, Krasner and Tate introduce analytic functions immediately even when interested in algebraic varieties, so that the name of the subject is rigid *analytic* geometry, but this is not necessary, a rigid algebraic geometry exists as well.

While we discussed o-minimality as an analogy, our real goal is a *reduction* of questions over valued fields to the o-minimal setting. The value group Γ of a valued field is o-minimal of a simple kind, where all definable objects are piecewise $\mathbb{Q}$-linear. Our main result is that for any quasi-projective variety V over K, $\widehat{V}$ admits a definable deformation retraction to a subset S, called a skeleton, which is definably homeomorphic to a space defined over Γ. There is a delicate point here: the definable homeomorphism is valid semi-algebraically, but if one stays in the (tropical) locally semi-linear setting, one must take into account subspaces of Γ_∞^n, where Γ_∞ is a partial completion of Γ by the addition of a point at ∞. The intersection of the space with the points at ∞ contains valuable additional information. In general, such a skeleton is non-canonical. At this point, o-minimal results such as triangulation can be quoted. As a corollary we obtain an equivalence of categories between the category of definable subsets of quasi-projective varieties over K, with homotopy classes of definable continuous maps $\widehat{U} \to \widehat{V}$ as morphisms $U \to V$, and a homotopy category of definable spaces over the o-minimal Γ.

In case the value group is $\mathbb{R}$, our results specialize to similar tameness theorems for Berkovich spaces. In particular we obtain local contractibility for Berkovich spaces associated to algebraic varieties, a result which was proved by Berkovich under smoothness assumptions [5], [6]. We also show that for projective varieties, the corresponding Berkovich space is homeomorphic to a projective limit of finite-dimensional simplicial complexes that are deformation retracts of itself. We further obtain finiteness statements that were not known classically; we refer to Chapter 14 for these applications.

We now present the contents of the chapters and a sketch of the proof of the main theorem.

Chapter 2 includes some background material on definable sets, definable types, orthogonality and domination, especially in the valued field context. In 2.11 we present the main result of [20] with a new insight regarding one point, that will be used in several critical points later in the paper. We know that every nonempty definable set over an algebraically closed substructure of a model of ACVF extends to a definable type. A definable type p can be decomposed into a definable type q on Γ^n, and a map f from this type to stably dominated definable types. In previous definitions of metastability, this decomposition involved an uncontrolled base change that prevented any canonicity. We note here that the q-germ of f is defined with no additional parameters, and that it is this germ that really determines p. Thus a general definable type is a function from a definable type on Γ^n to stably dominated definable types.

In Chapter 3 we introduce the space $\widehat{V}$ of stably dominated types on a definable set V. We show that $\widehat{V}$ is pro-definable; this is in fact true in any NIP theory, and not just in ACVF. We further show that $\widehat{V}$ is strict pro-definable, i.e. the image of $\widehat{V}$ under any projection to a definable set is definable. This uses metastability, and also a classical definability property of irreducibility in algebraically closed fields. In the case of curves, we note later that $\widehat{V}$ is in fact definable; for many purposes strict pro-definable sets behave in the same way. Still in Chapter 3, we define a topology on $\widehat{V}$, and study the connection between this topology and V. Roughly speaking, the topology on $\widehat{V}$ is generated by $\widehat{U}$, where U is a definable set cut out by *strict* valuation inequalities. The space V is a dense subset of $\widehat{V}$, so a continuous map $\widehat{V} \to \widehat{U}$ is determined by the restriction to V. Conversely, given a definable map $V \to \widehat{U}$, we explain the conditions for extending it to $\widehat{V}$. This uses the interpretation of $\widehat{V}$ as a set of definable types. We determine the Grothendieck topology on V itself induced from the topology on $\widehat{V}$; the closure or continuity of definable subsets or of functions on V can be described in terms of this Grothendieck topology without reference to $\widehat{V}$, but we will see that this viewpoint is more limited.

In Chapter 4 we define the central notion of definable compactness; we give a general definition that may be useful whenever one has definable topologies with enough definable types. The o-minimal formulation regarding limits of curves is replaced by limits of definable types. We relate definable compactness to being closed and bounded. We show the expected properties hold, in particular the image of a definably compact set under a continuous definable map is definably compact.

The definition of $\widehat{V}$ is a little abstract. In Chapter 5 we give a concrete representation of $\widehat{\mathbb{A}^n}$ in terms of spaces of semi-lattices. This was already alluded to in the first paragraphs of the introduction.

A major issue in this paper is the frontier between the definable and the topological categories. In o-minimality automatic continuity theorems play a role. Here we did not find such results very useful. At all events in 6.2 we characterize topologically those subspaces of $\widehat{V}$ that can be definably parameterized by Γ^n. They turn out to be o-minimal in the topological sense too. We use here in an essential way the construction of $\widehat{V}$ in terms of spaces of semi-lattices, and the characterization in [19] of definable maps from Γ into such spaces. We shall prove that our retractions provide skeleta lying in the subspace $V^\#$ of $\widehat{V}$ of *strongly stably dominated types* introduced in 8.1. This is another canonical space associated with V, ind-definable this time, admitting a natural continuous map into $\widehat{V}$ which restricts to a topological embedding on definable subsets. We study it further in Chapter 8; our uniformity results for $\widehat{V}$ depend on it.

Chapter 7 is concerned with the case of curves. We show that $\widehat{C}$ is definable (and not just pro-definable) when C is a curve. The case of $\mathbb{P}^1$ is elementary, and in equal characteristic zero it is possible to reduce everything to this case. But in general we use model-theoretic methods. We construct a definable deformation retraction from $\widehat{C}$ into a Γ-internal subset. We consider relative curves too, i.e. varieties V with maps $f : V \to U$, whose fibers are of dimension one. In this case we prove the existence of a deformation retraction of all fibers that is globally continuous and takes $\widehat{C}$ into a Γ-internal subset for almost all fibers C, i.e. all outside a proper subvariety of U. On curves lying over this variety, the motions on nearby curves do not converge to any continuous motion.

Chapter 9 contains some algebraic criteria for the verification of continuity. For the Zariski topology on algebraic varieties, the valuative criterion is useful: a constructible set is closed if it is invariant under specializations. Here we are led to doubly valued fields. These can be obtained from valued fields *either* by adding a valued field structure to the residue field, *or* by enriching the value group with a new convex subgroup. The functor $\widehat{X}$ is meaningful for definable sets of this theory as well, and interacts well with the various specializations. These criteria are used in Chapter 10 to verify the continuity of the relative homotopies of Chapter 7.

Chapter 10 includes some additional material on homotopies. In particular, for a smooth variety V, there exists an "inflation" homotopy, taking a simple point to the generic type of a small neighborhood of that point. This homotopy has an image that is properly a subset of $\widehat{V}$, and cannot be understood directly in terms of definable subsets of V. The image of this homotopy retraction has the merit of being contained in $\widehat{U}$ for any dense Zariski open subset U of V.

Chapter 11 contains the statement and proof of the main theorem. For any quasi-projective algebraic variety V, we prove the existence of a definable homotopy retraction from $\widehat{V}$ to an o-minimal subspace of the type described in 6.2. After some preliminary reductions, we may assume V fibers over a variety U of lower dimension and the fibers are curves. On each fiber, a homotopy retraction can be constructed with o-minimal image, as in Chapter

7; above a certain Zariski open subset U_1 of U, these homotopies can be viewed as the fibers of a single homotopy h_1. We require however a global homotopy. The homotopy h_1 itself does not extend to the complement of U_1; but in the smooth case, one can first apply an inflation homotopy whose image lies in $\widehat{V_1}$, where V_1 is the pullback of U_1. If V has singular points, a more delicate preparation is necessary. Let S_1 be the image of the homotopy h_1. Now a relative version of the results of 6.2 applies (Theorem 6.4.2); after pulling back the situation to a finite covering U' of U, we show that S_1 embeds topologically into $U' \times \Gamma_\infty^N$. Now any homotopy retraction of $\widehat{U}$, lifting to $\widehat{U'}$ and fixing certain functions into Γ^m, can be extended to a homotopy retraction of S_1 (Theorem 6.4.4). Using induction on dimension, we apply this to a homotopy retraction taking U to an o-minimal set; we obtain a retraction of V to a subset S_2 of S_1 lying over an o-minimal set, hence itself o-minimal. At this point o-minimal topology as in [9] applies to S_2, and hence to the homotopy type of $\widehat{V}$. In 11.7 we give a uniform version of Theorem 11.1.1 with respect to parameters. In Chapter 12 we examine the simplifications occuring in the proof of the main theorem in the smooth case and in Chapter 13 we deduce an equivalence of categories between a certain homotopy category of definable subsets of quasi-projective varieties over a given valued field and a suitable homotopy category of definable spaces over the o-minimal Γ.

Chapter 14 contains various applications to classical Berkovich spaces. Let V be a quasi-projective variety over a field F endowed with a non-archimedean norm and let V^{an} be the corresponding Berkovich space. We deduce from our main theorem several new results on the topology of V^{an} which were not known previously in such a level of generality. In particular we show that V^{an} admits a strong deformation retraction to a subspace homeomorphic to a finite simplicial complex and that V^{an} is locally contractible. We prove a finiteness statement for the homotopy type of fibers in families. We also show that if V is projective, V^{an} is homeomorphic to a projective limit of finite-dimensional simplicial complexes that are deformation retracts of V^{an}.

We do not assume any previous knowledge of Berkovich spaces, but highly recommend the survey [13], as well as [14] for an introduction to the model-theoretic viewpoint, and a sketch of proof of Theorem 11.1.1.

We are grateful to Vladimir Berkovich, Antoine Chambert-Loir, Zoé Chatzidakis, Antoine Ducros, Martin Hils, Dugald Macpherson, Kobi Peterzil, Anand Pillay, and Sergei Starchenko for their very useful comments. We address special thanks to Antoine Chambert-Loir for his invaluable help in preparing the final version of the text. The paper has also benefited greatly from highly extensive and thorough comments by anonymous referees, and we are very grateful to them.

During the preparation of this paper, the research of the authors has been partially supported by the following grants: E. H. by ISF 1048/07 and the European Research Council under the European Union's Seventh Framework Programme (FP7/2007-2013)/ERC Grant Agreement No. 291111; F.L. by ANR-06-BLAN-0183 and the European Research Council under the European Union's Seventh Framework Programme (FP7/2007-2013)/ERC Grant Agreement No. 246903/NMNAG.

⋆ ⋆ ⋆

Chapter Two

Preliminaries

Summary. In 2.1-2.6 we recall some model theoretic notions we shall use in an essential way in this work: definable, pro-definable and ind-definable sets, definable types, orthogonality to a definable set, stable domination. In 2.7-2.9 we consider more specifically these concepts in the framework of the theory ACVF of algebraically closed valued fields and recall in particular some results of [19] and [20] we rely on. In 2.10 we describe the definable types concentrating on a stable definable V as an ind-definable set. In 2.11, we prove a key result allowing us to view definable types as integrals of stably dominated types along some definable type on the value group sort. Finally, in 2.12 we discuss the notion of pseudo-Galois coverings that we shall use in Chapter 6.

We will rapidly recall the basic model theoretic notions of which we make use, but we recommend to the non-model theoretic reader an introduction such as [34] (readers seeking a more comprehensive text on stability may also consult [33]).

2.1 DEFINABLE SETS

Let us fix a first order language $\mathcal{L}$ and a complete theory T over $\mathcal{L}$. The language $\mathcal{L}$ may be multisorted. If $\mathcal{S}$ is a sort, and A is an $\mathcal{L}$-structure, we denote by $\mathcal{S}(A)$, the part of A belonging to the sort $\mathcal{S}$. For C a set of parameters in a model of T and x any set of variables, we denote by $\mathcal{L}_C$ the language $\mathcal{L}$ with symbols of constants for element of C added and by $S_x(C)$ the set of types over C in the variables x. Thus, $S_x(C)$ is the Stone space of the Boolean algebra of formulas with free variables contained in x and parameters from C up to equivalence over T. If A is a tuple or a set of parameters and B is a set of parameters, we shall denote by $\mathrm{tp}(A/B)$ the type of A over B. We write $\mathrm{tp}(A/B) \vdash \mathrm{tp}(A/BC)$ to mean that $\mathrm{tp}(A/B)$ implies $\mathrm{tp}(A/BC)$, i.e. $\mathrm{tp}(A/BC) = \mathrm{tp}(A'/BC)$ whenever $\mathrm{tp}(A/B) = \mathrm{tp}(A'/B)$.

We shall work in a large saturated model $\mathbb{U}$ (a universal domain for T). More precisely, we shall fix some uncountable cardinal κ larger than any cardinality of interest, and consider a model $\mathbb{U}$ of cardinality κ such that for every $A \subset U$ of cardinality $< \kappa$, every p in $S_x(A)$ is realized in $\mathbb{U}$, for x any set of variables of cardinality $< \kappa$. Such a $\mathbb{U}$ is unique up to isomorphism. Set theoretic issues involved in the choice of κ turn out to be unimportant and resolvable in numerous ways; cf. [7] or [21], Appendix A.

All sets of parameters A we shall consider will be *small* subsets of $\mathbb{U}$, that is of cardinality $< \kappa$, and all models M of T we shall consider will be elementary substructures of $\mathbb{U}$ with cardinality $< \kappa$. By a *substructure* of $\mathbb{U}$ we shall generally mean a small definably closed subset of $\mathbb{U}$.

If φ is a formula in $\mathcal{L}_C$, involving some sorts $\mathcal{S}_i$ with arity n_i, for every small model M containing C, one can consider the set $Z_\varphi(M)$ of tuples a in the cartesian product of the $\mathcal{S}_i(M)^{n_i}$ such that $M \models \varphi(a)$. One can view Z_φ as a functor from the category of models and elementary embeddings, to the category of sets. Such functors will be called *definable sets* over C. Note that a definable set X is completely determined by the (large) set $X(\mathbb{U})$, so we may identify definable sets with subsets of cartesian products of sets $\mathcal{S}_i(\mathbb{U})^{n_i}$. Definable sets over C form a category Def_C in a natural way. Under the previous identification a definable morphism between definable sets $X_1(\mathbb{U})$ and $X_2(\mathbb{U})$ is a function $X_1(\mathbb{U}) \to X_2(\mathbb{U})$ whose graph is definable.

By a definable set, we mean definable over some C. When C is empty one says $\varnothing$-definable or 0-definable. A subset of a given definable set X which is an intersection of $< \kappa$ definable subsets of X is said to be ∞-*definable.*

When the theory T has quantifier elimination, any definable set can be defined by a quantifier-free formula, and in any place where it matters we will always suppose that it is so defined.

Sets of $\mathbb{U}$-points of definable sets satisfy the following form of compactness: if X is a definable set such that $X(\mathbb{U}) = \bigcup_{i \in I} X_i(\mathbb{U})$, with $(X_i)_{i \in I}$ a small family of definable sets, then $X = \bigcup_{i \in A} X_i$ with A a finite subset of I.

Recall that if C is a subset of a model M of T, by the *algebraic closure* of C, denoted by $\mathrm{acl}(C)$, one denotes the subset of those elements c of M, such that, for some formula φ over C with one free variable, $Z_\varphi(M)$ is finite and contains c. The *definable closure* $\mathrm{dcl}(C)$ of C is the subset of those elements c of M, such that, for some formula φ over C with one free variable, $Z_\varphi(M) = \{c\}$.

If X is a C-definable set and $C \subset B$, we write $X(B)$ for $X(\mathbb{U}) \cap \mathrm{dcl}(B)$.

2.2 PRO-DEFINABLE AND IND-DEFINABLE SETS

We define the category ProDef_C of *pro-definable* sets over C as the category of pro-objects in the category Def_C indexed by a small directed partially ordered set. Thus, if $X = (X_i)_{i \in I}$ and $Y = (Y_j)_{i \in J}$ are two objects in ProDef_C,

$$\mathrm{Hom}_{\mathrm{ProDef}_C}(X, Y) = \varprojlim_j \varinjlim_i \mathrm{Hom}_{\mathrm{Def}_C}(X_i, Y_j).$$

Elements of $\mathrm{Hom}_{\mathrm{ProDef}_C}(X, Y)$ will be called C-pro-definable morphisms (or C-definable for short) between X and Y.

By a result of Kamensky [29], the functor of "taking $\mathbb{U}$-points" induces an equivalence of categories between the category ProDef_C and the sub-category of the category of sets whose objects and morphisms are inverse limits of $\mathbb{U}$-points of definable sets indexed by a small directed partially ordered set (here the word "co-filtering" is also used, synonymously with "directed"). By pro-definable, we mean pro-definable over some C. In this paper we shall freely identify a pro-definable set $X = (X_i)_{i \in I}$ with the set $X(\mathbb{U}) = \varprojlim_i X_i(\mathbb{U})$. For any set B with $C \subset B \subset \mathbb{U}$, we set $X(B) = X(\mathbb{U}) \cap \mathrm{dcl}(B) = \varprojlim_i X_i(B)$. Pro-definable is thus the same as $*$-definable in the sense of Shelah, that is, a small projective limit of definable subsets. One defines similarly the category IndDef_C of *ind-definable* sets over C for which a similar equivalence holds.

Let X be a pro-definable set. We shall say it is *strict* pro-definable if it may be represented as a pro-object $(X_i)_{i \in I}$, with surjective transition morphisms $X_j(\mathbb{U}) \to X_i(\mathbb{U})$. Equivalently, it is a $*$-definable set, such that the projection to any finite number of coordinates is definable.

Dual definitions apply to ind-definable sets; thus "strict" means that the maps are injective: in $\mathbb{U}$, a small union of definable sets is a strict ind-definable set.

By a morphism from an ind-definable set $X = \varinjlim_i X_i$ to a pro-definable one $Y = \varprojlim_j Y_j$, we mean a compatible family of morphisms $X_i \to Y_j$. A morphism $Y \to X$ is defined dually; it is always represented by a morphism $Y_j \to X_i$, for some j, i.

REMARK 2.2.1. Any strict ind-definable set X with a definable point admits a bijective morphism to a strict pro-definable set. On the other hand, if Y is strict pro-definable and X is strict ind-definable, a morphism $Y \to X$ always has definable image.

Proof. Fix a definable point p in X. If $f : X_i \to X_j$ is injective, define $g : X_j \to X_i$ by setting it equal to f^{-1} on $\mathrm{Im}(f)$, constant equal to p outside $\mathrm{Im}(f)$. The second statement is clear by compactness. □

DEFINITION 2.2.2. Let $Y = \varprojlim_i Y_i$ be pro-definable. Assume given, for each i, $X_i \subset Y_i$ such that the transition maps $Y_i \to Y_{i'}$, for $i \geq i'$, restrict to maps $X_i \to X_{i'}$ and set $X = \varprojlim_i X_i$.

(1) If each X_i is definable and, for some i_0, the maps $X_i \to X_{i'}$ are bijections for all $i \geq i' \geq i_0$, we say X is *iso-definable.*

(2) If each X_i is ∞-definable and, for some i_0, the maps $X_i \to X_{i'}$ are bijections for all $i \geq i' \geq i_0$, we say X is *iso-∞-definable.*

(3) If there exists a definable set W and a pro-definable morphism $g : W \to Y$ such that for each i, the composition of g and the projection $Y \to Y_i$ has image X_i, we say X is *definably parameterized.*

In Example 6.1.1 we shall give an example, for the spaces we will consider, of a definably parameterized subset which is not iso-definable. In Question

7.1.4 we formulate an open problem about inverse images of iso-definable subsets under finite morphisms. We now give two conditions under which definably parameterized sets are iso-definable.

LEMMA 2.2.3. *Let W be a definable set, Y a pro-definable set, and let $f : W \to Y$ be a pro-definable map. Then the image of W in Y is strict pro-definable. If f is injective, or more generally if the equivalence relation on W defined by $f(w) = f(w')$ is definable, then $f(W)$ is iso-definable.*

Proof. Say $Y = \varprojlim_i Y_i$. Let f_i be the composition $W \to Y \to Y_i$. Then f_i is a function whose graph is ∞-definable. By compactness there exists a definable function $F : W \to Y_i$ whose graph contains f_i; but then clearly $F = f_i$ and so the image $X_i = f_i(W)$ and f_i itself are definable. Now $f(W)$ is the projective limit of the system (X_i), with maps induced from (Y_i); the maps $X_i \to X_j$ are surjective for $i > j$, since $W \to X_j$ is surjective. Now assume there exists a definable equivalence relation E on W such that $f(w) = f(w')$ if and only if $(w, w') \in E$. If $(w, w') \in W^2 \smallsetminus E$ then w and w' have distinct images in some X_i. By compactness, for some i_0, if $(w, w') \in W^2 \smallsetminus E$ then $f_{i_0}(w) \neq f_{i_0}(w')$. So for any i mapping to i_0 the map $X_i \to X_{i_0}$ is injective. □

COROLLARY 2.2.4. *Let Y be pro-definable and let $X \subset Y$ be a pro-definable subset. Then X is iso-definable if and only if X is in (pro-definable) bijection with a definable set.* □

LEMMA 2.2.5. *Let Y be pro-definable, X an iso-definable subset. Let G be a finite group acting on Y, and leaving X invariant. Let $f : Y \to Y'$ be a map of pro-definable sets, whose fibers are exactly the orbits of G. Then $f(X)$ is iso-definable.*

Proof. Let U be a definable set, and $h : U \to X$ a pro-definable bijection. Define $g(u) = u'$ if $gh(u) = h(u')$. This induces a definable action of G on U. We have $f(h(u)) = f(h(u'))$ iff there exists g such that $gu = u'$. Thus the equivalence relation $f(h(u)) = f(h(u'))$ is definable; by Lemma 2.2.3, the image is iso-definable. □

We shall call a subset X of a pro-definable set Y *relatively definable* in Y if X is cut out from Y by a single formula. More precisely, if $Y = \varprojlim_i Y_i$ is pro-definable, X will be relatively definable in Y if there exists some index i_0 and a definable subset Z of Y_{i_0}, such that, denoting by X_i the inverse image of Z in Y_i for $i \geq i_0$, $X = \varprojlim_{i \geq i_0} X_i$. A subset of a pro-definable set Y is called *relatively ∞-definable* in Y if it is the intersection of a small family of relatively definable subsets of Y.

Iso-definability and relative definability are related somewhat as finite dimension is related to finite codimension; so they rarely hold together. In this terminology, a semi-algebraic subset of $\widehat{V}$, that is, a subset of the form

$\widehat{X}$, where X is a definable subset of V, will be relatively definable, but most often not iso-definable.

LEMMA 2.2.6. (1) *Let X be pro-definable, and assume that the equality relation Δ_X is a relatively definable subset of X^2. Then X is iso-∞-definable.*

(2) *A pro-definable subset of an iso-∞-definable set is iso-∞-definable.*

Proof. (1) X is the projective limit of an inverse system $\{X_i\}$, with maps $f_i : X \to X_i$. We have $(x, y) \in \Delta_X$ if and only if $f_i(x) = f_i(y)$ for each i. It follows that for some i, $(x, y) \in \Delta_X$ if and only if $f_i(x) = f_i(y)$. For otherwise, for any finite set I_0 of indices, we may find $(x, y) \notin \Delta_X$ with $f_i(x) = f_i(y)$ for every $i \in I_0$. But then by compactness, and using the relative definability of (the complement of) Δ_X, there exist $(x, y) \in X^2 \smallsetminus \Delta_X$ with $f_i(x) = f_i(y)$ for all i, a contradiction. Thus the map f_i is injective. (2) follows from (1), or can be proved directly. □

LEMMA 2.2.7. *Let $f : X \to Y$ be a morphism between pro-definable sets. If Y is (isomorphic to) a definable set, then* Imf *is ∞-definable. In general* Imf *is pro-definable.*

Proof. Follows easily from compactness. □

LEMMA 2.2.8. *Let $f : X \to Y$ be a morphism between strict pro-definable sets. Then* Imf *is strict pro-definable, as is the graph of f.*

Proof. We can represent X and Y as respectively projective limit of definable sets X_i and Y_j with surjective transition mappings and f by $f_j : X_{c(j)} \to Y_j$, for some function c between the index sets. The projection of Imf to Y_j is the same as the image of f_j, using the surjectivity of the maps between the sets $X_{c(j)}$ and $f_j(X_{c(j)})$. The graph of f is the image of $\mathrm{Id} \times f : X \to (X \times Y)$. □

REMARK 2.2.9 (On terminology). We often have a set $D(A)$ depending functorially on a structure A. We say that D is pro-definable if there exists a pro-definable set D' such that $D'(A)$ and $D(A)$ are in canonical bijection; in other words D and D' are isomorphic functors.

In practice we have in mind a choice of D' arising naturally from the definition of D; usually various interpretations are possible, but all are isomorphic as pro-definable sets. Once D' is specified, so is, for any pro-definable W and any A, the set of A-definable maps $W \to D'$. If worried about the identity of D', it suffices to specify what we mean by an A-definable map $W \to D$. Then Yoneda's lemma ensures the uniqueness of a pro-definable set D' compatible with this notion.

The same applies for ind. For instance, let $\mathrm{Fn}(V, V')(A)$ be the set of A-definable functions between two given $\varnothing$-definable sets V and V'. Then

$\mathrm{Fn}(V, V')$ is an ind-definable set. The representing ind-definable set is clearly determined by the description.

To avoid all doubts, we specify that $\mathrm{Fn}(U, \mathrm{Fn}(V, V')) = \mathrm{Fn}(U \times V, V')$.

2.2.10 Maps from ind-definable to pro-definable sets

Let $X = \varinjlim_i X_i$ be an ind-definable set, and let $Y = \varprojlim_j Y_j$ be a pro-definable set. Recall that $\mathrm{Hom}(X, Y) = \varprojlim_{i,j} \mathrm{Hom}(X_i, Y_j)$, where one denotes by $\mathrm{Hom}(X_i, Y_j)$ the set of definable maps $X_i \to Y_j$. Clearly, if $f \in \mathrm{Hom}(X, Y)$ then f induces a map $f_M : X(M) \to Y(M)$, for any model M. In case X is strict ind-definable, we call f *injective* if in any model, f_M is injective. If X is strict ind-definable and f is represented by $(f_{i,j})$, then f is injective iff for each i, for some j, f_{ij} is injective; since if for arbitrarily large j there exist distinct $x, x' \in X_i$ with $f_{ij}(x) = f_{ij}(x')$, then by compactness we can find a pair $x \neq x' \in X_i$ such that for all j, $f_{ij}(x) = f_{ij}(x')$.

DEFINITION 2.2.11. Let X be a subset of a pro-definable set. By a *strict ind-definable structure* on X we shall mean a strict ind-definable set Z together with an injective morphism $g : Z \to Y$ with image X. Two such data $g : Z \to Y$ and $g' : Z' \to Y$ will be considered to induce the same structure if there exists an isomorphism $h : Z \to Z'$ of ind-definable sets with $g = h \circ g'$.

We will say that "X is strict ind-definable" if a strict ind-definable structure is fixed. In this situation we will view X itself as being ind-definable, and can apply any notion appropriate for ind-definable sets. Notably we can speak of *definable* subsets of X; these are iso-definable, but in general an iso-definable subset of a strict ind-definable set need not be definable in the sense of the given structure.

LEMMA 2.2.12. *Let Y be pro-definable. Assume $W \subset Y$ admits a strict ind-definable structure $f : X \to W$, such that for each definable $X' \subset X$, for some definable quotient $\pi : Y \to Y'$, the restriction $\pi|f(X)$ is injective above $\pi(f(X'))$. Then W has a unique such ind-definable structure, i.e. if $W = f'(X')$ with the same property, then there exists an isomorphism $g : X \to X'$ of ind-definable sets with $f = g \circ f'$.*

Proof. Let W be strict ind-definable via $f : X \to Y$ and via $f' : X' \to Y$ having the above properties. We need to show that $f^{-1} \circ f' : X' \to X$ is an isomorphism of ind-definable sets. As $f^{-1} \circ f'$ is a bijection on points, and since the restriction of the graph of this bijection to any product $U \times U'$ of definable subsets of X and X' respectively is ∞-definable, it suffices to show that $(f^{-1} \circ f')(U')$ is contained in a definable subset of X', for any definable $U' \subset X'$ (and vice versa). Let $\pi : Y \to D$ be a morphism to a definable set D, such that $\pi \circ f'$ is injective above $\pi(f'(U'))$. Now $U' \subset \bigcup_U ((f')^{-1} \circ f)(U)$, where U ranges over all definable subsets of X, defined over a given set of definition for X. For $u \in U, u' \in U'$, we have $u' = ((f')^{-1} \circ f)(u)$ iff

$f(u) = f'(u')$ iff $\pi \circ f(u) = \pi \circ f'(u')$; this is a definable condition. So $((f')^{-1} \circ f)(U)$ is definable. By compactness, U' is contained in a finite union of sets $((f')^{-1} \circ f)(U)$; as the union of finitely many definable subsets of X is definable, it is contained in such a set. □

Let Y be pro-definable, and consider an injective morphism f from an ind-definable set $X = \varinjlim_i X_i$ to Y. Then $f(X)$ is strict pro-definable assuming that the equivalence relation E_i on X_i defined by $f(x) = f(x')$ be definable; for then $f_{\mathbb{U}}(X(\mathbb{U})) = g_{\mathbb{U}}(\mathbf{X}(\mathbb{U}))$, where $\mathbf{X} = \varinjlim_i (X_i/E_i)$, $\pi : X \to \mathbf{X}$ is the natural quotient, and g is the map such that $f = g \circ \pi$; note that $\mathbf{X}$ is strict ind-definable.

From this, and the fact that strict ind-definable sets are closed under disjoint unions, we obtain:

LEMMA 2.2.13. *Let Y be pro-definable. Let $S_k \subset Y$ admit a strict ind-definable structure, via ind-definable sets X_k and injective morphisms f_k with $f_k(X_k) = S_k$. Assume the pullback to $S_k \times S_{k'}$ of the diagonal $\Delta_Y \subset Y \times Y$ is piecewise definable; i.e. $(f_k \times f_{k'})^{-1}(\Delta_Y) \cap (X \times X')$ is definable, for any definable $X \subset X_k, X' \subset X_{k'}$. Then $\cup_k S_k$ admits a strict ind-definable structure.* □

2.3 DEFINABLE TYPES

For any set z of variables, we shall denote by $\mathcal{L}_z$ the set of $\mathcal{L}$-formulas in variables in z up to equivalence in the theory T. A type $p(x)$ in variables $x = (x_1, \dots, x_n)$ can be viewed as a Boolean homomorphism from $\mathcal{L}_x$ to the two-element Boolean algebra. Now consider variables y_i running through all finite products of sorts. A *0-definable type* $p(x)$ is defined to be a function $d_p x : \mathcal{L}_{x,y_1,\dots,} \to \mathcal{L}_{y_1,\dots,}$, such that for any finite $y = (y_1, \dots, y_n)$, $d_p x$ restricts to a Boolean retraction $\mathcal{L}_{x,y} \to \mathcal{L}_y$. An A-definable type p is a 0-definable type of the theory T_A obtained by naming constants for the elements of the substructure A. Sometimes we shall also say p is *based on* A. By a *definable type* we mean a $\mathbb{U}$-definable type. The image of a formula $\phi(x, y)$ under $d_p x$ is called the ϕ-definition of p. Note that this definition makes sense for any, possibly infinite, set of variables x. When there is no risk of confusion, we sometimes will write d_p instead of $d_p x$.

Given such a retraction, and given any model M of T, one obtains a type over M, namely

$$p|M := \{\varphi(x, b_1, \dots, b_n) : M \models (d_p x)(\varphi)(b_1, \dots, b_n)\}.$$

This type over M determines p; this explains the use of the term definable type. However viewed as above, a definable type is really not a type but a different kind of object. We will often identify p with the type $p|\mathbb{U}$ which is $\mathrm{Aut}(\mathbb{U})$-invariant, and determines p. For any $B \subset \mathbb{U}$, we denote by $p|B$ the restriction of $p|\mathbb{U}$ to B. Similarly, for any $C \subset \mathbb{U}$, replacing $\mathcal{L}$ by $\mathcal{L}_C$ one gets the notion of C-definable type. If p is C-definable, then the type $p|\mathbb{U}$ is $\mathrm{Aut}(\mathbb{U}/C)$-invariant.

If p is a definable type and X is a definable set, or a pro-definable set, one says p is on X if all realizations of $p|\mathbb{U}$ lie in X. One denotes by $S_{def,X}$ the set of definable types on X. Let $f : X \to Z$ be a definable map between definable sets, or a pro-definable map between pro-definable sets. For p in $S_{def,X}$ one denotes by $f_*(p)$ the definable type defined by $(d_{f_*(p)}z)(\varphi(z,y)) = (d_p x)(\varphi(f(x),y))$. This gives rise to a mapping $f_* : S_{def,X} \to S_{def,Z}$.

For a $\varnothing$-definable set V, let L_V denote the Boolean algebra of $\varnothing$-definable subsets of V. Then a type on V corresponds to an element of $\mathrm{Hom}(L_V, 2)$ and a definable type on V is the same as a compatible family of elements of $\mathrm{Hom}_W(L_{V\times W}, L_W)$, for W running over the $\varnothing$-definable sets, where Hom_W denotes the set of Boolean homomorphisms h such that $h(V \times X) = X$ for $X \subset W$. Let U be a pro-definable set. By a *definable function* $U \to S_{def,V}$, we mean a compatible family of elements of $\mathrm{Hom}_{W\times U}(L_{V\times W\times U}, L_{W\times U})$ for W running over the $\varnothing$-definable sets. Any element $u \in U$ gives a Boolean retraction $L_{W\times U} \to L_W(u)$ by $Z \mapsto Z(u) = \{z : (z,u) \in Z\}$, with $L_W(u)$ the Boolean algebra of u-definable subsets of W. So a definable function $U \to S_{def,V}$ gives indeed a U-parameterized family of definable types on V.

Let p be a partial $\mathbb{U}$-type. Let us say p is *definably generated* over A if it is generated by a partial type of the form $\bigcup_{(\phi,\theta)\in S} P(\phi,\theta)$, where S is a set of pairs of formulas $(\phi(x,y), \theta(y))$ over A, and $P(\phi,\theta) = \{\phi(x,b) : \mathbb{U} \models \theta(b)\}$.

LEMMA 2.3.1. *Let p be a type over $\mathbb{U}$. If p is definably generated over A, then p is A-definable.*

Proof. This follows from Beth's theorem: if one adds a predicate for the p-definitions of all formulas $\phi(x,y)$, with the obvious axioms, there is a unique interpretation of these predicates in $\mathbb{U}$, hence they must be definable.

Alternatively, let $\phi(x,y)$ be any formula. From the fact that p is definably generated it follows easily that $\{b : \phi(x,b) \in p\}$ is a strict ind-definable set over A. Indeed, $\phi(x,b) \in p$ if and only if for some $(\phi_1,\theta_1),\dots,(\phi_m,\theta_m) \in S$, $(\exists c_1,\dots,c_m)(\theta_i(c_i) \wedge (\forall x)(\bigwedge_i \phi_i(x,c) \implies \phi(x,b)))$. Applying this to $\neg\phi$, we see that the complement of $\{b : \phi(x,b) \in p\}$ is also strict ind-definable. Hence $\{b : \phi(x,b) \in p\}$ is A-definable. □

COROLLARY 2.3.2. *Let $f : X \to Y$ be an A-definable (or pro-definable) function. Let q be an A-definable type on Y. Let p_A be a type over A such that $f_* p_A = q|A$ and such that, for any B containing A there exists a unique type p_B over B such that p_B contains p_A, and $f_* p_B = q|B$. Here $f_* p_B$ denotes the type generated by $\mathcal{L}(B)$-formulas $\varphi(y)$ such that $\varphi(f(x))$ belongs to p_B. Then there exists a unique A-definable type p such that for all B, $p|B = p_B$.* □

DEFINITION 2.3.3. In the situation of Corollary 2.3.2, p is said to be *dominated by q via f*.

Let us recall that a theory T is said to have *elimination of imaginaries* if, for any $M \models T$, any collection $\mathcal{S}_1, \ldots, \mathcal{S}_k$ of sorts in T, and any $\varnothing$-definable equivalence relation E on $\mathcal{S}_1(M) \times \cdots \times \mathcal{S}_k(M)$, there is a $\varnothing$-definable function f from $\mathcal{S}_1(M) \times \cdots \times \mathcal{S}_k(M)$ into a product of sorts of M, such that for any $a, b \in \mathcal{S}_1(M) \times \cdots \times \mathcal{S}_k(M)$, we have Eab if and only if $f(a) = f(b)$. Given a complete theory T, it is possible to extend it to a complete theory T^{eq} over a language $\mathcal{L}^{eq}$ by adjoining, for each collection $\mathcal{S}_1, \ldots, \mathcal{S}_k$ of sorts and $\varnothing$-definable equivalence relation E on $\mathcal{S}_1 \times \cdots \times \mathcal{S}_k$, a sort $(\mathcal{S}_1 \times \cdots \times \mathcal{S}_k)/E$, together with a function symbol for the natural map $a \mapsto a/E$. Any model M of T can be canonically extended to a model of T^{eq}, denoted M^{eq}. We shall refer to the new sorts of T^{eq} as imaginary sorts, and to elements of them as *imaginaries*.

Suppose that D is a definable set in $M \models T$, defined say by the formula $\phi(x, a)$. There is a $\varnothing$-definable equivalence relation $E_\phi(y_1, y_2)$, where $E_\phi(y_1, y_2)$ holds if and only if $\forall x(\phi(x, y_1) \leftrightarrow \phi(x, y_2))$. Now a/E_ϕ is identifiable with an element of an imaginary sort; it is determined uniquely (up to interdefinability over $\varnothing$) by D, and will often be referred to as a *code* for D, and denoted $\ulcorner D \urcorner$. We prefer to think of $\ulcorner D \urcorner$ as a fixed object (e.g. as a member of $\mathbb{U}^{eq}$) rather than as an equivalence class of M; for viewed as an equivalence class it is formally a different set (as is D itself) in elementary extensions of M.

LEMMA 2.3.4. *Assume the theory T has elimination of imaginaries. Let $f : X \to Y$ be a C-definable mapping between C-definable sets. Assume f has finite fibers, say of cardinality bounded by some integer m. Let p be a C-definable type on Y. Then, any global type q on X such that $f_*(q) = p|\mathbb{U}$ is* $\mathrm{acl}(C)$*-definable.*

Proof. Let $\mathbf{p} = p|\mathbb{U}$. The partial type $\mathbf{p}(f(x))$ admits at most m distinct extensions $q_1, \ldots, q_\ell$ to a complete type. Choose $C' \supset C$ such that all $q_i|C'$ are distinct. Certainly the q_i are C'-invariant. It is enough to prove they are C'-definable, since then, for every formula φ, the $\mathrm{Aut}(\mathbb{U}/C)$-orbit of $d_{q_i}(\varphi)$ is finite, hence $d_{q_i}(\varphi)$ is equivalent to a formula in $\mathcal{L}(\mathrm{acl}(C))$. To prove q_i is C'-definable note that

$$p(f(x)) \cup (q_i|C')(x) \vdash q_i(x).$$

Thus, there is a set A of formulas $\varphi(x, y)$ in $\mathcal{L}$, a mapping $\varphi(x, y) \to \vartheta_\varphi(y)$ assigning to formulas in A formulas in $\mathcal{L}(C')$ such that q_i is generated by $\{\varphi(x, b) : \mathbb{U} \models \vartheta_\varphi(b)\}$. It then follows from Lemma 2.3.1 that q_i is indeed C'-definable. □

2.4 STABLE EMBEDDEDNESS

A C-definable set D in $\mathbb{U}$ is *stably embedded* if, for any definable set E (with parameters a from $\mathbb{U}$) and $r > 0$, $E \cap D^r$ is definable over $C \cup D$. To state a more explicit version that does not use $\mathbb{U}$: for any formula $\phi(x, y)$ there is a formula $\psi(x, z)$ such that for all a there is a sequence d from D such that

$$\{x \in D^r : \models \phi(x, a)\} = \{x \in D^r : \models \psi(x, d)\}.$$

For more on stably embedded sets, we refer to the Appendix of [8].

LEMMA 2.4.1. *Let T be a complete theory in a language L and D a stably embedded sort. Let L_D be the restriction of L to D and L^*_D any enrichment of L_D. Let T_D be the restriction of T to D and let T^*_D be any expansion of T_D to a complete theory in L^*_D. Let $T^\sharp_D$ be the relativization of T^*_D to D, i.e the theory that states that $D \models T^*_D$. Let $L^* = L \cup L^*_D$ and let $T^* = T \cup T^\sharp_D$. Then T^* is complete, D is stably embedded in T^*, and the L^*_D-type of a tuple b of elements of D implies its L^*-type. Moreover, assume T and T^*_D admit quantifier elimination, and for any tuple a in a model of T, $\mathrm{dcl}(a) \cap \mathrm{dcl}(D) = \mathrm{dcl}((f_i(a)))$ where (f_i) enumerates term functions with values in D. Then T^* admits quantifier elimination.*

Proof. Let M^*, N^* be two saturated models of T^* of the same cardinality. To prove completeness, we must show that $M^* \cong N^*$. To prove stable embeddedness, we must show that any isomorphism $f : D^*_M \to D^*_N$ extends to an isomorphism $M^* \to N^*$. But D is stably embedded with respect to L, so f extends to an L-isomorphism $M \to N$, which is by definition also an L^*-isomorphism. This proves both stable embeddedness and completeness; completeness also follows since by completeness of T^*_D, we do have $D^*_M \cong D^*_N$. The statement about the type of a tuple b follows from the completeness result applied to T and T^*_D, each expanded by constants for b.

To prove the "moreover" statement, we must show that if a, b are tuples from M^* respectively with the same quantifier-free type, then there exists an automorphism of M^* with $a \mapsto b$. Let $c = (f_i(a))$ and $d = (f_i(b))$ where (f_i) enumerates term functions with values in D. Then c and d have the same quantifier-free type in $D(M^*)$ so there exists an automorphism of D as an L^*-structure with $a \mapsto b$. As above this automorphism extends to M^*; so we may assume it is the identity. Now in the restriction M of M^* to L, we have $\mathrm{tp}(a/D) = \mathrm{tp}(b/D)$ so there exists an automorphism σ of M fixing D pointwise with $a \mapsto b$; and as it fixes D, σ is also an L^*-automorphism. □

2.5 ORTHOGONALITY TO A DEFINABLE SET

Let Q be a fixed $\varnothing$-definable set. We give definitions of orthogonality to Q that are convenient for our purposes, and are equivalent to the usual ones when Q is stably embedded and admits elimination of imaginaries; this is the only case we will need.

Let A be a substructure of $\mathbb{U}$. A type $p = \mathrm{tp}(c/A)$ is said to be *almost orthogonal* to Q if $Q(A(c)) = Q(A)$. Here $A(c)$ is the substructure generated by c over A, and $Q(A) = Q \cap \mathrm{dcl}(A)$ is the set of points of Q definable over A.

An A-definable type p is said to be *orthogonal* to Q, and one writes $p \perp Q$, if $p|B$ is almost orthogonal to Q for any substructure B containing A. Equivalently, for any B and any B-definable function f into Q the push-forward $f_*(p)$ is a type concentrating on one point, i.e. including a formula of the form $y = \gamma$.

Let us recall that for F a structure containing C, $\mathrm{Fn}(W, Q)(F)$ denotes the family of F-definable functions $W \to Q$ and that $\mathrm{Fn}(W, Q) = \mathrm{Fn}(W, Q)(\mathbb{U})$ is an ind-definable set.

Let V be a C-definable set. Let p be a definable type on V, orthogonal to Q. Any $\mathbb{U}$-definable function $f : V \to Q$ is generically constant on p. Equivalently, any C-definable function $f : V \times W \to Q$ (where W is some C-definable set) depends only on the W-argument, when the V-argument is a generic realization of p. More precisely, we have a mapping

$$p_*^W : \mathrm{Fn}(V \times W, Q) \longrightarrow \mathrm{Fn}(W, Q)$$

(denoted by p_* when there is no possibility of confusion) given by $p_*(f)(w) = \gamma$ if $(d_p v)(f(v, w) = \gamma)$ holds in $\mathbb{U}$.

Uniqueness of γ is clear for any definable type. Orthogonality to Q is precisely the statement that for any f, $p_*(f)$ is a function on W, i.e. for any w, such an element γ exists. The advantage of the presentation $f \mapsto p_*(f)$, rather than the two-valued $\phi \mapsto p_*(\phi)$, is that it makes orthogonality to Q evident from the very data.

Let $S^Q_{def,V}(A)$ denote the set of A-definable types on V orthogonal to Q. It will be useful to note the (straightforward) conditions for pro-definability of $S^Q_{def,V}$. Given a function $g : S \times W \to Q$, we let $g_s(w) = g(s, w)$, thus viewing it as a family of functions $g_s : W \to Q$.

LEMMA 2.5.1. *Assume the theory T eliminates imaginaries, and that for any formula $\phi(v, w)$ without parameters, there exists a formula $\theta(w, s)$ without parameters such that for any $p \in S^Q_{def,V}$, for some e,*

$$\phi(v, c) \in p \iff \theta(c, e).$$

Then $S^Q_{def,V}$ is pro-definable, i.e. there exists a canonical pro-definable Z and a canonical bijection $Z(A) = S^Q_{def,V}(A)$ for every A.

Proof. We first extend the hypothesis a little. Let $f : V \times W \to Q$ be $\varnothing$-definable. Then there exists a $\varnothing$-definable $g : S \times W \to Q$ such that for any

$p \in S^Q_{def,V}$, for some $s \in S$, $p_*(f) = g_s$. Indeed, let $\phi(v, w, q)$ be the formula $f(v, w) = q$ and let $\theta(w, q, s)$ be the corresponding formula provided by the hypothesis of the lemma. Let S be the set of all s such that for any $w \in W$ there exists a unique $q \in Q$ with $\theta(w, q, s)$. Now, by setting $g(s, w) = q$ if and only if $\theta(w, q, s)$ holds, one gets the more general statement.

Let $f_i : V \times W_i \to Q$ be an enumeration of all $\varnothing$-definable functions $f : V \times W \to Q$, with i running over some index set I. Let $g_i : S_i \times W_i \to Q$ be the corresponding functions provided by the previous paragraph. Elimination of imaginaries allows us to assume that s is a canonical parameter for the function $g_{i,s}(w) = g_i(s, w)$, i.e. for no other s' do we have $g_{i,s} = g_{i,s'}$. We then have a natural map $\pi_i : S^Q_{def,V} \to S_i$, namely $\pi_i(p) = s$ if $p_*(f_i) = g_{i,s}$. Let $\pi = \Pi_i \pi_i : S^Q_{def,V} \to \Pi_i S_i$ be the product map. Now $\Pi_i S_i$ is canonically a pro-definable set, and the map π is injective. So it suffices to show that the image is relatively ∞-definable in ΠS_i. Indeed, $s = (s_i)_i$ lies in the image if and only if for each finite tuple of indices $i_1, \dots, i_n \in I$ (allowing repetitions),

$$(\forall w_{i_1} \in W_{i_1}) \cdots (\forall w_{i_n} \in W_{i_n})(\exists v \in V) \bigwedge_{j=1}^{n} f_{i_j}(v, w_{i_j}) = g_{i_j}(s_{i_j}, w_{i_j}).$$

For given this consistency condition, there exists $a \in V(\mathbb{U}')$ for some $\mathbb{U} \prec \mathbb{U}'$ such that $f_i(a, w) = g_i(s, w)$ for all $w \in W_i$ and all i. It follows immediately that $p = \mathrm{tp}(a/\mathbb{U})$ is definable and orthogonal to Q, and $\pi(p) = s$. Conversely if $p \in S^Q_{def,V}(\mathbb{U})$ and $a \models p|\mathbb{U}$, for any $w_1 \in W_1(\mathbb{U}), \dots, w_n \in W_n(\mathbb{U})$, the element a witnesses the existence of v as required. So the image is cut out by a set of formulas concerning the s_i. □

If Q is a two-element set, any definable type is orthogonal to Q, and $\mathrm{Fn}(V, Q)$ can be identified with the algebra of formulas on V, via characteristic functions. The presentation of definable types as a Boolean retraction from formulas on $V \times W$ to formulas on W can be generalized to definable types orthogonal to Q, for arbitrary Q. An element p of $S^Q_{def,V}(A)$ yields a compatible system of retractions $p^W_* : \mathrm{Fn}(V \times W, Q) \longrightarrow \mathrm{Fn}(W, Q)$. These retractions are also compatible with definable functions $g : Q^m \to Q$, namely $p_*(g \circ (f_1, \dots, f_m)) = g \circ (p_* f_1, \dots, p_* f_m)$. One can restrict attention to $\varnothing$-definable functions $Q^m \to Q$ along with compositions of the following form: given $F : V \times W \times Q \to Q$ and $f : V \times W \to Q$, let $F \circ' f(v, w) = F(v, w, f(v, w))$. Then $p_*(F \circ' f) = p_*(F) \circ' p_*(f)$. It can be shown that any compatible system of retractions compatible with these compositions arises from a unique element p of $S^Q_{def,V}(A)$. This can be shown by the usual two way translation between sets and functions: a set can be coded by a function into a two-element set (in case two constants are not available, one can add variables x, y, and consider functions whose values are among the variables). On the other hand a function can be coded by a set, namely its graph. This characterization will not be used, and we will leave the details to the reader. It does give a slightly different way to see the ∞-definability of the image in Lemma 2.5.1.

2.6 STABLE DOMINATION

We shall assume from now on that the theory T has elimination of imaginaries.

DEFINITION 2.6.1. A C-definable set D in $\mathbb{U}$ is said to be *stable* if the structure with domain D, when equipped with all the C-definable relations, is stable.

One considers the multisorted structure St_C whose sorts D_i are the C-definable, stable and stably embedded subsets of $\mathbb{U}$. For each finite set of sorts D_i, all the C-definable relations on their union are considered as $\emptyset$-definable relations R_j. The structure St_C is stable by Lemma 3.2 of [20].

For any $A \subset \mathbb{U}$, one sets $\mathrm{St}_C(A) = \mathrm{St}_C \cap \mathrm{dcl}(CA)$.

LEMMA 2.6.2. *For any subsets A, B and C of $\mathbb{U}$ the following conditions are equivalent:*

(1) $\mathrm{tp}(B/C\,\mathrm{St}_C(A)) \vdash \mathrm{tp}(B/CA)$;

(2) $\mathrm{tp}(A/C\,\mathrm{St}_C(B)) \vdash \mathrm{tp}(A/CB)$;

(3) $\mathrm{tp}(A/C\,\mathrm{St}_C(A)) \vdash \mathrm{tp}(A/C\,\mathrm{St}_C(A)B)$.

Proof. The equivalence of (1) and (2) is Lemma 3.8 (i) of [20]. The equivalence of (1) and (3) follows from the elementary fact that $\mathrm{tp}(A/D) \vdash \mathrm{tp}(A/DB)$ is always equivalent to $\mathrm{tp}(B/D) \vdash \mathrm{tp}(B/DA)$, cf. [20] p. 29. $\square$

REMARK 2.6.3. For any subsets A, B and C of $\mathbb{U}$, it is a consequence of stable embeddedness that $\mathrm{tp}(A/C\,\mathrm{St}_C(A)) \vdash \mathrm{tp}(A/C\,\mathrm{St}_C(A)\,\mathrm{St}_C(B))$, as explained in Remark 3.7 of [20].

DEFINITION 2.6.4. A type $\mathrm{tp}(A/C)$ is *stably dominated* if, for any B such that $\mathrm{St}_C(A) \downarrow_{\mathrm{St}_C(C)} \mathrm{St}_C(B)$, the conditions in Lemma 2.6.2 are satisfied.

By [20] 3.13, if $\mathrm{tp}(a/C)$ is stably dominated, then it has an $\mathrm{acl}(C)$-definable extension p to $\mathbb{U}$; this definable type will also be referred to as stably dominated. In fact our focus is really on stably dominated *definable types*, and we will rarely refer to *types* as stably dominated.

The unique stably dominated extension of $\mathrm{tp}(a/\mathrm{acl}(C))$ will be denoted by $\mathrm{tp}(a/\mathrm{acl}(C))|\mathbb{U}$; more generally, for any B with $\mathrm{acl}(C) \leq B \leq \mathbb{U}$, write $p|B = \mathrm{tp}(a/\mathrm{acl}(C))|B$.

We say that a stably dominated type $\mathrm{tp}(a/C)$ is *stationary* if it has a C-definable extension p to $\mathbb{U}$. Equivalently,

$$\mathrm{dcl}(Ca) \cap \mathrm{acl}(C) = \mathrm{dcl}(C).$$

One direction of the equivalence follows from the remark immediately following, applied to $N = \mathbb{U}$; for the other see, e.g., [24] Lemma 2.9.

For any $|C|^+$-saturated, $|C|^+$-homogenous extension N of C, $p|N$ is the unique $\mathrm{Aut}(N/\mathrm{acl}(C))$-invariant extension of $\mathrm{tp}(a/\mathrm{acl}(C))$. We will need a slight extension of this:

LEMMA 2.6.5. *Let $p = \mathrm{tp}(a/C)$ be a stably dominated C-definable type, $C = \mathrm{acl}(C)$. Let $C \subset B = \mathrm{dcl}(B)$. Assume that, for any $b \in \mathrm{St}_C(B) \smallsetminus C$, there exists $b' \in B$, $b' \neq b$, such that b and b' are $\mathrm{Aut}(B/C)$-conjugate. Then $p|B$ is the unique $\mathrm{Aut}(B/C)$-invariant extension of $\mathrm{tp}(a/C)$.*

Proof. By hypothesis, p is stably dominated via some C-pro-definable function h into St_C. Let q be an $\mathrm{Aut}(B/C)$-invariant extension of $\mathrm{tp}(a/C)$, say $q = \mathrm{tp}(d/B)$. Let $h_*q = \mathrm{tp}(h(d)/\mathrm{St}_C(B))$ be its pushforward. Let b enumerate the canonical base of h_*q, so that h_*q is the unique nonforking extension to $\mathrm{St}_C(B)$ of $h_*q|C(b)$, and any automorphism fixing q also fixes b. As q is C-invariant, any automorphism of B/C fixes b. But by assumption, if $b \notin C$ there is an elementary permutation σ of $\mathrm{St}_C(B)$ over C with $\sigma(b) \neq b$; it follows that b is contained in C. Thus h_*q does not fork over C, so $h_*q = h_*p$. By definition of stable domination, it follows that $q = p$. □

PROPOSITION 2.6.6 ([20] Proposition 6.11). *Assume the types $\mathrm{tp}(a/C)$ and $\mathrm{tp}(b/aC)$ are stably dominated, then $\mathrm{tp}(ab/C)$ is stably dominated.*

A formula $\varphi(x, y)$ is said to *shatter* a subset W of a model of T if for any two finite disjoint subsets U, U' of W there exists b with $\phi(a, b)$ for $a \in U$, and $\neg\phi(a', b)$ for $a' \in U'$. Shelah says that a formula $\varphi(x, y)$ has the *independence property* if it shatters arbitrarily large finite sets; otherwise, it has NIP. Finally, T has NIP if every formula has NIP. Stable and o-minimal theories are NIP, as is ACVF.

If $\varphi(x, y)$ has NIP then there exists a positive integer k, such that for any finite (or infinite) indiscernible sequence $(a_1, \dots, a_n)$ and any b in a model of T, $\{i : \phi(a_i, b)\}$ is the union of $\leq k$ convex segments. If $\{a_1, \dots, a_n\}$ is an indiscernible set, i.e. the type of $(a_{\sigma(1)}, \dots, a_{\sigma(n)})$ does not depend on $\sigma \in \mathrm{Sym}(n)$, it follows that $\{i : \phi(a_i, b)\}$ has size $\leq k$, or else the complement has size $\leq k$.

DEFINITION 2.6.7. If T is a NIP-theory, and p is an $\mathrm{Aut}(\mathbb{U}/C)$-invariant type over $\mathbb{U}$, one says that p is *generically stable* over C if it is C-definable and finitely satisfiable in any model containing C (that is, for any formula $\varphi(x)$ in p and any model M containing C, there exists c in M such that $\mathbb{U} \models \varphi(c)$).

In general, when $p(x), q(y)$ are $\mathrm{Aut}(\mathbb{U}/C)$-invariant types, there exists a unique $\mathrm{Aut}(\mathbb{U}/C)$-invariant type $r(x, y)$, such that for any $C' \supset C$, $(a, b) \models r(x, y)$ if and only if $a \models p|C$ and $b \models q|C(a)$. This type is denoted $p(x) \otimes$

$q(y)$. In general $\otimes$ is associative but not necessarily symmetric. We define p^n by $p^{n+1} = p^n \otimes p$.

The following characterization of generically stable types in NIP theories from [27] will be useful:

LEMMA 2.6.8 ([27] Proposition 3.2). *Assume T has* NIP. *An* $\mathrm{Aut}(\mathbb{U}/C)$-*invariant type $p(x)$ is generically stable over C if and only if p^n is symmetric with respect to permutations of the variables $x_1, \ldots, x_n$.*

For any formula $\varphi(x, y)$, there exists a natural number n such that whenever p is generically stable and $(a_1, \ldots, a_N) \models p^N|C$ with $N > 2n$, for every c in $\mathbb{U}$, $\varphi(x, c) \in p$ if and only if $\mathbb{U} \models \bigvee_{i_0 < \cdots < i_n} \varphi(a_{i_0}, c) \wedge \cdots \wedge \varphi(a_{i_n}, c)$.

The second part of the lemma is an easy consequence of the definition of a NIP formula, or rather the remark on indiscernible sets just below the definition.

We remark that Proposition 2.6.6 also follows from the characterization of generically stable definable types in NIP theories as those with symmetric tensor powers in Lemma 2.6.8, cf. [27].

We also recall the notion of a *strongly stably dominated type* from [23]. These are the stably dominated types that are dominated within a single formula, rather than a type. The distinction is analogous to that between ω-stability and stability, or regular and strongly regular types in stability theory.

DEFINITION 2.6.9. Say $\mathrm{tp}(a/C)$ is *strongly stably dominated* if there exists $\phi(x) \in \mathrm{tp}(a/\mathrm{St}_C(a))$ such that for any tuple b with $\mathrm{St}_C(a) \downarrow_{\mathrm{St}_C(C)} \mathrm{St}_C(b)$, ϕ implies $\mathrm{tp}(a/\mathrm{St}_C(a)b)$. If h is a C-definable function and $h(a)$ are the parameters of ϕ, we say $\mathrm{tp}(a/C)$ is strongly stably dominated via ϕ and h.

We say that a definable type p is *strongly stably dominated* if for some $A = \mathrm{acl}(A)$ such that p is A-definable, $p|A$ is strongly stably dominated. Note that it follows that p is stably dominated.

REMARK 2.6.10. Assume $\mathrm{tp}(a/C)$ is stably dominated. Then $\mathrm{tp}(a/C)$ is strongly stably dominated iff $\mathrm{tp}(a/C')$ is isolated for some (or any) C' with $\mathrm{St}_C(a) \subset C' \subset \mathrm{St}_C$. Indeed, by Remark 2.6.3, $\mathrm{tp}(a/\mathrm{St}_C(a)) \vdash \mathrm{tp}(a/C')$ for any C' with $\mathrm{St}_C(a) \subset C' \subset \mathrm{St}_C$.

For part (3) of the following proposition, we will need a special hypothesis (we refer to the beginning of 2.8 for the notion of internality):

(Sp) There exists a sort (or union of sorts) $\mathbf{k}$ such that $\mathbf{k}$ is ω-stable, and for any A, St_A is $\mathbf{k}$-internal, i.e. any A-definable stable and stably embedded is $\mathbf{k}$-internal.

Note that (Sp) holds in ACVF with $\mathbf{k}$ the residue field sort. This follows from Lemma 2.7.4 since, as recalled in Proposition 2.7.1, in this case the induced structure on $\mathbf{k}$ is that of a pure algebraically closed field, which is ω-stable.

LEMMA 2.6.11. *Assume* (Sp) *holds. Then, for any A, any $c \in \mathrm{St}_A$, and any A-definable set D containing* $\mathbf{k}$, *or union of such sets,* $\mathrm{tp}(c/A \cup D)$ *is isolated.*

Proof. To see this let P be the solution set of $\mathrm{tp}(c/A \cup D)$. Then P is $\mathbf{k}$-internal, so the automorphism group $G = \mathrm{Aut}(P)$ is an ∞-definable group internal to $\mathbf{k}$ by Theorem B.1′ (1) in [21]. Since $\mathbf{k}$ is ω-stable, it follows then from Corollaire 5.19 in [37] that G is definable. Thus P, being a G-orbit, is a definable set. □

PROPOSITION 2.6.12. *Let p be a strongly stably dominated definable type, based on $A = \mathrm{acl}(A)$.*

(1) *$p|A$ is strongly stably dominated.*

(2) *If f is a definable function such that p belongs to its domain, then $f_*(p)$ is strongly stably dominated.*

(3) *Assume* (Sp) *holds. Let $b \models p|A$, and let* $\mathrm{tp}(c/\mathrm{acl}(Ab))$ *be strongly stably dominated. Then so is* $\mathrm{tp}(c/A)$.

Proof. (1) If p is based on A and on $A' = \mathrm{acl}(A')$, we have to show that $p|A$ is strongly stably dominated iff $p|A'$ is strongly stably dominated. We may assume here that $A \subset A'$.

Let us show $p|A'$ is strongly stably dominated, assuming the same for $p|A$. Let $a \models p|A'$. Now $p|A = \mathrm{tp}(a/A)$ is strongly stably dominated, say via $\phi(x,e)$ with $e \in \mathrm{St}_A(a)$; $e = h(a)$ for some A-definable function h. Now if b is such that $e \downarrow_{\mathrm{St}_{A'}(A')} \mathrm{St}_{A'}(b)$, we have to show that $\phi(x,e)$ implies $\mathrm{tp}(a/A'be)$. Since $e \models h_*p|A'$, and $e \downarrow_{\mathrm{St}_{A'}(A')} \mathrm{St}_{A'}(b)$, we have $e \models h_*p|A'b$. In particular, $e \models h_*p|\,\mathrm{St}_A(A'b)$ so $e \downarrow_{\mathrm{St}_A(A)} \mathrm{St}_A(A'b)$, i.e. $e \downarrow_{\mathrm{St}_A(A)} \mathrm{St}_A(A'b)$. By stable domination, it follows that $\mathrm{tp}(a/Ae)$ implies $\mathrm{tp}(a/A'be)$; but $\phi(x,e)$ implies $\mathrm{tp}(a/Ae)$, so $\phi(x,e)$ implies $\mathrm{tp}(a/A'be)$.

Assume now that $p|A'$ is strongly stably dominated. We have to show that $\mathrm{tp}(a/E)$ is isolated where $E = \mathrm{St}_A(a)$. Let $E' = \mathrm{dcl}(A' \cup E)$, so $\mathrm{St}_{A'}(a) \subset E'$ by [20] 6.10 (iv). Then $\mathrm{tp}(a/E')$ is isolated, i.e. $\mathrm{tp}(a/EA')$ is isolated, say by $\psi(x,e,a')$. But $\mathrm{tp}(a/E)$ implies $\mathrm{tp}(a/Ea')$. So some $\theta(x,e) \in \mathrm{tp}(a/E)$ implies $\psi(x,e,a')$, and this $\theta(x,e)$ clearly isolates $\mathrm{tp}(a/E)$.

(2) Say p and f are defined over A. Let $c \models p|A$. Then $\mathrm{tp}(c/\mathrm{St}_A(c))$ is isolated, so $\mathrm{tp}(f(c)/\mathrm{St}_A(c))$ is isolated. Since $\mathrm{St}_A(f(c)) \subset \mathrm{St}_A(c) \subset \mathrm{St}_A$, and $\mathrm{tp}(f(c)/\mathrm{St}_A(f(c)) \vdash \mathrm{tp}(f(c)/\mathrm{St}_A))$, as noted above it follows that $\mathrm{tp}(f(c)/A)$ is strongly stably dominated.

(3) We have $\mathrm{tp}(c/\mathrm{St}_{\mathrm{acl}(Ab)})$ isolated by some formula over $Ab'c'$, where $b' \in \mathrm{acl}(Ab)$ and $c' \in \mathrm{St}_{Abb'}(c)$. In particular $\mathrm{tp}(c/bb'c'\,\mathrm{St}_A)$ is isolated. By Lemma 2.6.11 $\mathrm{tp}(c'/\mathrm{St}_{Abb'})$ is also isolated. By transitivity of stable domination, Proposition 2.6.6, $\mathrm{tp}(bb'c/A)$ is stably dominated. Moreover

tp($bb'c/\operatorname{St}_A$) is isolated, since tp(b/St_A), tp($b'/b,\operatorname{St}_A$), tp($c'/\operatorname{St}_{Abb'}$) and tp($c/bb'c'\operatorname{St}_A$) are all isolated. By Remark 2.6.10, tp($bb'c/A$) is strongly stably dominated, and by (2) so is tp(c/A). □

2.7 REVIEW OF ACVF

A valued field consists of a field K together with a homomorphism v from the multiplicative group to an ordered abelian group Γ, such that $v(x+y) \geq \min(v(x), v(y))$, for every x and y in $K^{\times}$. In this paper we shall write the law on Γ additively. We shall write Γ_∞ for Γ with an element ∞ added with usual conventions, namely ∞ is larger than any element of Γ and is absorbing for the addition. In particular we extend v to $K \to \Gamma_\infty$ by setting $v(0) = \infty$. We denote by $\mathcal{O}$ or by R the valuation ring, by $\mathcal{M}$ the maximal ideal and by k the residue field.

Now assume K is algebraically closed and v is surjective. The value group Γ is then divisible and the residue field k is algebraically closed. We shall denote by ACVF of algebraically closed valued fields with nontrivial valuation. By a classical result of A. Robinson, the completions of ACVF are the theories ACVF_{p_1,p_2} of nontrivially valued algebraically closed fields of characteristic p_1 and residue characteristic p_2. Several quantifier elimination results hold for ACVF. In particular ACVF admits quantifier elimination in the 3-sorted language $\mathcal{L}_{k,\Gamma}$, with sorts VF, Γ and $\mathbf{k}$ for the valued field, value group and residue field sorts, with respectively the ring, ordered abelian group and ring language, and additional symbols for the valuation v and the map $\operatorname{Res} : \mathrm{VF}^2 \to k$ sending (x, y) to the residue of xy^{-1} if $v(x) \geq v(y)$ and $y \neq 0$ and to 0 otherwise (cf. [19] Theorem 2.1.1). Sometimes we shall also write val instead of v for the valuation. In this paper we shall use the extension $\mathcal{L}_{\mathcal{G}}$ of $\mathcal{L}_{k,\Gamma}$ considered in section 3.1 of [19] for which elimination of imaginaries holds. In addition to sorts VF, Γ and $\mathbf{k}$, there are *geometric sorts* S_n and T_n, $n \geq 1$. The sort S_n is the collection of all codes for free rank n R-submodules of K^n. For $s \in S_n$, we denote by red(s) the reduction modulo the maximal ideal of the lattice $\Lambda(s)$ coded by s. This has $\varnothing$-definably the structure of a rank n k-vector space. We denote by T_n the set of codes for elements in $\cup_{s\in S_n}\{\operatorname{red}(s)\}$. Thus an element of T_n is a code for the coset of some element of $\Lambda(s)$ modulo $\mathcal{M}\Lambda(s)$. For each $n \geq 1$, we have symbols τ_n for the functions $\tau_n : T_n \to S_n$ defined by $\tau_n(t) = s$ if t codes an element of red(s). We shall set $\mathcal{S} = \cup_{n\geq 1} S_n$ and $\mathcal{T} = \cup_{n \geq 1} T_n$. The main result of [19] is that ACVF admits elimination of imaginaries in $\mathcal{L}_{\mathcal{G}}$. It is also proved in [19] that ACVF admits elimination of quantifiers in $\mathcal{L}_{\mathcal{G}}$.

With our conventions, if $C \subset \mathbb{U}$, we write $\Gamma(C)$ for $\operatorname{dcl}(C) \cap \Gamma$ and $\mathbf{k}(C)$ for $\operatorname{dcl}(C) \cap \mathbf{k}$. If K is a subfield of $\mathbb{U}$, one denotes by Γ_K the value group, thus $\Gamma(K) = \mathbb{Q} \otimes \Gamma_K$. If the valuation induced on K is nontrivial, then the model theoretic algebraic closure acl(K) is a model of ACVF. In particular the structure $\Gamma(K)$ has definable choice, hence is Skolemized, being an expansion by constants of a divisible ordered abelian group (cf. Proposition 2.7.1).

We shall denote in the same way a finite cartesian product of sorts and the corresponding definable set. For instance, we shall denote by Γ the definable set which to any model K of ACVF assigns $\Gamma(K)$ and by $\mathbf{k}$ the definable set which to K assigns its residue field. We shall also sometimes write K for the sort VF.

For a field F, we denote by F^{alg} an algebraic closure of F. By an algebraic variety over a field F (or variety for short), we shall always mean a reduced and separated scheme of finite type over F.

The following follows from the different versions of quantifier elimination (cf. [19] Proposition 2.1.3):

PROPOSITION 2.7.1.

(1) *The definable set Γ is o-minimal in the sense that every definable subset of Γ is a finite union of intervals.*

(2) *Any K-definable subset of k is finite or cofinite (uniformly in the parameters), i.e. k is strongly minimal.*

(3) *The definable set Γ is stably embedded.*

(4) *If $A \subset K$, then $\mathrm{acl}(A) \cap K$ is equal to the field algebraic closure of A in K.*

(5) *If $S \subset k$ and $\alpha \in k$ belongs to $\mathrm{acl}(S)$ in the K^{eq} sense, then α belongs to the field algebraic closure of S.*

(6) *The definable set k is stably embedded.*

In fact, Γ is endowed with the structure of a pure divisible ordered abelian group and k with the structure of a pure algebraically closed field.

LEMMA 2.7.2 ([19] Lemma 2.1.7). *Let C be an algebraically closed valued field, and let $s \in S_n(C)$, with $\Lambda = \Lambda_s \subset K^n$ the corresponding lattice. Then Λ is C-definably isomorphic to R^n, and the torsor $\mathrm{red}(s)$ is C-definably isomorphic to k^n.*

A C-definable set D is called *k-internal* if there exists a finite $F \subset \mathbb{U}$ such that $D \subset \mathrm{dcl}(k \cup F)$ (this is a special case of the more general definition given at the beginning of section 2.8).

We have the following characterizations of k-internal sets:

LEMMA 2.7.3 ([19] Lemma 2.6.2). *Let D be a C-definable set. Then the following conditions are equivalent:*

(1) *D is k-internal;*

(2) *for any* $m \geq 1$*, there is no surjective definable map from* D^m *to an infinite interval in* Γ*;*

(3) D *is finite or, up to permutation of coordinates, is contained in a finite union of sets of the form* $\mathrm{red}(s_1) \times \cdots \times \mathrm{red}(s_m) \times F$*, where* $s_1, \ldots, s_m$ *are* $\mathrm{acl}(C)$*-definable elements of* $\mathcal{S}$ *and* F *is a* C*-definable finite set of tuples from* $\mathcal{G}$*.*

For any parameter set C, let $\mathrm{VC}_{k,C}$ be the many-sorted structure whose sorts are k-vector spaces $\mathrm{red}(s)$ with s in $\mathrm{dcl}(C) \cap \mathcal{S}$. Each sort $\mathrm{red}(s)$ is endowed with a k-vector space structure. In addition, as its $\varnothing$-definable relations, $\mathrm{VC}_{k,C}$ has all C-definable relations on products of sorts.

By Proposition 3.4.11 of [19], we have:

Lemma 2.7.4 ([19] Proposition 3.4.11). *Let* D *be a* C*-definable set of* K^{eq}*. Then the following conditions are equivalent:*

(1) D *is* k*-internal;*

(2) D *is stable and stably embedded;*

(3) D *is contained in* $\mathrm{dcl}(C \cup \mathrm{VC}_{k,C})$*.*

By combining Proposition 2.7.1, Lemma 2.7.2 and Lemma 2.7.4, one sees that (over a model) the ϕ-definition of a stably dominated type factors through some function into k^n, where k is the residue field.

Corollary 2.7.5. *Let* C *be a model of* ACVF*, let* V *be a* C*-definable set and let* $a \in V$*. Assume* $p = \mathrm{tp}(a/C)$ *is a stably dominated type. Let* $\phi(x, y)$ *be a formula over* C*. Then there exists a* C*-definable map* $g : V \to k^n$ *and a formula* θ *over* C *such that, if* $g(a) \downarrow_{\mathbf{k}(C)} \mathrm{St}_C(b)$*, then* $\phi(a, b)$ *holds if and only if* $\theta(g(a), b)$*.*

The following lemma from [19] is also useful:

Lemma 2.7.6 ([19] Lemma 3.4.12). *If* $B = \mathrm{acl}(B)$*, then, for any* $\alpha \in \Gamma$*,* $\mathrm{acl}(B\alpha) = \mathrm{dcl}(B\alpha)$*.*

2.8 Γ-INTERNAL SETS

Let Q be an F-definable set. An F-definable set X is *Q-internal* if there exists $F' \supset F$, and an F'-definable surjection $h : Y \to X$, with Y an F'-definable subset of Q^n for some n. When Q is stably embedded and eliminates imaginaries, as is the case of Γ in ACVF, we can take h to be a bijection, by factoring out the kernel. If one can take $F' = F$ we say that X is *directly Q-internal.* We shall say an iso-definable subset of a pro-definable set is Q-internal if it is pro-definably isomorphic to some Q-internal set.

In the case of $Q = \Gamma$ in ACVF, we mention some equivalent conditions.

LEMMA 2.8.1. *Let X be an F-definable set. The following conditions are equivalent:*

(1) *X is Γ-internal;*

(2) *X is internal to some o-minimal definable linearly ordered set;*

(3) *X admits a definable linear ordering;*

(4) *every stably dominated type on X (over any base set) is constant (i.e. contains a formula $x = a$);*

(5) *there exists an $\mathrm{acl}(F)$-definable injective map $h : X \to \Gamma^n$ for some $n \geq 0$.*

Proof. The fact that (2) implies (3) follows easily from elimination of imaginaries in ACVF: by inspection of the geometric sorts, the only o-minimal one is Γ itself. Condition (3) clearly implies (4) by the symmetry property of generically stable types p: $p(x) \otimes p(y)$ has $x \leq y$ if and only if $y \leq x$, hence $x = y$. We now prove that (4) implies (5) using elimination of imaginaries in ACVF, and inspection of the geometric sorts. Namely, let $A = \mathrm{acl}(F)$ and let $c \in Y$. Assuming (4), let us show that $c \in \mathrm{dcl}(A \cup \Gamma)$. This reduces to the case that $\mathrm{tp}(c/A)$ is unary in the sense of section 2.3 of [19]; for if $c = (c_1, c_2)$ and the implication holds for $\mathrm{tp}(c_2/A)$ and for $\mathrm{tp}(c_1/A(c_2))$ we obtain $c_2 \in \mathrm{acl}(A, \Gamma, c_1)$; it follows that (4) holds for $\mathrm{tp}(c_1/A)$, so $c_1 \in \mathrm{dcl}(A, \gamma)$ and the result follows since $\mathrm{acl}(A, \gamma) = \mathrm{dcl}(A, \gamma)$ for $\gamma \in \Gamma^m$ by Lemma 2.7.6. So assume $\mathrm{tp}(c/A)$ is unary, i.e. it is the type of a sub-ball b of a free $\mathcal{O}$-module M. The radius γ of b is well-defined. Now $\mathrm{tp}(c/A, \gamma(b))$ is a type of balls of constant radius; if $c \notin \mathrm{acl}(A, \gamma(b))$ then there are infinitely many balls realizing this type, and their union fills out a set containing a larger closed sub-ball. In this case the generic type of the closed sub-ball induces a stably dominated type on a subset of $\mathrm{tp}(c/A, \gamma(b))$, contradicting (4). Thus $c \in \mathrm{acl}(A, \gamma(b)) = \mathrm{dcl}(A, \gamma(b))$. This provides an $\mathrm{acl}(F)$-definable surjection from a definable subset of some Γ^n to X. Using definable Skolem functions, one obtains a definable injection from X to some Γ^n.

It remains to prove that (1) implies (2) and (5) implies (1), which is clear. □

Let U and V be definable sets. A definable map $f : U \to V$ with all fibers Γ-internal is called a *Γ-internal cover.* If $f : U \to V$ is an F-definable map, such that for every $v \in V$ the fiber is $F(v)$-definably isomorphic to a definable set in Γ^n, then by compactness and stable embeddedness of Γ, U is isomorphic over V to a fiber product $V \times_{g,h} Z$, where $g : V \to Y \subset \Gamma^m$, and $Z \subset \Gamma^n$, and $h : Z \to Y$. We call such a cover *directly Γ-internal.*

Any finite cover of V is Γ-internal, and so is any directly Γ-internal cover.

LEMMA 2.8.2. *Let V be a definable set in* ACVF_F. *Then any Γ-internal cover $f : U \to V$ is isomorphic over V to a finite disjoint union of sets which are a fiber product over V of a finite cover and a directly Γ-internal cover.*

Proof. It suffices to prove this at a complete type $p = \mathrm{tp}(c/F)$ of U, since the statement will then be true (using compactness) above a (relatively) definable neighborhood of $f_*(p)$, and so (again by compactness, on V) everywhere. Let $F' = F(f(c))$. By assumption, $f^{-1}(f(c))$ is Γ-internal. So over F' there exists a finite definable set H, for $t \in H$ an $F'(t)$-definable bijection $h_t : W_t \to U$, with $W_t \subset \Gamma^n$, and $c \in \mathrm{Im}(h_t)$. We can assume H is an orbit of $G = \mathrm{Aut}(\mathrm{acl}(F')/F')$. In this case, since Γ is linearly ordered, W_t cannot depend on t, so $W_t = W$. Similarly let $G_c = \mathrm{Aut}(\mathrm{acl}(F)(c)/F(c)) \leq G$. Then the element $h_t^{-1}(c)$ of W depends only on the G_c-orbit of h_t. Let H_c be such an orbit (defined over $F(c)$), and set $h^{-1}(c) = h_t^{-1}(c)$ for t in this orbit and some $h \in H_c$. Then H_c has a canonical code $g_1(c)$, and we have $g_1(c) \in \mathrm{acl}(F(f(c)))$, and $c \in \mathrm{dcl}(F(f(c), g_1(c), h^{-1}(c)))$. Let $g(c) = (f(c), g_1(c))$. Then $\mathrm{tp}(g(c)/F)$ is naturally a finite cover of $\mathrm{tp}(f(c)/F)$, and $\mathrm{tp}(f(c), h^{-1}(c)/F)$ is a directly Γ-internal cover. □

We write VF^* for VF^n when we do not need to specify n (similarly for $\mathrm{VF}^* \times \Gamma^*$).

LEMMA 2.8.3. *Let F be a definably closed substructure of $\mathrm{VF}^* \times \Gamma^*$, let $B \subset \mathrm{VF}^m$ be ACVF_F-definable, and let B' be a definable set in any sorts (including possibly imaginaries). Let $g : B' \to B$ be a definable map with finite fibers. Then there exists a definable $B'' \subset \mathrm{VF}^{m+\ell}$ and a definable bijection $B' \to B''$ over B.*

Proof. By compactness, working over $F(b)$ for $b \in B$, this reduces to the case that B is a point. So B' is a finite ACVF_F-definable set, and we must show that B' is definably isomorphic to a subset of VF^ℓ. Now we can write $F = F_0(\gamma)$ for some $\gamma \in \Gamma^*$ with $F_0 = F \cap \mathrm{VF}$. By Lemma 2.7.6, $\mathrm{acl}(F) = \mathrm{dcl}(\mathrm{acl}(F_0)(\gamma))$. So $B' = \{f(\gamma) : f \in B''\}$ where B'' is some finite F_0-definable set of functions on Γ. Replacing F by F_0 and B' by B'', we may assume F is a field.

CLAIM. $\mathrm{acl}(F) = \mathrm{dcl}(F^{\mathrm{alg}})$.

Proof of the claim. This is clear if F is not trivially valued since then F^{alg} is an elementary substructure of $\mathbb{U}$.

When F is trivially valued, suppose $e \in \mathrm{acl}(F)$; we wish to show that $e \in \mathrm{dcl}(F^{\mathrm{alg}})$; we may assume $F = F^{\mathrm{alg}}$. The easiest proof is by inspection of the geometric imaginaries: the only F-algebraic sublattice of K^n is $\mathcal{O}^n$, and the elements of the sort T_n above it are indexed by k^n. (Here is a sketch of a more direct proof: let t and t' be elements with $0 < \mathrm{val}(t) \ll \mathrm{val}(t')$. Then

$e \in \mathrm{dcl}(F(t)^{\mathrm{alg}})$ and $e \in \mathrm{dcl}(F(t')^{\mathrm{alg}})$ by the nontrivially valued case. But by the stationarity lemma ([20] 8.11), $\mathrm{tp}((e,t)/F) \cup \mathrm{tp}((e,t')/F)$ generates $\mathrm{tp}((e,t),(e,t')/F)$, forcing $e \in \mathrm{dcl}(F)$.) □

Now we have $B' \subset \mathrm{acl}(F) = \mathrm{dcl}(F^{\mathrm{alg}})$. Using induction on $|B'|$ we may assume B' is irreducible, and also admits no nonconstant ACVF_F-definable map to a smaller definable set. If B' admits a nonconstant definable map into VF then it must be 1-1 and we are done. Let $b \in B'$ and let $F' = \mathrm{Fix}(\mathrm{Aut}(F^{\mathrm{alg}}/F(b)))$. Then F' is a field, and if $d \in F' \setminus F$, then $d = h(b)$ for some definable map h, which must be nonconstant since $d \notin F$. If $F' = F$ then by Galois theory, $b \in \mathrm{dcl}(F)$, so again the statement is clear. □

Note that the last part of the argument is valid in any expansion of the theory of fields: if C is definably closed and $F \subset C \subset \mathrm{dcl}(F')$, with F' an algebraic extension of F, then $C = \mathrm{dcl}(C \cap F')$.

COROLLARY 2.8.4. *The composition of two definable maps with Γ-internal fibers also has Γ-internal fibers. In particular if f has finite fibers and g has Γ-internal fibers then $g \circ f$ and $f \circ g$ have Γ-internal fibers.*

Proof. As pointed out by a referee this follows from characterization (4) in Lemma 2.8.1, which is clearly closed under towers. Let us also give a direct proof. We may work over a model A. By Lemma 2.8.2 and the definition, the class of Γ-internal covers is the same as compositions $g \circ f$ of definable maps f with finite fibers, and g with directly Γ-internal covers. Hence to show that this class is closed under composition it suffices to show that if f has finite fibers and g has directly Γ-internal covers, then $f \circ g$ has Γ-internal fibers; in other words that if $b \in \mathrm{acl}(A(a))$, $a \in \mathrm{dcl}(A \cup \{\gamma\})$ with γ a tuple from Γ, then $(a,b) \in \mathrm{dcl}(A \cup \Gamma)$. But $\mathrm{acl}(A,\gamma) = \mathrm{dcl}(A,\gamma)$ for $\gamma \in \Gamma^m$ by Lemma 2.7.6, so $(a,b) \in \mathrm{dcl}(A \cup \Gamma)$. □

WARNING 2.8.5. The corollary applies to definable maps between definable sets, hence also to iso-definable sets. However if $f : X \to Y$ is a map between pro-definable sets and U is a Γ-internal, iso-definable subset of Y, we do not know if $f^{-1}(U)$ must be Γ-internal, even if f is ≤ 2-to-one.

2.9 ORTHOGONALITY TO Γ

Let A be a substructure of $\mathbb{U}$.

PROPOSITION 2.9.1. (a) *Let p be an A-definable type. The following conditions are equivalent:*

(1) *p is stably dominated;*

(2) *p is orthogonal to Γ;*

(3) *p is generically stable.*

(b) *A type p over A extends to at most one generically stable A-definable type.*

Proof. The equivalence of (1) and (2) follows from [20] 10.7 and 10.8. Using Proposition 10.16 in [20], and [27], Proposition 3.2(v), we see that (2) implies (3). (In fact (1) implies (3) is easily seen to be true in any theory, in a similar way.) To see that (3) implies (2) (again in any theory), note that if p is generically stable and f is a definable function, then f_*p is generically stable (by any of the criteria of [27] 3.2, say the symmetry of indiscernibles). Now a generically stable definable type on a linearly ordered set must concentrate on a single point: a two-element Morley sequence (a_1, a_2) based on p will otherwise consist of distinct elements, so either $a_1 < a_2$ or $a_1 > a_2$, neither of which can be an indiscernible set. The statement on unique extensions follows from [27], Proposition 3.2(v). □

We shall use the following statement, Theorem 12.18 from [20]:

THEOREM 2.9.2.

(1) *Suppose that $C \leq L$ are valued fields with C maximally complete, $\mathbf{k}(L)$ is a regular extension of $\mathbf{k}(C)$ and Γ_L/Γ_C is torsion free. Let a be a sequence in $\mathbb{U}$, $a \in \mathrm{dcl}(L)$. Then $\mathrm{tp}(a/C \cup \Gamma(Ca))$ is stably dominated.*

(2) *Let C be a maximally complete algebraically closed valued field, and a be a sequence in $\mathbb{U}$. Then $\mathrm{tp}(\mathrm{acl}(Ca)/C \cup \Gamma(Ca))$ is stably dominated.*

We use this especially when C is algebraically closed, so that the conditions on regularity and torsion-freeness are redundant.

In particular, if $C = \mathrm{acl}(C)$ and $\Gamma(C) = \mathbb{R}$, every type of elements of Γ over M is definable, so every type over C is definable. This is relevant to Berkovich spaces. We note another instance of this, when the value group is extended only by infinite or infinitesimal elements.

LEMMA 2.9.3. *Let A be a divisible abelian group. Let B be an extension of A containing no proper extension of A in which A is order-dense. Then every type realized in B over A is definable.*

Proof. Indeed, let B be a finitely generated extension of A. We show that B/A is definable by induction on $\mathrm{rk}(B/A)$. If there are any positive elements $b \in B$ with $b < a$ for any $0 < a \in A$, one can find such a b with smallest archimedean class; so any element b' of B with $0 < b' < b$ has the form αb, $\alpha \in \mathbb{Q}$. Let $B' = \{b' \in B : b \ll |b'|\}$. Let $B'' = \{b'' \in B : (\exists n \in \mathbb{N})(|b''| < nb)\}$. Then $B \cong B'' \oplus B'$, by induction B'/A is definable, and as B''/B' is definable by inspection, the result follows. Similarly, though slightly less canonically, if there are any $b \in B$ with $b > A$, find such a b with maximal archimedean class. Pick a maximal set of $\mathbb{Q}$-linearly independent elements b_i in the same archimedean class as b. Let $B' = \{b' \in B : |b'| \ll b\}$. Then again $B = B' \oplus \oplus \mathbb{Q} b_i$, $\mathrm{tp}(b_1, \dots, b_m)/B'$ is definable, and the result follows. Finally, if there are no infinitesimal nor any infinite elements in B over A, then by assumption we have $A = B$, and certainly B/A is definable. □

2.10 $\widehat{V}$ FOR STABLE DEFINABLE V

We end with a description of the set $\widehat{V}$ of definable types concentrating on a stable definable V, as an ind-definable set. The notation $\widehat{V}$ is compatible with the one that will be introduced in greater generality in 3.1. Such a representation will not be possible for algebraic varieties V in ACVF and so the picture here is not at all suggestive of the case that will mainly interest us, but it is simpler and will be lightly used at one point.

A family X_a of definable sets is said to be *uniformly definable* in the parameter a if there exists a definable X such that $X_a = \{x : (a, x) \in X\}$. An ind-definable set X_a depending on a parameter a is said to be uniformly definable in a if it can be presented as the direct limit of a system $X_{a,i}$, with each $X_{a,i}$ and the morphisms $X_{a,i} \to X_{a,j}$ definable uniformly in a. If U is a definable set, and $X_u = \varinjlim_i X_{u;i}$ is (strict) ind-definable uniformly in u, then the disjoint union of the X_u is clearly (strict) ind-definable too.

Recall $\mathbf{k}$ denotes the residue field sort. Given a Zariski closed subset $W \subset \mathbf{k}^n$, define $\deg(W)$ to be the degree of the Zariski closure of W in projective n-space. Let $\mathrm{ZC}_d(\mathbf{k}^n)$ be the family of Zariski closed subsets of degree $\leq d$ and let $\mathrm{IZC}_d(\mathbf{k}^n)$ be the sub-family of absolutely irreducible varieties. It is well-known that $\mathrm{IZC}_d(\mathbf{k}^n)$ is definable (cf., for instance, chapter17 of [17]). These families are invariant under $\mathrm{GL}_n(\mathbf{k})$, hence for any definable $\mathbf{k}$-vector space V of dimension n, we may consider their pullbacks $\mathrm{ZC}_d(V)$ and $\mathrm{IZC}_d(V)$ to families of subsets of V, under a $\mathbf{k}$-linear isomorphism $V \to \mathbf{k}^n$. Then $\mathrm{ZC}_d(V)$ and $\mathrm{IZC}_d(V)$ are definable, uniformly in any definition of V.

LEMMA 2.10.1. *If V is a finite-dimensional $\mathbf{k}$-space, then $\widehat{V}$ is strict ind-definable.*

The disjoint union D_{st} of the $\widehat{V_\Lambda}$ with $V_\Lambda = \Lambda/\mathcal{M}\Lambda$ and where Λ ranges over the definable family S_n of lattices in K^n is also strict ind-definable.

Proof. Since $\widehat{V}$ can be identified with the limit over all d of $\mathrm{IZC}_d(V)$, it is strict ind-definable uniformly in V. The family of lattices Λ in K^n is a definable family, so the disjoint union of $\widehat{V_\Lambda}$ over all such Λ is strict ind-definable. □

If K is a valued field, we set $\mathrm{RV} = K^\times/1 + \mathcal{M}$ and denote by rv the canonical morphism $K^\times \to \mathrm{RV}$. So we have an exact sequence of abelian groups $0 \to k^\times \to \mathrm{RV} \to \Gamma \to 0$. For $\gamma \in \Gamma$, denote by $V_\gamma^\times$ the preimage of γ in RV. It is a principal homogeneous space for $k^\times$. It becomes a k-vector space V_γ after adding an element 0 and defining addition in the obvious way. For $m \geq 0$, we denote by $\widehat{\mathrm{RV}^m}$ the set of stably dominated types on RV^m.

LEMMA 2.10.2. *For $m \geq 0$, $\widehat{\mathrm{RV}^m}$ is strict ind-definable.*

Proof. Note that RV is the union over $\gamma \in \Gamma$ of the principal homogeneous spaces $V_\gamma^\times$. For $\overline{\gamma} = (\gamma_1, \dots, \gamma_n) \in \Gamma^n$, let $V_{\overline{\gamma}} = \Pi_{i=1}^n V_{\gamma_i}$. Since the image of a stably dominated type on RV^m under the morphism $\mathrm{RV}^m \to \Gamma^m$ is constant, any stably dominated type must concentrate on a finite product $V_{\overline{\gamma}}$. Thus it suffices to show, uniformly in $\overline{\gamma} \in \Gamma^n$, that $\widehat{V_{\overline{\gamma}}}$ is strict ind-definable. Indeed $\widehat{V_{\overline{\gamma}}}$ can be identified with the limit over all d of $\mathrm{IZC}_d(V_{\overline{\gamma}})$. □

REMARK 2.10.3. By the above proof, the function dim on $\mathrm{IZC}_d(V_{\overline{\gamma}})$ induces a constructible function on $\widehat{\mathrm{RV}^m}$, that is, having definable fibers on each definable piece of $\widehat{\mathrm{RV}^m}$.

2.11 DECOMPOSITION OF DEFINABLE TYPES

We seek to understand a definable type in terms of a definable type q on Γ^n, and the germ of a definable map from q to stably dominated types.

Let us start by recalling the notion of an A-definable germ, cf. Definition 6.1 in [20]. Let p be an A-definable type on some A-definable set X. Let $\varphi(x, y, b)$ be a formula defining a function $f_b(x)$ whose domain contains all realizations of p. The *germ of f_b on p*, or *p-germ of f_b*, is the equivalence class of b under the equivalence relation $\sim$, where $b \sim b'$ if the formula $f_b(x) = f_{b'}(x)$ is in p. Equivalently, $b \sim b'$ if and only if for any $a \models p|Abb'$, $f_b(a) = f_{b'}(a)$. As p is A-definable, $\sim$ is also A-definable, and the germ of f_b on p is a definable object.

Now assume $Y = \varprojlim_i Y_i$ is an A-pro-definable set and let h and h' be two pro-definable maps over $B \supset A$ taking values in Y whose domain contains all realizations of p. We say h and h' have the same *p-germ* if $h(e) = h'(e)$ when $e \models p|B$. The p-germ of h is the equivalence class of h. Thus, h and h' have the same p-germ if and only if for every i the maps h_i and h'_i given by composing h and h' with the projection to Y_i have the same p-germ; and the p-germ of h is determined by the sequence of p-germs of the h_i.

Let p be an A-definable type. Define $\mathrm{rk}_\Gamma(p) = \mathrm{rk}_\mathbb{Q}\Gamma(M(c))/\Gamma(M)$, where $A \leq M \models \mathrm{ACVF}$ and $c \models p|M$. Since p is definable, this rank does not depend on the choice of M, but for the present discussion it suffices to take M somewhat saturated, to make it easy to see that $\mathrm{rk}_\Gamma(p)$ is well-defined.

If p has rank r, then there exists a definable function to Γ^r whose image is not contained in a smaller dimensional set. We show first that at least the germ of such a function can be chosen A-definable.

LEMMA 2.11.1. *Let p be an A-definable type and set $r = \mathrm{rk}_\Gamma(p)$. Then there exists a nonempty A-definable set Q'' and for $b \in Q''$ a function*

$\gamma_b = ((\gamma_b)_1, \ldots, (\gamma_b)_r)$ from a definable set containing p into Γ^r, definable uniformly in b, such that

(1) *If $b \in Q''$ and $c \models p|A(b)$ then the image of $\gamma_b(c)$ in $\Gamma(A(b,c))/\Gamma(A(b))$ is a $\mathbb{Q}$-linearly independent r-tuple;*

(2) *if $b, b' \in Q''$ and $c \models p|A(b,b')$ then $\gamma_b(c) = \gamma_{b'}(c)$.*

Proof. Take M sufficiently saturated and consider an M-definable function $\gamma = (\gamma_1, \ldots, \gamma_r)$ into Γ^r, such that if $c \models p|M$ then $\gamma_1(c), \ldots, \gamma_r(c)$ have $\mathbb{Q}$-linearly independent images in $\Gamma(M(c))/\Gamma(M)$. Say $\gamma = \gamma_a$ is defined over $A(a)$ with a a finite tuple and let $Q = \mathrm{tp}(a/A)$. If $b \in Q$ there exist a unique $N(a,b) \in \mathrm{GL}_r(\mathbb{Q})$ and $\gamma' = \gamma'(a,b) \in \Gamma(M)^r$ such that for $c \models p|M(b)$, $\gamma_b(c) = N(a,b)\gamma_a(c) + \gamma'$. By compactness and because p is A-definable, as b varies the matrices $N(a,b)$ vary among a finite number of possibilities $N_1, \ldots, N_k$ and there exist finitely many A-definable functions $\gamma'_i : Q \times Q \to \Gamma^r$ such that whenever $a', b \in Q$, then for some i, and for any $c \models p|M(a',b)$, $\gamma_{a'}(c) = N_i\gamma_b(c) + \gamma'_i(a',b)$. By compactness again, the same is true for some A-definable set Q' containing Q.

Consider the A-definable equivalence relation E on Q' defined by $b'Eb$ if $(d_p x)(\gamma_{b'}(x) = \gamma_b(x))$. Then by the above discussion, $Q'/E \subset \mathrm{dcl}(A(a), \Gamma)$ (in particular Q'/E is Γ-internal, cf. 2.8). By Lemma 2.8.1 it follows that $Q'/E \subset \mathrm{acl}(A, \Gamma)$, and there exists a definable map $g : Q'/E \to \Gamma^\ell$ with finite fibers.

We can consider the following partial orderings on Q': $b' \leq_i b$ if and only if $(d_p x)((\gamma_{b'})_i(x) \leq (\gamma_b)_i(x))$. These induce partial orderings on Q'/E, such that if $x \neq y$ then $x <_i y$ or $y <_i x$ for some i. This permits a choice of an element from any given finite subset of Q'/E; thus the map g admits a definable section e.

It follows in particular there exists a nonempty A-definable subset $Y \subset \Gamma^\ell$ and for $y \in Y$ an element $e(y) \in Q'/E$. If Y has an A-definable element then there exists an A-definable E-class in Q'/E; let Q'' be this class. This is always the case unless $\Gamma(A) = (0)$, and $Y \subset (0)^{\ell_1} \times (0,\infty)^{\ell_2} \times (-\infty, 0)^{\ell_3}$, with $\ell_2 + \ell_3 > 0$; but we give another argument that works in general.

For $y \in Y$ we have a p-germ of a function $\gamma[y]$ into Γ^r, and the germs of $y, y' \in Y$ differ by an element $(M(y,y'), d(y,y'))$ of $\mathrm{GL}_r(\mathbb{Q}) \ltimes \Gamma^r$. It is easy to cut down Y so that $M(y,y') = 1$ for all y, y'. Indeed, let q be any definable type on Y; then for some $M_0 \in \mathrm{GL}_r(\mathbb{Q})$, for $y \models q$ and $y' \models q|y$ we have $M(y,y') = M_0$. It follows that $M_0^2 = M_0$ so that $M_0 = 1$, hence we may impose that $(d_q y')(M(y,y') = 1)$ holds on Y. Now $d(y,y'') = d(y,y') + d(y',y'')$. Pick $a \in Y$, and let $d_a(y) = d(y,a)$; then we have $d(y,y') = d_a(y) - d_a(y')$. Let $\gamma_0 \in \Gamma_\infty$ be some A-definable limit point of Y. (Such a point exists by induction on dimension; consider the boundary.) Then d_a has a finite number of limit values at γ_0, being piecewise linear; let c_a be the smallest of them. So $d'_a = d_a - c_a$ still satisfies $d(y,y') = d'_a(y) - d'_a(y')$, and now 0 is a limiting value of d'_a. Any conjugate $d'_{a'}$ of d'_a differs from d'_a by a constant, and only finitely many constants are possible (since both functions have 0 as a limit value at γ_0). Thus d'_a has only finitely many conjugates, so it is acl(A)-definable; as above it follows that it is A-definable. Set $d' = d'_a$ and replace each germ $\gamma[y]$ by $\gamma[y] - d'(y)$. The result is another family of germs with $M(y, y') = 1$ and $d(y, y') = 0$. This means that the germ does not depend on the choice of $y \in Y$. □

LEMMA 2.11.2. *Let p be an A-definable type on some A-definable set V and set $r = \mathrm{rk}_\Gamma(p)$. There exists an A-definable germ of maps $\delta : p \to \Gamma^r$ of maximal rank. Furthermore for any such δ the definable type $\delta_*(p)$ is A-definable.*

Proof. The existence of the germ δ follows from Lemma 2.11.1. It is clear that any two such germs differ by composition with an element of $\mathrm{GL}_r(\mathbb{Q}) \ltimes \Gamma(A)^r$. So, if one fixes such a germ, it is represented by any element of the A-definable family $(\gamma_a : (a \in Q''))$ in Lemma 2.11.1. The definable type $\delta_*(p)$ on Γ^r does not depend on the choice of δ within this family, hence $\delta_*(p)$ is an A-definable type. □

In the remainder of this section, we will use the notation $\widehat{V}$ for the space of stably dominated types on V, for V an A-definable set, introduced in 3.1. In Theorem 3.1.1 we prove that $\widehat{V}$ can be canonically identified with a strict A-pro-definable set. More generally, if $V = \varprojlim_i V_i$ is an A-pro-definable set, we denote by $\widehat{V}$ the set of stably dominated types on V. Note that $\widehat{V}$ is canonically isomorphic to $\varprojlim_i \widehat{V_i}$, hence is A-pro-definable.

LEMMA-DEFINITION 2.11.3. *Let V and W be A-definable sets, or A-pro-definable sets. If q is an A-definable type on V and $h : V \to \widehat{W}$ is an A-pro-definable map, there exists a unique A-definable type r on W such that for any model M containing A, if $e \models q|M$ and $b \models h(e)|Me$ then $b \models r|M$. We refer to r as the integral $\int_q h$ of h along q. As by definition r depends only on the q-germ $\underline{h}$ of h, we set $\int_q \underline{h} := \int_q h$.*

Note that for h as above, if the q-germ $\underline{h}$ is A-definable (equivalently $\mathrm{Aut}(\mathbb{U}/A)$-invariant), then so is r; again the definition of r depends only on $\underline{h}$ hence if $\underline{h}$ is $\mathrm{Aut}(\mathbb{U}/A)$-invariant then so is r (even if h is not).

REMARK 2.11.4. One can consider the space $\widehat{\widehat{V}}$ of stably dominated types on the strict pro definable set $\widehat{V}$, for V a definable set. There is a canonical pro-definable map $\vartheta : \widehat{\widehat{V}} \to \widehat{V}$ sending a stably dominated type q on $\widehat{V}$ to $\vartheta(q) = \int_q \mathrm{id}_{\widehat{V}}$. So $\vartheta(q)$ is a definable type, and by Proposition 2.6.6 it is stably dominated.

The following key Theorem 2.11.5 states that any definable type may be viewed as an integral of stably dominated types along some definable type on Γ^r. The proposition states the existence of certain A-definable germs of pro-definable functions; there may be no A-definable function with this germ.

THEOREM 2.11.5. *Let p be an A-definable type on some A-definable set V and let $\delta : p \to \Gamma^r$ be as in* Lemma 2.11.2. *Let $s = \delta_* p$. There exists an A-definable s-germ $f : s \to \widehat{V}$ such that $p = \int_s f$.*

Proof. Let M be a maximally complete model, and let $c \models p|M$, $t = \delta(c)$. Then $\Gamma(M(c))$ is generated over $\Gamma(M)$ by $\delta(c)$. By [19], Corollary 3.4.3 and Theorem 3.4.4, $M(t) := \mathrm{dcl}(M \cup \{t\})$ is algebraically closed. By Theorem 2.9.2 $\mathrm{tp}(c/M(t))$ is stably dominated, hence extends to a unique element $f_M(t)$ of $\widehat{V}(M(t))$. We will show that f_M does not depend on M.

Let $M \leq N \models$ ACVF, with N large and saturated. We first show that $f_M = f_N$. Let $c \models p|N$, $t = \delta(c)$; so $t \models s|N$. We will show that the homogeneity hypotheses of Lemma 2.6.5 hold. Consider an element b of $N(t) \smallsetminus M(t)$; it has the form $h(e,t)$ with $e \in N$. Let $\bar{e}$ be the class of e modulo the definable equivalence relation: $x \sim x'$ if $(d_s t)(h(x,t) = h(x',t))$. Since b is not $M(t)$-definable, $\bar{e} \notin M$. Hence there exists $e' \in N$ with $\mathrm{tp}(e'/M) = \mathrm{tp}(e/M)$, but $e' \not\sim e$; and there exists an automorphism of N over M, taking e to e' which may be extended to an automorphism of $N(t)/M(t)$, taking b to $b' = h(e',t) \neq b$ with $\mathrm{tp}(b'/M(t)) = \mathrm{tp}(b/M(t))$. Since $\mathrm{tp}(c/N(t))$ is $\mathrm{Aut}(N(t)/M(t))$-invariant, by Lemma 2.6.5, $\mathrm{tp}(c/N(t)) = f_M(t)|N(t)$. Hence $f_M(t)|N(t) = f_N(t)|N(t)$; but as above $N(t)$ is algebraically closed, so two stably dominated types based on $N(t)$ and with the same restriction to $N(t)$ must be equal; hence $f_M(t) = f_N(t)$, so $f_M = f_N$.

Given two maximally complete fields M and M' we see by choosing N containing both that $f_M(t)|N(t) = f_{M'}(t)|N(t)$; another use of Lemma 2.6.5, this time over $N(t)$ and with $\mathbb{U}$ as the homogeneous larger model, gives $f_M(t) = f_{M'}(t)$. So $f_M(t)$ does not depend on M and can be denoted $f(t)$. We obtain a pro-definable map $f : P \to \widehat{V}$, where $P = \mathrm{tp}(t/A)$. The $\delta_*(p)$-germ of this function f does not depend on the choice of δ. It follows that the germ is $\mathrm{Aut}(\mathbb{U}/A)$-invariant, hence A-definable; and by construction we have $p = \int_{\delta_*(p)} f$. □

2.12 PSEUDO-GALOIS COVERINGS

We finally recall a notion of Galois covering at the level of points; it is essentially the notion of a Galois covering in the category of varieties in which radicial morphisms (EGA I, (3.5.4)) are viewed as invertible. Recall a morphism of schemes $V \to W$ is *radicial* if for every field K, the morphism $V(K) \to W(K)$ is injective.

Following [40] p. 52, we call a finite surjective morphism $Y \to X$ of integral separated noetherian schemes a *pseudo-Galois covering* if the field extension $F(Y)/F(X)$ is normal and the canonical group homomorphism $\mathrm{Aut}_X(Y) \to \mathrm{Gal}(F(Y), F(X))$ is an isomorphism, where $\mathrm{Gal}(F(Y), F(X))$ is by definition the group $\mathrm{Aut}_{F(X)}(F(Y))$. Injectivity follows from the irreducibility of Y and the separateness assumption.

Let V be a normal irreducible variety over a field F. Recall that by a variety over F we mean a reduced and separated scheme of finite type over F. We take normality to imply irreducibility, but sometimes will repeat the irreducibility condition. Let K' be a finite, normal field extension of $F(V)$. Then the normalization V' of V in K' is a pseudo-Galois covering since the canonical morphism $\mathrm{Aut}_V(V') \to G = \mathrm{Gal}(K', F(V))$ is an isomorphism. This is a special case of the functoriality in K' of the map taking K' to the normalization of V in K'. The action of $g \in G$ on V' may be described as follows. To g corresponds a rational map $V' \to V'$; let W_g be the graph of this map, a closed subvariety of $V' \times V'$. Each of the projections $W_g \to V'$ is birational, and finite. Since V' is normal, these projections are isomorphisms, so g is the graph of an isomorphism $V' \to V'$.

As observed in [40] p. 53, if $Y \to X$ is a pseudo-Galois covering and X is normal, for any morphism $X' \to X$ with X' an integral noetherian scheme, the Galois group $G = \mathrm{Gal}(F(Y), F(X))$ acts transitively on the components of $X' \times_X Y$. Here is a brief argument. Note that if X' is the normalization of X in a finite purely inseparable extension K' of its function field $F(X)$, the morphism $X' \to X$ is radicial. Indeed one may assume $X = \mathrm{Spec}\ A$, $X' = \mathrm{Spec}\ A'$ and the characteristic is p. For some integer h, $F(X)$ contains K'^{p^h} and an element x of K' lies in A' if and only if $x^{p^h} \in A$. It follows that the morphism $Y/G \to X$ is radicial, hence G is transitive on fibers of Y/X. So there are no proper G-invariant subvarieties of Y mapping onto X. It is clear from Galois theory that G acts transitively on the components of $X' \times_X Y$ mapping dominantly to X'; it follows that the union of these components is a $\mathrm{Gal}(F(Y), F(X))$-invariant subset mapping onto X', hence is all of $X' \times_X Y$. So there are no other components.

If Y is a finite disjoint union of nonempty integral noetherian schemes Y_i, we say a finite surjective morphism $Y \to X$ is a pseudo-Galois covering if each restriction $Y_i \to X$ is a pseudo-Galois covering. Also, if X is a finite disjoint union of nonempty integral noetherian schemes X_i, we shall say $Y \to X$ is a pseudo-Galois covering if its pullback over each X_i is a pseudo-Galois covering.

Chapter Three

The space $\widehat{V}$ of stably dominated types

Summary. The core of this chapter is the study of the space $\widehat{V}$ of stably dominated types on a definable set V. It is endowed with a canonical structure of a (strict) pro-definable set in 3.1. Some examples of stably dominated types are given in 3.2. Then, in 3.4 we endow it with the structure of a definable topological space, a notion defined in 3.3. The properties of that definable topology are discussed in 3.5. In 3.6 we study the canonical embedding of V in $\widehat{V}$ as the set of simple points. An essential feature in our approach is the existence of a canonical extension for a definable function on V to $\widehat{V}$. This is discussed in 3.8 where continuity criteria are given. They rely on the notion of v-, g-, and v+g-continuity introduced in 3.7. In 3.9 we introduce basic notions of (generalized) paths and homotopies. In the remaining 3.10-3.12 we introduce notions of use in later chapters: good metrics, Zariski topology, schematic distance.

3.1 $\widehat{V}$ AS A PRO-DEFINABLE SET

We shall now work in a big saturated model $\mathbb{U}$ of ACVF in the language $\mathcal{L}_{\mathcal{G}}$. We fix a substructure C of $\mathbb{U}$. If X is an algebraic variety defined over the valued field part of C, we can view X as embedded as a constructible in affine n-space, via some affine chart. Alternatively we could make new sorts for $\mathbb{P}^n$, and consider only quasi-projective varieties. At all events we will treat X as we treat the basic sorts. By a "definable set" we mean: a definable subset of some product of sorts (and varieties), unless otherwise specified.

For a C-definable set V, and any substructure F containing C, we denote by $\widehat{V}(F)$ the set of F-definable stably dominated types p on V (that is such that $p|F$ contains the formulas defining V).

We will now construct the fundamental object of the present work, initially as a pro-definable set. We will later define a topology on $\widehat{V}$.

We show that there exists a canonical pro-definable set E and a canonical identification $\widehat{V}(F) = E(F)$ for any F. We will later denote E as $\widehat{V}$. We call $\widehat{V}$ the *stable completion* of V. Here "stable" makes reference to the notion of stably dominated or generically stable type, and "completion" refers to the density of simple points, cf. Lemma 3.6.1.

THEOREM 3.1.1. *Let V be a C-definable set. Then there exists a canonical strict C-pro-definable set E and a canonical identification $\widehat{V}(F) = E(F)$ for* *any F. Moreover, if $f : V \to W$ is a morphism of C-definable sets, the induced map $\widehat{V} \to \widehat{W}$ is a morphism of C-pro-definable sets.*

REMARK 3.1.2. (This very formal remark can be skipped with no loss of understanding.) The canonical pro-definable set E described in the proof will be denoted as $\widehat{V}$ throughout the rest of the paper.

If one wishes to bring the choice of E out of the proof and into a formal definition, a Grothendieck-style approach can be adopted. The pro-definable structure of E determines in particular the notion of a pro-definable map $U \to E$, where U is any pro-definable set. We thus have a functor from the category of pro-definable sets to the category of sets, $U \mapsto E(U)$, where $E(U)$ is the set of (pro-)definable maps from U to $\widehat{V}$. This includes the functor $F \mapsto E(F)$ considered above: in case U is a complete type associated with an enumeration of a structure A, then $\widehat{V}(U)$ can be identified with $\widehat{V}(A)$. Now instead of describing E we can explicitly describe this functor. Then the representing object E is uniquely determined, by Yoneda's lemma, and can be called $\widehat{V}$. Yoneda's lemma also automatically yields the functoriality of the map $V \mapsto \widehat{V}$ from the category of C-definable sets to the category of C-pro-definable sets.

In the present case, any reasonable choice of pro-definable structure satisfying the theorem will be pro-definably isomorphic to the E we chose, so the more category-theoretic approach does not appear to us necessary. As usual in model theory, we will say "Z is pro-definable" to mean that Z can be canonically identified with a pro-definable set E, where no ambiguity regarding E is possible.

One more remark before beginning the proof. Suppose Z is a strict ind-definable set of pairs (x, y), and let $\pi(Z)$ be the projection of Z to the x-coordinate. If $Z = \cup Z_n$ with each Z_n definable, then $\pi(Z) = \cup \pi(Z_n)$. Hence $\pi(Z)$ is naturally represented as an ind-definable set (and is itself strict).

Proof of Theorem 3.1.1. A definable type p is stably dominated if and only if it is generically stable (Proposition 2.9.1). The definition of $\phi(x, c) \in p$ stated in Lemma 2.6.8 clearly runs over a uniformly definable family of formulas. Hence by Lemma 2.5.1, $\widehat{V}$ is pro-definable.

To show strict pro-definability, let $f : V \times W \to \Gamma_\infty$ be a definable function. Write $f_w(v) = f(v, w)$, and define $p_*(f) : W \to \Gamma_\infty$ by $p_*(f)(w) = p_*(f_w)$. Let $Y_{W,f}$ be the subset of $\mathrm{Fn}(W, \Gamma_\infty)$ consisting of all functions $p_*(f)$, for p varying in $\widehat{V}(\mathbb{U})$. By the proof of Lemma 2.5.1 it is enough to prove that $Y_{W,f}$ is definable. Since by pro-definability of $\widehat{V}$, $Y_{W,f}$ is ∞-definable, it remains to show that it is ind-definable.

Set $Y = Y_{W,f}$ and consider the set Z of quadruples (g, h, q, L) such that:

(1) $L = k^n$ is a finite-dimensional $\mathbf{k}$-vector space;

(2) $q \in \widehat{L}$;

(3) h is a definable function $V \to L$ (with parameters);

(4) $g : W \to \Gamma_\infty$ is a function satisfying: $g(w) = \gamma$ if and only if

$$(d_q\bar{v})((\exists v \in V)(h(v) = \bar{v}) \wedge (\forall v \in V)(h(v) = \bar{v} \implies f(v,w) = \gamma))$$

i.e. for $\bar{v} \models q$, $h^{-1}(\bar{v})$ is nonempty, and for any $v \in h^{-1}(\bar{v})$, $g(w) = f(v,w)$.

Let Z_1 be the projection of Z to the first coordinate. Note that Z is strict ind-definable by Lemma 2.10.1 and hence Z_1 is also strict ind-definable.

Let us prove $Y \subset Z_1$. Take p in $\widehat{V}(\mathbb{U})$, and let $g = p_*(f)$. We have to show that $g \in Z_1$. Say $p \in \widehat{V}(C')$, with C' a model of ACVF and let $a \models p|C'$. By Corollary 2.7.5 there exists a C'-definable function $h : V \to L = k^n$ and a formula θ over C' such that if $C' \subset B$ and $b, \gamma \in B$, if $h(a) \downarrow_{\mathbf{k}(C')} \mathrm{St}_{C'}(B)$, then $f(a,b) = \gamma$ if and only if $\theta(h(a), b, \gamma)$. Let $q = \mathrm{tp}(h(a)/C')$. Then (1-4) hold and (g, h, q, L) lies in Z.

Conversely, let $(g, h, q, L) \in Z$; say they are defined over some base set M; we may take M to be a maximally complete model of ACVF. Let $\bar{v} \models q|M$, and pick $v \in V$ with $h(v) = \bar{v}$. Let $\bar{\gamma}$ generate $\Gamma(M(v))$ over $\Gamma(M)$. By Theorem 2.9.2 $\mathrm{tp}(v/M(\bar{\gamma}))$ is stably dominated. Let $M' = \mathrm{acl}(M(\bar{\gamma}))$ (actually $\mathrm{dcl}(M(\bar{\gamma}))$ is algebraically closed). Let p be the unique element of $\widehat{V}(M')$ such that $p|M' = \mathrm{tp}(v/M')$. We need not have $p \in \widehat{V}(M)$, i.e. p may not be M-definable, but since $\mathbf{k}$ and Γ are orthogonal and k is stably embedded, $h_*(p)$ is M-definable. Thus $h_*(p)$ is the unique M-definable type whose restriction to M is $\mathrm{tp}(\bar{v}/M)$, i.e. $h_*(p) = q$. By definition of Z it follows that $p_*(f) = g$. Thus $Y = Z_1$ and $Y_{W,f}$ is strict ind-definable, hence C-definable.

Now let $f : V \to W$ be a morphism of C-definable sets and denote by $\widehat{f} : \widehat{V} \to \widehat{W}$ the corresponding map. For any definable map $g : W \times Z \to \Gamma_\infty$, let $\tilde{g} := g \circ (f \times \mathrm{Id}_Z)$. Since for any $p \in \widehat{V}$ we have $p_*(\tilde{g}) = \widehat{f}(p)_*(g)$, there is a definable inclusion $Y_{Z,\tilde{g}} \hookrightarrow Y_{Z,g}$. The composition of $\widehat{f}$ with the projection $\widehat{W} \to Y_{Z,g}$ factors through that inclusion, and it follows that $\widehat{f}$ is a morphism of C-pro-definable sets. $\square$

If $f : V \to W$ is a morphism of definable sets, we shall denote by $\widehat{f} : \widehat{V} \to \widehat{W}$ the corresponding morphism of pro-definable sets. Sometimes we shall write f instead of $\widehat{f}$.

3.2 SOME EXAMPLES

EXAMPLE 3.2.1. If b is a closed ball in $\mathbb{A}^1$, let $p_b \in \widehat{\mathbb{A}^1}$ be the generic type of b: it can be defined by $(p_b)_*(f) = \min\{\mathrm{val}(f(x)) : x \in b\}$, for any polynomial f. This applies even when b has valuative radius ∞, i.e. consists of a single point. The generic type of a finite product of closed balls is defined by exactly the same formula. If b and b' are (finite products of) closed balls, in the notation of Remark 3.6.3, $p_{b\times b'} = p_b \otimes p_{b'}$. By [19], 2.3.6, 2.3.8, and 2.5.5, $\widehat{\mathbb{A}^1}$ is equal to the set of all generic types of closed balls of valuative radius in Γ_∞. As a set, $\widehat{\mathbb{P}^1}$ consists of the disjoint union of $\widehat{\mathbb{A}^1}$ and the definable type concentrating on the point ∞.

EXAMPLE 3.2.2. Let us give examples of a more exotic nature. Let F be a field and set $K = F(t)$ with valuation trivial on F and $\mathrm{val}(t) = 1$. Let $\varphi = \sum_{i=0}^{\infty} a_i x^i$ be a formal series with coefficients $a_i \in F$. Assume φ is not algebraic. Let $p_0(x; y)$ consist of all formulas over K of the form

$$\mathrm{val}(y - \sum_{i=0}^{n} a_i (tx)^i) \geq n+1.$$

Then $p_0(x; y) + (p_{\mathcal{O}}|\mathbb{U})(x)$ generates a complete type p_φ which is a stably dominated type. However, $p_\varphi \in \widehat{\mathbb{A}^2}$ is not strongly stably dominated in the sense of Definition 2.6.9.

Proof. Let M be any valued field extension of K^{alg} such that $\mathbb{Z}$ is cofinal in $\Gamma(M)$. For a series $\beta = \sum_{i=0}^{\infty} b_i z^i$, $b_i \in \mathcal{O}_M$, define $p_{0,\beta}$ to consist of all formulas

$$\mathrm{val}(y - \sum_{i=0}^{n} b_i (xt)^i) \geq n+1.$$

Let $c \models p_{\mathcal{O}}|M$. First suppose $p_{0,\beta}(c; 0)$ holds. Then

$$\min_{i\leq n}(\mathrm{val}(b_i) + i) = \mathrm{val}(\sum_{i=0}^{n} b_i (ct)^i) \geq n+1,$$

since $c \models p_{\mathcal{O}}|M$. So $\mathrm{val}(b_i) \geq n+1-i$. Letting $n \to \infty$ we see that $b_i = 0$; so $\beta = 0$.

Next suppose just that $p_0(c; d)$ holds for some $d \in M(c)^{\mathrm{alg}}$. So $Q(c, d) = 0$ for some polynomial $Q \in \mathcal{O}_M[x, y]$. Let $\varphi' = Q(x, \varphi(tx))$ be the power series obtained by substituting $\varphi(tx)$ for y. Then $p_{0,\varphi'}(c; 0)$ holds. Hence by the previous paragraph, $\varphi' = 0$, so $\varphi(tx)$ is algebraic, and φ is also algebraic.

Thus, $p_0(c; y)$ defines an infinite intersection b of balls over $M(c)$, with no algebraic point. Hence b contains no nonempty $M(c)$-definable subset. So $p_0 + \mathrm{tp}(c/M)$ generates a complete type p_φ over $M(c)$. Now assume M is maximally complete and let $(c, d) \models p_\varphi|M$. Since $\Gamma(M(c, d)) = \Gamma(M)$, it follows from Theorem 2.9.2 that $\mathrm{tp}((c, d)/M)$ is stably dominated. One has $\mathrm{trdeg}_M M(c, d) = 2$, while the corresponding residue field extension has transcendence degree 1 by Lemma 2.5.5 of [19]. By Proposition 8.1.2 it follows that p_φ is not strongly stably dominated. □

EXAMPLE 3.2.3. By Example 13.1 in [20], which is rather similar to Example 3.2.2, over any valued field K, there exist points p of $\widehat{\mathbb{A}^2}$ defined over some extension M of K such that if $c \models p|M$, then $\mathrm{trdeg}_M M(c) = 2$ while the residue field extension has transcendence degree 1. By Proposition 8.1.2 such points are not strongly stably dominated in the sense of Definition 2.6.9.

3.3 THE NOTION OF A DEFINABLE TOPOLOGICAL SPACE

We will consider topologies on definable and pro-definable sets X. With the formalism of the universal domain $\mathbb{U}$, we can view these as certain topologies on $X(\mathbb{U})$, in the usual sense.

If M is a model, the space $X(M)$ will not be a subspace of $X(\mathbb{U})$; indeed in the case of an order topology, or any Hausdorff Ziegler topology in the sense defined below, the induced topology from a saturated model on a small set is always discrete. Instead we define $X(M)$ to be the topological space whose underlying set is $X(M)$, and whose topology is generated by sets $U(M)$ with U an M-definable open set.

We will not have occasion to consider $X(A)$ when A is a substructure, which is not a model. We remark however that if $\mathrm{acl}(A) = M$ is a model, then the induced topology on $X(A)$ from $X(M)$ is induced by the A-definable open sets. Indeed if $p \in X(A)$ and $p \in U$ with U definable over M, let $\mathbf{U}$ be the intersection of all $\mathrm{Aut}(M/A)$-conjugates of U; then $\mathbf{U}$ is open, A-definable, and $p \in \mathbf{U} \subset U$.

We will say that a topological space X is *definable in the sense of Ziegler* if the underlying set X is definable, and there exists a definable family B of definable subsets of X forming a neighborhood basis at each point. This allows for a good topological logic, see [42]. But it is too restrictive for our purposes. An algebraic variety with the Zariski topology is not a definable space in this sense; nor is the topology even generated by a definable family.

Let X be an A-definable, resp. pro-definable, set. Let $\mathcal{T}$ be a topology on $X(\mathbb{U})$, and let $\mathcal{T}_d$ be the intersection of $\mathcal{T}$ with the class of relatively $\mathbb{U}$-definable subsets of X. We will say that $\mathcal{T}$ is an *A-definable topology* if it is generated by $\mathcal{T}_d$, and for any A-definable family $\mathcal{W} = (W_u : u \in U)$ of relatively definable subsets of X, $\mathcal{W} \cap \mathcal{T}$ is ind-definable over A. The second condition is equivalent to the statement that $\{(x, W) : x \in W, W \subset X, W \in \mathcal{W} \cap \mathcal{T}\}$ is ind-definable over A. An equivalent definition is that the topology is generated by an ind-definable family of relatively definable sets over A. We will also say that $(X, \mathcal{T})$ is a definable space over A, resp. a pro-definable space over A, or just that X is a definable, resp. pro-definable, space over A when there is no ambiguity about $\mathcal{T}$. We say X is a (pro-)definable space if it is a (pro-)A-definable space for some small A. As usual the smallest such A may be recognized Galois theoretically.

If $\mathcal{T}_0$ is any ind-definable family of relatively definable subsets of X, the set $\mathcal{T}_1$ of finite intersections of elements of $\mathcal{T}_0$ is also ind-definable. Let $\mathcal{T}$ be the family of subsets of $X(\mathbb{U})$ that are unions of sets $Z(\mathbb{U})$, with $Z \in \mathcal{T}_1$. Then $\mathcal{T}$ is a topology on $X(\mathbb{U})$, generated by the relatively definable sets within it. By compactness, a relatively definable set $Y \subset X$ is in $\mathcal{T}$ if and only if for some definable $T' \subset \mathcal{T}_1$, Y is a union of sets $Z(\mathbb{U})$ with $Z \in T'$. It follows that the topology $\mathcal{T}$ generated by $\mathcal{T}_0$ is a definable topology. In the above situation, note also that if Y is A-relatively definable, then Y is an A-definable union of relatively definable open sets from T'. Indeed, let $Y' = \{Z \in T' : Z \subset Y\}$, then $Y = \cup_{Z \in Y'} Z$. In general Y need not be a union of sets from $\mathcal{T}_1(A)$, for any small A.

As is the case with groups, the notion of a pro-definable space is more general than that of pro-(definable spaces). However the spaces we will consider will be pro-(definable spaces).

When Y is a definable topological space, and A a base substructure, the set $Y(A)$ is topologized using the family of A-definable open subsets of Y. We do not use externally definable open subsets (i.e. A'-definable for larger A) to define the topology on $Y(A)$; if we did, we would obtain the discrete topology on $Y(A)$ whenever Y is Hausdorff. The same applies in the pro-definable case; thus in the next section we shall topologize $\widehat{X}(K)$ using the K-definable open subsets of $\widehat{X}$, restricted to $\widehat{X}(K)$.

When we speak of the topology of Y without mention of A, we mean to take $A = \mathbb{U}$, the universal domain; often, any model will also do.

3.4 $\widehat{V}$ AS A TOPOLOGICAL SPACE

Assume that V comes with a definable topology $\mathcal{T}_V$, and an ind-definable sheaf $\mathcal{O}$ of definable functions into Γ_∞. We define a topology on $\widehat{V}$ as follows. A pre-basic open set has the form: $\{p \in \widehat{O} : p_*(\phi) \in U\}$, where $O \in \mathcal{T}_V$, $U \subset \Gamma_\infty$ is a definable set, open for the order topology, and $\phi \in \mathcal{O}(O)$. A basic open set is by definition a finite intersection of pre-basic open sets.

When V is an algebraic variety, we take the topology to be the Zariski topology, and the sheaf to be the sheaf of regular functions composed with val.

When X is a definable subset of a given algebraic variety V, we give $\widehat{X}$ the subspace topology.

3.5 THE AFFINE CASE

Assume V is a definable subset of some affine variety. Let $\mathrm{Fn}_r(V, \Gamma_\infty)$ denote the functions of the form $\mathrm{val}(F)$, where F is a regular function on the Zariski closure of V. By quantifier elimination any definable function in $\mathrm{Fn}(V, \Gamma_\infty)$ is piecewise a difference of the form $\frac{1}{n}f - \frac{1}{m}g$ with f and g in Fn_r and n and m positive integers. Moreover, by piecewise we mean sets cut out by Boolean combinations of sets of the form $f \leq g$, where $f, g \in \mathrm{Fn}_r(V, \Gamma_\infty)$. It follows that if p is a definable type and $p_*(f)$ is defined for $f \in \mathrm{Fn}_r(V, \Gamma_\infty)$, then p is stably dominated, and determined by $p_*|\mathrm{Fn}_r(V \times W, \Gamma_\infty)$ for all W. A *basic open* set is defined by finitely many strict inequalities $p_*(f) < p_*(g)$, with $f, g \in \mathrm{Fn}_r(V, \Gamma_\infty)$. (In case $f = \mathrm{val}(F)$ and $g = \mathrm{val}(G)$ with $G = 0$, this is the same as $F \neq 0$.) It is easy to verify that the topology generated by these basic open sets coincides with the definition of the topology on $\widehat{V}$ above, for the Zariski topology and the sheaf of functions $\mathrm{val}(f)$, f regular.

Note that if $F_1, \dots, F_n$ are regular functions on V, and each $p \mapsto p_*(f_i)$ is continuous, with $f_i = \mathrm{val}(F_i)$, then $p \mapsto (p_*(f_1(x)), \dots, p_*(f_n(x)))$ is continuous. Thus the topology on $\widehat{V}$ is the coarsest one such that all $p \mapsto p_*(f)$ are continuous, for $f \in \mathrm{Fn}_r(V, \Gamma_\infty)$. So the basic open sets with f or g constant suffice to generate the topology.

The topology on $\widehat{V}$ is strict pro-definably generated in the following sense: for each definable set W, one endows $\mathrm{Fn}(W, \Gamma_\infty)$ with the product topology induced by the order topology on Γ_∞. Now for a definable function $f : V \times W \to \Gamma_\infty$ the topology induced on the definable set $Y_{W,f}$ is generated by a definable family of definable subsets of $Y_{W,f}$ (recall that $Y_{W,f}$ is the subset of $\mathrm{Fn}(W, \Gamma_\infty)$ consisting of all functions $p_*(f)$, for p varying in $\widehat{V}(\mathbb{U})$). By definition, the pullbacks to $\widehat{V}$ of the definable open subsets of the $\mathrm{Fn}(W, \Gamma_\infty)$ generate the topology on $\widehat{V}$. In particular, $\widehat{V}$ is a pro-definable space in the sense of 3.3.

When V is a definable subset of an algebraic variety over VF, the topology on $\widehat{V}$ can also be defined by gluing the affine pieces. It is easy to check that this is consistent (if V' is an affine open of the affine V, obtained say by inverting g, then any function $\mathrm{val}(f/g)$ can be written $\mathrm{val}(f) - \mathrm{val}(g)$, hence is continuous on $\widehat{V'}$ in the topology induced from $\widehat{V}$). Moreover, this coincides with the topology defined via the sheaf of regular functions.

For any definable set X, we have an embedding $X \to \widehat{X}$, taking a point x to the definable type $\mathrm{tp}(x/\mathbb{U})$ concentrating on x.

LEMMA 3.5.1. *If X is a definable subset of Γ_∞^n then $X = \widehat{X}$ canonically. More generally if U is a definable subset of VF^n or a definable subset of an algebraic variety over* VF *and W is a definable subset of Γ_∞^m, then the canonical map $\widehat{U} \times W \to \widehat{U \times W}$ is a bijection.*

Proof. Let $f : U \times W \to U$ and $g : U \times W \to W$ be the projections. If $p \in \widehat{U \times W}$ we saw that $g_*(p)$ concentrates on some $a \in W$; so $p = f_*(p) \times g_*(p)$ (i.e. $p(u, w)$ is generated by $f_*(p)(u) \cup g_*(p)(w)$). □

If U is a definable subset of an algebraic variety over VF, we endow $\widehat{U} \times \Gamma_\infty^m \simeq \widehat{U \times \Gamma_\infty^m}$ with the quotient topology for the surjective mapping $\widehat{U \times \mathbb{A}^m} \to \widehat{U \times \Gamma_\infty^m}$ induced by $\mathrm{id} \times \mathrm{val}$.

We will see below (as a special case of Lemma 3.5.3) that the topology on $\Gamma_\infty = \widehat{\Gamma_\infty}$ is the order topology, and the topology on $\widehat{\Gamma_\infty^m} = \Gamma_\infty^m$, is the product topology.

For $\gamma = (\gamma_1, \dots, \gamma_n) \in \Gamma_\infty^n$, let $b(\gamma) = \{x = (x_1, \dots, x_n) \in \mathbb{A}_1^n : \mathrm{val}(x_i) \geq \gamma_i, i = 1, \dots, n\}$. We set $p_\gamma = p_{b(\gamma)}$ with the notation from Example 3.2.1.

LEMMA 3.5.2. *The map* $j : \widehat{\mathbb{A}^n} \times \Gamma_\infty \to \widehat{\mathbb{A}^{n+1}}$, $(q, \gamma) \mapsto q \otimes p_\gamma$ *is continuous for the product topology of* $\widehat{\mathbb{A}^n}$ *with the order topology on* Γ_∞.

Proof. We have to show that for each polynomial $f(x_1, \ldots, x_n, y)$ with coefficients in VF, the map $(p, \gamma) \mapsto j(p, \gamma)_*(\mathrm{val}(f))$ is continuous. The functions min and + extend naturally to continuous functions $\Gamma_\infty^2 \to \Gamma_\infty$. Now if $f(x_1, \ldots, x_n, y)$ is a polynomial with coefficients in VF, there exists a function $P(\gamma_1, \ldots, \gamma_n, \tau)$ obtained by composition of min and +, and polynomials h_i such that

$$\min_{\mathrm{val}(y)=\alpha} \mathrm{val}(f(x_1, \ldots, x_n, y)) = P(\mathrm{val}(h_1(x)), \ldots, \mathrm{val}(h_d(x)), \alpha),$$

namely, $\min_{\mathrm{val}(y)=\alpha} \mathrm{val}(\sum h_i(x) y^i) = \min_i(\mathrm{val}(h_i(x)) + i\alpha)$. So $P : \Gamma_\infty^{n+1} \to \Gamma_\infty$ is continuous. Hence $j(p, \gamma)_* f = P(p_*(h_1), \ldots, p_*(h_d), \gamma)$. Continuity follows, by composition. □

LEMMA 3.5.3. *If* U *is a definable subset of* $\mathbb{A}^n \times \Gamma_\infty^\ell$ *and* W *is a definable subset of* Γ_∞^m, *the induced topology on* $\widehat{U \times W} = \widehat{U} \times W$ *coincides with the product topology.*

Proof. We have seen that the natural map $\widehat{U \times W} \to \widehat{U} \times W$ is bijective; it is clearly continuous, where $\widehat{U} \times W$ is given the product topology. To show that it is closed, it suffices to show that the inverse map is continuous, and we may take $U = \mathbb{A}^n$ and $W = \Gamma_\infty^m$. By factoring $\widehat{U \times \Gamma_\infty^m} \to \widehat{U \times \Gamma_\infty^{m-1}} \times \Gamma_\infty \to \widehat{U} \times \Gamma_\infty^{m-1} \times \Gamma_\infty$, we may assume $m = 1$. Having said this, by pulling back to $\mathbb{A}^{n+\ell}$ we may assume $\ell = 0$. The inverse map is equal to the composition of j as in Lemma 3.5.2 with a projection, hence is continuous. □

Let U be a definable subset of V, over a structure A. Say $\widehat{U}(A)$ is *explicitly* A*-open* if for any $p \in \widehat{U}(A)$, there exists a Zariski open V' with $p \in \widehat{V'}$, and regular functions $G_1, \ldots, G_n$ on V', $g_i = \mathrm{val}(G_i) : V' \to \Gamma_\infty$ and open neighborhoods E_i of $p_*(g_i)$, all defined over A, such that $\cap_i g_i^{-1}(E_i) \subset \widehat{U}$.

The following lemma will be used in Chapter 14 for structures of the form $\mathbf{F} = (F, \mathbb{R})$.

LEMMA 3.5.4. *Let* $\mathbf{F}$ *be any structure consisting of field points and* Γ*-points including at least one positive element of* Γ. *Let* V *be a variety defined over* $\mathbf{F}$ *and let* U *be an* $\mathbf{F}$ *definable subset of* V. *If* $\widehat{U}$ *is open in* $\widehat{V}$, *then* $\widehat{U}(\mathbf{F})$ *is explicitly* $\mathbf{F}$*-open.*

Proof. Covering V by affine subsets, we may assume V is affine. Let $F = \mathrm{VF}(\mathbf{F})$ be the field points.

We first show that if the statement holds for $(F^{\mathrm{alg}}, \Gamma(F^{\mathrm{alg}}))$, then it holds for $\mathbf{F} = (F, \Gamma(F))$. Note that it is enough to show it holds for $(F, \Gamma(F^{\mathrm{alg}}))$

since $g_i^{-1}((\alpha, \beta)) = (ng_i)^{-1}((n\alpha, n\beta))$. Let $p \in \widehat{U}(F)$. There exist regular functions $G_1, \ldots, G_n$ over F^{alg}, and intervals I_j of Γ_∞, defined over $\Gamma(F)$, such that $p \in \cap_j \widehat{g_j}^{-1}(I_j) \subset \widehat{U}$, with $g_j = \mathrm{val}(G_j)$. So it suffices to show, for each j, that the intersection of the Galois conjugates of $\widehat{g_j}^{-1}(I_j)$ contains an open neighborhood of p in $\widehat{V}(F)$. Let $G = G_j$, $g = g_j$ and $I = I_j$, and let G^ν be the Galois conjugates of G over F, $g^\nu = \mathrm{val}(G^\nu)$.

Let $b \models p$. Then the G^ν are Galois conjugate over $F(b)$, p being F-definable. The elements $c_\nu = G^\nu(b)$ are Galois conjugate over $F(b)$; they are the roots of a polynomial $H(b, y) = \Pi_\nu(y - G^\nu(b)) = \sum_m h_\mu(b) y^m$. For all b' in some F-definable Zariski open set U' containing b, the set of roots of $H(b', y)$ is equal to $\{G^\nu(b')\}$. Within U', the set of b' such that, for all ν, $g^\nu(b') \in I$ can therefore be written in terms of the Newton polygon of $H(b', y)$, i.e. in terms of certain inequalities between convex expressions in $\mathrm{val}(h_k(b'))$. This shows that the intersection of Galois conjugates of $\widehat{G}^{-1}(I)$ contains an open neighborhood of p.

This permits us to assume F is algebraically closed, as we will do from now on.

Assume first $\mathbf{F} \subset \mathrm{dcl}(F)$. In particular, by assumption, F is not trivially valued and since $\mathbf{F} = \mathrm{acl}(F)$ is an elementary submodel, the statement is clear.

We now have to deal with the case that $\mathbf{F}$ is bigger than F; we may assume $\mathbf{F}$ is generated over F by finitely many elements of Γ, and indeed, adding one element at a time, that $\mathbf{F} = F(\gamma)$ for some $\gamma \in \Gamma$. Let c be a field element with $\mathrm{val}(c) = \gamma$; it suffices to show that if U is open over $F(c)$, then it is over F too. Let $G(x, c) = \sum G_i(x) c^i$ be a polynomial (where $x = (x_1, \ldots, x_n), V \leq \mathbb{A}^n$). Let $g(p, c)$ be the generic value of $\mathrm{val}(G(x, c))$ at p and $g_i(p)$ the one of $\mathrm{val}(G_i)$. Then $g(p, c) = \min_i g_i(p) + i\gamma$. From this the statement is clear. □

When $\Gamma(\mathbf{F}) = (0)$, Lemma 3.5.4 is not true as stated. Here is a counterexample: Let $V = \mathbb{A}^2$, $U = \{(x, y) \in V : \mathrm{val}(x) < \mathrm{val}(y)\}$, let p be the generic type of $\mathcal{O} \times \{0\}$. If $G(x, y)$ is any polynomial over $F = \mathcal{O}_F$, then $p_*(G) = 0$ unless $y | G$ and then $p_*(G) = \infty$. The only conditions about G one can form around p over F are, in case $G = yG_1$, that $p_*(G) > 0$. So no F-explicit open set can be contained in U, since one can always take $0 \ll \mathrm{val}(y) < \mathrm{val}(x)$ to satisfy $p_*(yG_1) > 0$. But nevertheless, we still have:

COROLLARY 3.5.5. *Let $\mathbf{F}$ be any structure consisting of field points and Γ-points. Let V be a variety defined over $\mathbf{F}$ and let U be an $\mathbf{F}$-definable subset of V. Let $p \in \widehat{U}(\mathbf{F})$. If $\widehat{U}$ is open in $\widehat{V}$, then there exists a definable function $\alpha : V \to \Gamma_\infty^n$, an open neighborhood E of $p_*(\alpha)$, and a Zariski open V' with $p \in \widehat{V'}$, all defined over $\mathbf{F}$, such that $\alpha^{-1}(E) \subset \widehat{U}$ is explicitly $\mathbf{F}$-open and α has the form $(\mathrm{val}(G_1), \ldots, \mathrm{val}(G_n))$ for some regular functions G_i on V'.*

Proof. This follows from Lemma 3.5.4 unless $\Gamma(\mathbf{F}) = (0)$. Assume therefore that $\Gamma(\mathbf{F}) = 0$, so that all elements of Γ of positive valuation have the same type over $\mathbf{F}$. Let γ be such an element. By Lemma 3.5.4, there exist $G_1, \ldots, G_n, V', E_\gamma$ as required but over $\mathbf{F}(\gamma)$. So $G_1, \ldots, G_n, V'$ are defined over $\mathbf{F}$; E_γ depends on γ. Let $E = \cup_{\gamma>0} E_\gamma$. Then clearly E is open and the statement holds. □

3.6 SIMPLE POINTS

Recall that for any definable set V, we have an embedding $V \to \widehat{V}$, taking a point x to the definable type $\mathrm{tp}(x/\mathbb{U})$ concentrating on x. The points of the image are said to be *simple.*

LEMMA 3.6.1. *Let X be a definable subset of* VF^n.

(1) *The set of simple points of $\widehat{X}$ (which we identify with X) is an iso-definable and relatively definable dense subset of $\widehat{X}$. If M is a model of* ACVF, *then $X(M)$ is dense in $\widehat{X}(M)$.*

(2) *The induced topology on X agrees with the valuation topology on X.*

Proof. (1) The fact that X is iso-definable in $\widehat{X}$ is clear. For relative definability, note that a point of $\widehat{X}$ is simple if and only if each of its projections to $\widehat{\mathbb{A}^1}$ is simple and that on $\mathbb{A}^1$, the points are a definable subset of the set of closed balls (cf. Example 3.2.1). For density, consider (for instance) $p \in \widehat{X}(M)$ with $p_*(f) > \alpha$. Then $\mathrm{val}(f(x)) > \alpha \wedge x \in X$ is satisfiable in M, hence there exists a simple point $q \in \widehat{X}(M)$ with $q_*(f) > \alpha$.

(2) Clear from the definitions. The basic open subsets of the valuation topology are of the form $\mathrm{val}(f(x)) > \alpha$ or $\mathrm{val}(f(x)) < \alpha$. □

LEMMA 3.6.2. *Let $f : U \to V$ be a definable map between definable subsets of* VF^*. *If f has finite fibers, then the preimage of a simple point of $\widehat{V}$ under $\widehat{f}$ is simple in $\widehat{U}$.*

Proof. It is enough to prove that if X is a finite definable subset of VF^n, then $X = \widehat{X}$, which is clear by (1) of Lemma 3.6.1. □

REMARK 3.6.3. The natural projection $S_{def,U\times V} \to S_{def,U} \times S_{def,V}$ induces a continuous map $\widehat{U \times V} \to \widehat{U} \times \widehat{V}$. On the other hand, it admits a natural section, namely $\otimes : S_{def,U} \times S_{def,V} \to S_{def,U\times V}$, which restricts to a section of $\widehat{U \times V} \to \widehat{U} \times \widehat{V}$. This map is not continuous in the logic topology, nor is its restriction to $\widehat{U} \times \widehat{V} \to \widehat{U \times V}$ continuous. Indeed when $U = V$ the pullback of the diagonal $\widehat{\Delta_U}$ consists of simple points on the diagonal $\Delta_{\widehat{U}}$. But over a model, the set of simple points is dense, and hence not closed.

3.7 V-OPEN AND G-OPEN SUBSETS, V+G-CONTINUITY

DEFINITION 3.7.1. Let V be an algebraic variety over a valued field F. A definable subset of V is said to be *v-open* if it is open for the valuation topology. It is called *g-open* if it is defined by a positive Boolean combination of Zariski closed and open sets, and sets of the form $\{u \in U : \mathrm{val}(f(u)) > \mathrm{val}(g(u))\}$, for f and g regular functions on a Zariski open set $U \subset V$. More generally, if V is a definable subset of an algebraic variety W, a definable subset of V is said to be v-open (resp. g-open) if it is of the form $V \cap O$ with O v-open (resp. g-open) in W. A definable subset of $V \times \Gamma_\infty^m$ is called v- or g-open if its pullback to $V \times \mathbb{A}^m$ via id $\times$ val is. The complement of a v-open (resp. g-open) subset is said to be *v-closed* (resp. *g-closed*).

REMARK 3.7.2. If X is A-definable, the regular functions f and g in the definition of g-openness are *not* assumed to be A-definable; in general when A consists of imaginaries, no such f, g can be found. However when $A = \mathrm{dcl}(F)$ with F a valued field, they may be taken to be F-definable, by Lemma 9.1.1.

PROPOSITION 3.7.3. *Let V be an affine variety and X be a definable subset that is both v-closed and g-closed. Then X may be defined by a positive Boolean combination of subvarieties and sets defined by weak valuation inequalities* $\mathrm{val}(f(x)) \leq \mathrm{val}(g(x))$, *where f, g are regular functions on V. A similar statement may be made for V projective, using homogeneous polynomials.*

Proof. We prove this by induction on $\dim(V)$; assume the statement holds for varieties of lower dimension. We may assume V is irreducible. As X is g-closed, it is defined by weak valuation inequalities along with algebraic equalities and inequalities; thus away from some proper subvariety V' of V, X coincides with a set X' cut out by the inequalities $\mathrm{val}(f_i) \leq \mathrm{val}(g_i)$, $i = 1, \dots, n$. Thus $X' \smallsetminus V' = X \smallsetminus V'$; by induction, $X \cap V'$ has the right form; if we also show that $X' \cap V' \subset X$, then $X = X' \cup (X \cap V')$ will have the promised form. Thus it suffices to show that $X' \subset X$. As X' is v-closed, this follows from Lemma 3.7.4. □

LEMMA 3.7.4. *Let V be an affine variety, let f_i and g_i, $1 \leq i \leq n$, be nonzero regular functions on V and let V' be a proper subvariety of V. Let Y be the subset of V defined by the inequalities* $\mathrm{val}(f_i) \leq \mathrm{val}(g_i)$, $i = 1, \dots, n$. *Then any point b of Y lies arbitrarily close to a point of $Y \smallsetminus V'$ in the valuation topology.*

Proof. Let $p : \widetilde{V} \to V$ be the result of blowing up the ideal (f_1, g_1) on V; let b' be a point of $\widetilde{V}$ lying above b, and let $\widetilde{V}'$ be an affine open of $\widetilde{V}$ containing b'. If we show the existence of points of $p^{-1}(Y) \cap \widetilde{V}'$ arbitrarily close to b', avoiding the exceptional divisor as well as $p^{-1}(V')$, then by continuity of p the claim will be proved. Now on $\widetilde{V}'$, there is a regular function u_1 such that

$f_1 = g_1 u_1$ or $f_1 u_1 = g_1$; so the inequality $\mathrm{val}(f_1) \leq \mathrm{val}(g_1)$ can be replaced by $\mathrm{val}(u_1) \leq 0$, or $\mathrm{val}(u_1) \geq 0$. Iterating this construction, we may assume Y is defined by a conjunction of inequalities $\mathrm{val}(u_i) \leq 0$ or $\mathrm{val}(u_i) \geq 0$ for some regular functions u_i, $1 \leq i \leq n$. Now if we take any point of V very close to b in the valuation topology (but avoiding V') these inequalities are preserved. □

DEFINITION 3.7.5. Let V be an algebraic variety over a valued field F or a definable subset of such a variety. A definable function $h : V \to \Gamma_\infty$ is called *v-continuous* (resp. *g-continuous*) if the pullback of any v-open (resp. g-open) set is v-open (resp. g-open). A function $h : V \to \widehat{W}$ with W an affine F-variety is called v-continuous (resp. g-continuous) if, for any regular function $f : W \to \mathbb{A}^1$, $\mathrm{val} \circ f \circ h$ is v-continuous (resp. g-continuous).

Note that the topology generated by v-open subsets on Γ_∞ is discrete on Γ, while the neighborhoods of ∞ in this topology are the same as in the order topology. The topology generated by g-open subsets is the order topology on Γ, with ∞ isolated. We also have the topology on Γ_∞ coming from its canonical identification with $\widehat{\Gamma_\infty}$, or the v+g-topology; this is the intersection of the two previous topologies, that is, the order topology on Γ_∞.

Let V be an algebraic variety over a valued field F and let X be a definable subset of $V \times \Gamma_\infty^m$. We say that X is *v+g-open* if it is both v-open and g-open. The complement of a v+g-open subset is said to be *v+g-closed.* If W has a definable topology, a definable function $X \to W$ is called *v+g-continuous* if the pullback of a definable open subset of W is both v- and g-open, and similarly for functions to V.

REMARK 3.7.6. Note that v, g and v+g-open sets are *definable* sets. Over any given model it is possible to extend v to a topology in the usual sense, the valuation topology, whose restriction to definable sets is the family of v-open sets. But this is not true of g and of v+g; in fact they are not closed under definable unions, as the example $\mathcal{O} = \cup_{a \in \mathcal{O}} a + \mathcal{M}$ shows.

Any g-closed subset W of an algebraic variety is defined by a disjunction $\bigvee_{i=1}^m (\neg H_i \wedge \phi_i)$, with ϕ_i a finite conjunction of weak valuation inequalities $v(f) \leq v(g)$ and equalities, and H_i defining a Zariski closed subset. If W is also v-closed, W is equal to the union of the v-closures of the sets defined by $\neg H_i \wedge \phi_i$, $1 \leq i \leq m$.

LEMMA 3.7.7. *Let X be a definable subset of a variety V over a valued field. Let W be a definable subset of X which is v+g-closed in X. Then $\widehat{W}$ is closed in $\widehat{X}$. More generally, if W is g-closed in X, then $cl(\widehat{W}) \cap \widehat{X} \subset \widehat{cl_v(W)} \cap \widehat{X}$, with cl and cl_v denoting respectively the closure and the v-closure.*

Proof. Let M be a model, $p \in \widehat{X}(M)$, with $p \in cl(\widehat{W}(M))$. We will show that $p \in \widehat{cl_v(W)}$. Let (p_i) be a net in $\widehat{W}(M)$ approaching p. Let $a_i \models p_i|M$.

Let $\mathrm{tp}(a/M)$ be a limit type in the logic topology (so a can be represented by an ultraproduct of the a_i). For each i we have $\Gamma(M(a_i)) = \Gamma(M)$, but $\Gamma(M(a))$ may be bigger.

Consider the subset C of $\Gamma(M(a))$ consisting of those elements γ such that $-\alpha < \gamma < \alpha$ for all $\alpha > 0$ in $\Gamma(M)$. Thus C is a convex subgroup of $\Gamma(M(a))$; let N be the valued field extension of M with the same underlying M-algebra structure as $M(a)$, obtained by factoring out C. Let $\bar{a}$ denote a as an element of N. We have $a_i \in W$, so $a \in W$; since W is g-closed in X it is clear that $\bar{a} \in W$. (This is the easy direction of Lemma 9.1.1.)

Let $b \models p|M$. For any regular function f in $M[U]$, with U Zariski open in V, we have: $(*)$ $\mathrm{val}(f(a_i)) \to \mathrm{val}(f(b))$ in $\Gamma_\infty(M)$ (since $p_i \to p$).

Let $R = \{x \in N : (\exists m \in M)(\mathrm{val}(x) \geq \mathrm{val}(m))\}$. Then R is a valuation ring of N over M. By $(*)$, for large enough i, $\mathrm{val}(f(a_i))$ is bounded below by some element of $\Gamma(M)$ (namely any element below $p_*(f)$). So $\mathrm{val}(f(a))$ and $\mathrm{val}(f(\bar{a}))$ must lie above the same element. Thus $\bar{a} \in R$. Also by $(*)$, if $\mathrm{val}(f(\bar{a})) = \infty$, or just if $\mathrm{val}(f(\bar{a})) > \mathrm{val}(M)$, then $f(b) = 0$. Thus we have a well-defined map from the residue field of R to $M(b)$, with res $\bar{a} \mapsto b$. Since $\bar{a} \in W$, it follows that $b \in cl_v(W)$ (cf. the last part of the proof of Lemma 9.2.1), hence $p \in \widehat{cl_v(W)}$. □

PROPOSITION 3.7.8. *Let V be an algebraic variety over a valued field F. Then $\widehat{V}$ is Hausdorff.*

Proof. Let us first consider the case when V is affine. In this case we may assume $V = \mathbb{A}^n$. Let $p \in \widehat{\mathbb{A}^n}$ and $p' \in \widehat{\mathbb{A}^n}$. Let M be a model of ACVF such that $p \in \widehat{\mathbb{A}^n}(M)$ and $p' \in \widehat{\mathbb{A}^n}(M)$. There exists a polynomial $F \in M[x_1, \dots, x_n]$ such that $\mathrm{val}(F(p)) \neq \mathrm{val}(F(p'))$. Thus, for some $\alpha \in \Gamma(M)$, the disjoint open sets defined by the conditions $\mathrm{val}(F) < \alpha$ and $\mathrm{val}(F) > \alpha$ will separate p and p'.

In general, since V is separated as an algebraic variety, V is Hausdorff for the valuation topology. Indeed, the diagonal Δ of V in $V \times V$ being Zariski closed, it is closed for the valuation topology, and the valuation topology on $V \times V$ is the product topology.

Let $p \in \widehat{V}$. Fix U an affine open subset of V such that $p \in \widehat{U}$. Let Z be the intersection of the Zariski closed subsets W of U such that $p \in \widehat{W}$. Choose a closed embedding of U in some affine space $\mathbb{A}^n$. There exists a definable v-closed subset C of Z such that $p \in \widehat{C}$. We may further assume C to be bounded, meaning that there exists some γ in Γ such that for any $1 \leq i \leq n$, $\mathrm{val}(x_i) \geq \gamma$ on C. We denote by $d : U \times U \to \Gamma_\infty$ the restriction of the function $\min_{1 \leq i \leq n}(\mathrm{val}(x_i - y_i))$ on $\mathbb{A}^n \times \mathbb{A}^n$. Now let $p' \in \widehat{V}$ with $p' \neq p$. Fix U' an affine open subset of V such that $p' \in \widehat{U'}$ and a closed embedding of U' in some affine space $\mathbb{A}^{n'}$. Define Z' and d' similarly as Z and d and fix a bounded definable v-closed subset C' of Z' such that $p' \in \widehat{C'}$.

If $Z \cap U' \neq \varnothing$, p and p' both lie in $\widehat{U'}$ and we are done by the affine case. Thus, we may assume $Z \cap U' = \varnothing$ and $Z' \cap U = \varnothing$. In particular, $Z \cap Z' = \varnothing$ and so $B \cap B' = \varnothing$. For $c \in U$ and $\alpha \in \Gamma$, let $B_\alpha(c) = \{x \in U : d(x, c) > \alpha\}$. One defines similarly $B'_\alpha(c')$ for $c' \in U'$. Since V is Hausdorff for the valuation topology, for any $(c, c') \in C \times C'$ there exists some $\alpha \in \Gamma$ such that $B_\alpha(c) \cap B'_\alpha(c') = \varnothing$.

Let us now prove we can take a single α to work for all $(c, c') \in C \times C'$. Assume this is not the case and fix a valued field extension $F \leq M$ with M a model of ACVF. Then, for any $\alpha \in \Gamma(M)$, there would exist $c_\alpha \in C(M)$, $c'_\alpha \in C'(M)$ and $w_\alpha \in (U \cap U')(M)$ such that $d(c_\alpha, w_\alpha) > \alpha$ and $d'(c'_\alpha, w_\alpha) > \alpha$. Take a non-principal ultrafilter on the set $\Gamma(M)$ and let $\widetilde{M}$ be the corresponding ultrapower. Set $R = \{x \in \widetilde{M} : (\exists \alpha \in \Gamma(M))(\mathrm{val}(x) > \alpha)\}$ and $I = \{x \in \widetilde{M} : (\forall \alpha \in \Gamma(M))(\mathrm{val}(x) > \alpha)\}$. The quotient $M' = R/I$ is an elementary extension of M. Denote by $\tilde{c}$, $\tilde{c}'$ and $\tilde{w}$ the class of (c_α), (c'_α) and (w_α) in $V(\widetilde{M})$. The boundedness assumptions imply that $\tilde{c} \in C \cap V(R)$, $\tilde{c}' \in C' \cap V(R)$, and $\tilde{w} \in U \cap U' \cap V(R)$, so we can consider their images c, c'' and w in $V(M')$. Since C and C' are v-closed, it follows from Lemma 9.2.1 that $c \in C(M')$ and $c' \in C'(M')$. But then $c = w = c'$ so C and C' are not disjoint.

Fix such an α and set $O = \{x \in U : (\exists c \in C)(d(x, c) > \alpha)\}$. Define similarly $O' \subset U'$. By construction, O and O' are disjoint, thus $\widehat{O}$ and $\widehat{O'}$ are disjoint. We have $p \in \widehat{O}$ and $p' \in \widehat{O'}$. Let us check that $\widehat{O}$ is open. Let $q \in \widehat{O}$ and M a valued field extension of F which is a model of ACVF. Let a such that $\mathrm{tp}(a/M) = q|M$. Since a belongs to O, it belongs to $B_\alpha(c)$ for some $c \in C$. Thus q belongs to $\widehat{B_\alpha(c)}$ which is open and contained in $\widehat{O}$. It follows that $\widehat{O}$ is open. For similar reasons $\widehat{O'}$ is open, and thus $\widehat{V}$ is Hausdorff. □

3.8 CANONICAL EXTENSIONS

Let V be a definable set over some A and let $f : V \to \widehat{W}$ be an A-pro-definable morphism, where W is an A-definable subset of $Z \times \Gamma_\infty^m$, with Z an algebraic variety defined over A. We can define a *canonical extension* to $F : \widehat{V} \to \widehat{W}$, as follows.

If $p \in \widehat{V}(M)$, say $p|M = \mathrm{tp}(c/M)$, let $d \models f(c)|M(c)$. By transitivity of stable domination, Proposition 2.6.6, $\mathrm{tp}(cd/M)$ is stably dominated, and hence so is $\mathrm{tp}(d/M)$. Let $F(c) \in \widehat{W}(M)$ be such that $F(c)|M = \mathrm{tp}(d/M)$; this does not depend on d. Moreover $F(c)$ depends only on $\mathrm{tp}(c/M)$, so we can let $F(p) = F(c)$. By Lemma 3.8.1, $F : \widehat{V} \to \widehat{W}$ is an A-pro-definable morphism. Sometimes the canonical extension F of f will be denoted by $\widehat{f}$ or even by f.

LEMMA 3.8.1. *Let $f : V \to \widehat{W}$ be an A-pro-definable map as above. Then the canonical extension $F : \widehat{V} \to \widehat{W}$ is an A-pro-definable morphism.*

Proof. Let $g : W \times Z \to \Gamma_\infty$ be a definable map and let $Y_{Z,g}$ be the corresponding definable set of definable functions $Z \to \Gamma_\infty$ considered in the proof of Theorem 3.1.1. The composition of f with the projection $\widehat{W} \to Y_{Z,g}$ yields a definable map $\bar{f} : V \to Y_{Z,g}$. Let $\bar{g} : V \times Z \to \Gamma_\infty$ be the definable map sending $(v, z) \in V \times Z$ to $\bar{f}(v)(z)$. For any $p \in \widehat{V}$ we have $p_*(\bar{g}) = F(p)_*(g)$, hence there is a definable inclusion $Y_{Z,\bar{g}} \hookrightarrow Y_{Z,g}$. Since the composition of F with the projection $\widehat{W} \to Y_{Z,g}$ factors through that inclusion, it follows that F is an A-pro-definable morphism. □

LEMMA 3.8.2. *Let $f : V \to \widehat{W}$ be a pro-definable morphism, where V is a definable subset of an algebraic variety and W is a definable subset of $\mathbb{P}^n \times \Gamma_\infty^m$. Let X be a definable subset of V. Assume f is g-continuous on V and v-continuous at each point of X; i.e. $f^{-1}(G)$ is g-open whenever G is open, and $f^{-1}(G)$ is v-open at x whenever G is open, for any $x \in X \cap f^{-1}(G)$. Then the canonical extension F is continuous at each point of $\widehat{X}$.*

Proof. The topology on $\widehat{\mathbb{P}^n}$ may be described as follows, cf. 5.2. It is generated by the preimages of open sets of Γ_∞^N under continuous definable functions $\mathbb{P}^n \to \Gamma_\infty^N$ of the form

$$[x_0 : \ldots : x_n] \longmapsto [\mathrm{val}(x_0^d) : \ldots : \mathrm{val}(x_n^d) : \mathrm{val}(h_1) : \ldots : \mathrm{val}(h_{N-n})]$$

for some homogeneous polynomials $h_i(x_0, \ldots, x_n)$ of degree d; where in $\Gamma_\infty^N \smallsetminus \{\infty\}^N$ we define $[u_0 : \ldots : u_m]$ to be $(u_0 - \min u_i, \ldots, u_m - \min u_i)$. Composing with such functions we reduce to the case of Γ_∞^m, and hence to the case of $f : V \to \Gamma_\infty$.

Let $U = f^{-1}(G)$ be the pullback of a definable open subset G of Γ_∞. Then $F^{-1}(G) = \widehat{U}$. Now U is g-open, and v-open at any $x \in X \cap U$. By Lemma 3.7.7 applied to the complement of U in V, it follows that $\widehat{U}$ is open at any $x \in \widehat{X}$. □

LEMMA 3.8.3. *Let K be a valued field and V be an algebraic variety over K. Let $f : I \times V \to \widehat{V}$ be a g-continuous K-pro-definable morphism, where $I = [a, b]$ is a closed interval. Let i_I denote one of a or b and e_I denote the remaining point. Let X be a K-definable subset of V. Assume f restricts to a definable morphism $g : I \times X \to \widehat{X}$ and that f is v-continuous at every point of $I \times X$. Then g extends uniquely to a continuous K-pro-definable morphism $G : I \times \widehat{X} \to \widehat{X}$. If moreover, for every $v \in X$, $g(i_I, v) = v$ and $g(e_I, v) \in Z$, with Z a Γ-internal iso-definable subset, then for every $x \in \widehat{X}$, $G(i_I, x) = x$, and $G(e_I, x) \in Z$.*

Proof. Since $\widehat{I \times V} = I \times \widehat{V}$ by Lemma 3.5.1, the first statement follows from Lemma 3.8.2, by considering the pullback of I in $\mathbb{A}^1$. The equation $G(i_I, x) = x$ extends by continuity from the dense set of simple points to $\widehat{X}$. We have by construction $G(e_I, x) \in Z$, using the fact that any stably dominated type on Z is constant. □

LEMMA 3.8.4. *Let K be a valued field and V be a definable subset of an algebraic variety over K. Let $f : V \to \widehat{W}$ be a K-pro-definable morphism, with W a K-definable subset of $\mathbb{P}^n \times \Gamma_\infty^m$. Assume f is v+g-continuous. Then $F : \widehat{V} \to \widehat{W}$ is continuous and it is the unique extension of f to a continuous K-pro-definable morphism $\widehat{V} \to \widehat{W}$.*

Proof. Let us prove the continuity of F. As in the proof of Lemma 3.8.2, it is enough to consider the case $W = \Gamma_\infty$ which follows directly from Lemma 3.7.7. There is clearly at most one continuous extension, because of the density in $\widehat{V}$ of the set of simple points $V(\mathbb{U})$, cf. Lemma 3.6.1. □

LEMMA 3.8.5. *Let K be a valued field and V be a definable subset of an algebraic variety over K. Let $f : I \times V \to \widehat{V}$ be a v+g-continuous K-pro-definable morphism, where $I = [a, b]$ is a closed interval. Let i_I denote one of a or b and e_I denote the remaining point. Then f extends uniquely to a continuous K-pro-definable morphism $F : I \times \widehat{V} \to \widehat{V}$. If moreover, for every $v \in V$, $g(i_I, v) = v$ and $g(e_I, v) \in Z$, with Z an iso-definable Γ-internal subset, then, for every $x \in \widehat{V}$, $G(i_I, x) = x$, and $G(e_I, x) \in Z$.*

Proof. Follows from Lemma 3.8.4 similarly as Lemma 3.8.3 follows from Lemma 3.8.2. □

3.9 PATHS AND HOMOTOPIES

By an *interval* we mean a sub-interval of Γ_∞. Note that intervals of different length are in general not definably homeomorphic, and that the gluing of two intervals (e.g. $[0, 1]$ coming to the right of $[0, \infty]$) may not result in an interval. We get around the latter issue by formally introducing a more general notion, that of a generalized interval.

Given an interval I in Γ_∞, we may consider it either with the induced order or with the opposite order. The choice of one of these orders will be called an orientation of I. Let $I_1, \ldots, I_n$ be oriented sub-intervals of Γ_∞. Assume I_1 is right-closed (i.e. contains its largest endpoint), I_n is left-closed (i.e. contains its smallest endpoint), and that each I_j is closed for $1 < j < n$. Then one may glue end-to-end the intervals I_i in a way respecting the orientations by identifying the largest endpoint of I_i with the smallest endpoint of I_{i+1} for $1 \leq i < n$, and obtain a definable space. Any definable space I that may be obtained this way will be called a *generalized interval.*

If the generalized interval I is closed, we denote by i_I the smallest element of I and by e_I its largest element. Note that if I is obtained by gluing intervals $I_1, \ldots, I_n$, a function $I \times V \to W$ is definable, resp. continuous, resp. v+g-continuous, if and only if it is obtained by gluing definable, resp. continuous, resp. v+g-continuous, functions $\varphi_i : I_i \times V \to W$.

Let V be a definable set. By a *path* on $\widehat{V}$ we mean a continuous definable map $I \to \widehat{V}$ with I some generalized interval.

DEFINITION 3.9.1. Let X be a pro-definable subset of $\widehat{V} \times \Gamma_\infty^n$. A *homotopy* is a continuous pro-definable map $h : I \times X \to X$ with I a closed generalized interval. The maps h_{i_I} and h_{e_I} are then said to be homotopic (one denotes by h_t the map sending $x \in X$ to $h(t,x)$). The homotopy h is called a *deformation retraction* to $A \subset \widehat{V}$ if $h_{i_I} = \mathrm{id}_X$, $h(t,a) = a$ for all t in I and a in A and furthermore $h_{e_I}(x) \in A$ for each x. (In the literature, this is sometimes referred to as a *strong* deformation retraction.) We say $A = h_{e_I}(X)$ is the *image* of h. If $\varrho = h_{e_I}$, we say that $(\varrho, \varrho(X))$ is a deformation retract. Sometimes, we shall also call ϱ or $\varrho(X)$ a deformation retract, the other member of the pair being understood implicitly.

If W is a definable subset of $V \times \Gamma_\infty^n$, we will also refer to a v+g-continuous pro-definable map $h_0 : I \times W \to \widehat{W}$ as a homotopy; by Lemma 3.8.5, h_0 extends uniquely to a homotopy $h : I \times \widehat{W} \to \widehat{W}$. One defines similarly a deformation retraction $h_0 : I \times W \to \widehat{W}$ and its image.

By Lemma 3.8.5 if h_0 is a deformation retraction with image an iso-definable Γ-internal subset then its canonical extension is a deformation retraction with the same image.

EXAMPLE 3.9.2. Generalized intervals may in fact be needed to connect points of $\widehat{V}$. For instance let V be a cycle of $2n$ copies of $\mathbb{P}^1$, with consecutive pairs meeting in a point. By gluing $2n$ copies of the homotopy $\psi_{\{0,\infty\}}$ as defined in 7.5, one gets a deformation retraction $[0,\infty] \times \widehat{V} \to \widehat{V}$ with image a cycle made of $2n$ copies of $[0,\infty] \subset \Gamma_\infty$. However it is impossible to connect two points at extreme ends of this topological circle without gluing together n intervals.

DEFINITION 3.9.3. Let X be a pro-definable subset of $\widehat{V} \times \Gamma_\infty^n$. A homotopy $h : I \times X \to X$ is said to satisfy condition $(*)$ if $h(e_I, h(t,x)) = h(e_I, x)$ for every t and x. One defines similarly condition $(*)$ for a homotopy $h_0 : I \times W \to \widehat{W}$ when W is a definable subset of $V \times \Gamma_\infty^n$. Note that h_0 satisfies $(*)$ if and only if its canonical extension does.

Let $h_1 : I_1 \times X \to X$ and $h_2 : I_2 \times X \to X$ two homotopies. Denote by $I_1 + I_2$ the (generalized) interval obtained by gluing I_1 and I_2 at e_{I_1} and i_{I_2}. Assume $h_2(i_{I_2}, h_1(e_{I_1}, x)) = h_1(e_{I_1}, x)$ for every x in X. Then one denotes by $h_2 \circ h_1$ the homotopy $(I_1 + I_2) \times X \to X$ given by $h_1(t,x)$ for $t \in I_1$ and by $h_2(t, h_1(e_{I_1}, x))$ for t in I_2, and one calls $h_2 \circ h_1$ the composition (or concatenation) of h_1 and h_2.

DEFINITION 3.9.4. Let X be a pro-definable subset of $\widehat{V} \times \Gamma_\infty^n$ and let X' be a pro-definable subset of $\widehat{V'} \times \Gamma_\infty^{n'}$. A pro-definable map $f : X \to X'$ is said to be *definably closed* if for any closed pro-definable subset Z of X, $f(Z)$ is closed in X'.

REMARK 3.9.5. Note that an injective pro-definable map f is definably closed if and only if it is closed, since in this case taking the image under f commutes with arbitrary intersections.

LEMMA 3.9.6. *Let V be an algebraic variety over a valued field, and let X and X_1 be pro-definable subsets of $\widehat{V} \times \Gamma_\infty^N$. Let $f : X_1 \to X$ be a continuous, definably closed and surjective pro-definable map. Let I be a closed generalized interval and $h_1 : I \times X_1 \to X_1$ be a homotopy. Assume h_1 respects the fibers of f, in the sense that $f(h_1(t,x))$ depends only on t and $f(x)$. Then h_1 descends to a homotopy of X.*

Proof. Define $h : I \times X \to X$ by $h(t, f(x)) = f(h_1(t,x))$ for $x \in X_1$; then h is well-defined and pro-definable. We denote the map $\mathrm{Id} \times f : I \times X_1 \to X$ by f_2. Clearly, f_2 is a continuous, definably closed and surjective pro-definable map (the topology on $I \times X_1$, $I \times X$ being the product topology). To show that h is continuous, it suffices therefore to show that $h \circ f_2$ is continuous. Since $h \circ f_2 = f \circ h_1$ this is clear. □

REMARK 3.9.7. In particular, let $f : V_1 \to V$ be a proper surjective morphism of algebraic varieties over a valued field. Let h_1 be a homotopy $h_1 : I \times \widehat{V_1} \to \widehat{V_1}$, and assume h_1 respects the fibers of $\widehat{f}$. Then $\widehat{f}$ is surjective by Lemma 4.2.6, and definably closed by Lemma 4.2.26; so h_1 descends to a homotopy of X.

DEFINITION 3.9.8. Let X be a pro-definable subset of $\widehat{V} \times \Gamma_\infty^n$ and let X' be a pro-definable subset of $\widehat{V'} \times \Gamma_\infty^{n'}$. A continuous pro-definable map $F : X \to X'$ is said to be a *homotopy equivalence* if there exists a continuous pro-definable map $G : X' \to X$ such that $G \circ F$ is homotopic to Id_X and $F \circ G$ is homotopic to $\mathrm{Id}_{X'}$.

3.10 GOOD METRICS

By a *definable metric* on an algebraic variety V over a valued field F, we mean an F-definable function $d : V^2 \to \Gamma_\infty$ which is v+g-continuous and such that

(1) $d(x,y) = d(y,x)$; $d(x,x) = \infty$;

(2) $d(x,z) \geq \min(d(x,y), d(y,z))$;

(3) if $d(x,y) = \infty$ then $x = y$.

Note that given a definable metric on V, for any $v \in V$, $B(v; d, \gamma) := \{y : d(v,y) \geq \gamma\}$ is a family of g-closed, v-clopen sets whose intersection is $\{v\}$. Iit follows by a definable compactness argument that d induces the v-topology on V; this is anyhow clear for the metrics we will use.

We call d a *good metric* if there exists a v+g-continuous F-definable function $\rho : V \to \Gamma$ (so $\rho(v) < \infty$), such that for any $v \in V$ and any $\alpha \geq \rho(v)$, $B(v; d, \alpha)$ is affine and has a unique generic type, i.e. a definable type p such that for any Zariski closed subset V' of V not containing $B(v; d, \alpha)$ and any regular f on $V \smallsetminus V'$, p concentrates on $B(v; d, \alpha) \smallsetminus V'$, and $p_*(f)$ attains the minimum valuation of f on $B(v; d, \alpha) \smallsetminus V'$. Such a type is orthogonal to Γ, hence stably dominated.

The continuity of ρ can be replaced by local boundedness in this definition, using Lemma 10.1.8.

LEMMA 3.10.1.

(1) $\mathbb{P}^n$ *admits a good metric, with* $\rho = 0$.

(2) *Let* F *be a valued field,* V *a quasi-projective variety over* F. *Then there exists a definable metric on* V.

Proof. Consider first the case of $\mathbb{P}^1 = \mathbb{A}^1 \cup \{\infty\}$. If $x, y \in \mathcal{O}$, set $d(x, y) = d(x^{-1}, y^{-1}) = \mathrm{val}(x - y)$ and let $d(x, y) = 0$ if $v(x)$ and $v(y)$ have different signs. This is easily checked to be consistent, and to satisfy the conditions (1-3). It is also clearly v-continuous. Let us now prove g-continuity. By Proposition 9.6.1, it is enough to check that if $F \leq K$ is a valued field extension, $\pi : \Gamma(K) \to \bar{\Gamma}$ a homomorphism of ordered $\mathbb{Q}$-spaces extending $\Gamma(F)$, and $\mathbf{K} = (K, \pi \circ v)$, then $\pi(d_K(x, y)) = d_{\mathbf{K}}(x, y)$. If $x, y \in \mathcal{O}_K$ then $x, y \in \mathcal{O}_{\mathbf{K}}$ and $\pi(d_K(x, y)) = \pi(v_K(x - y)) = d_{\mathbf{K}}(x, y)$. Similarly for x^{-1}, y^{-1}. If $v(x) < 0 < v(y)$, then $v(x - y) < 0$ so $\pi(v(x - y)) \leq 0$, hence $d_{\mathbf{K}}(x, y) = 0 = d_K(x, y)$. This proves g-continuity. It is clear that the metric is good, with $\rho = 0$.

Now consider $\mathbb{P}^n$ with homogeneous coordinates $[X_0 : \ldots : X_n]$. For $0 \leq i \leq n$ denote by U_i the subset $\{x \in \mathbb{P}^n : X_i \neq 0 \wedge \inf \mathrm{val}(X_j/X_i) \geq 0\}$. If x and y belong both to U_i, one sets $d(x, y) = \inf \mathrm{val}(X_j/X_i - Y_j/Y_i)$. If $x \in U_i$ and $y \notin U_i$, one sets $d(x, y) = 0$. One checks that this definition is unambiguous and reduces to the former one when $n = 1$. The proof it is v+g-continuous is similar to the case $n = 1$ and the fact it is good with $\rho = 0$ is clear. This metric restricts to a metric on any subvariety of $\mathbb{P}^n$. □

3.11 ZARISKI TOPOLOGY

We shall occasionally use the Zariski topology on $\widehat{V}$. If V is an algebraic variety over a valued field, a subset of $\widehat{V}$ of the form $\widehat{F}$ with F Zariski closed, resp. open, in V is said to be *Zariski closed*, resp. *open*. Similarly, a subset E of $\widehat{V}$ is said to be *Zariski dense* in $\widehat{V}$ if $\widehat{V}$ is the only Zariski closed set containing E. For $X \subset \widehat{V}$, the Zariski topology on X is the one induced from the Zariski topology on $\widehat{V}$.

3.12 SCHEMATIC DISTANCE

Let $f(x_0, \ldots, x_m)$ be a homogenous polynomial with coefficients in the valuation ring $\mathcal{O}_F$ of a valued field F. One defines a function $\mathrm{val}(f) : \mathbb{P}^m \to [0, \infty]$ by $\mathrm{val}(f)([x_0 : \ldots : x_m]) = \mathrm{val}(f(x_0/x_i, \ldots, x_m/x_i))$ for any i such that $\mathrm{val}(x_i) = \min_j(\mathrm{val}(x_j))$.

Now let V be a projective variety over a valued field F and let Z be a closed F-subvariety of V. Fix an embedding $\iota : V \hookrightarrow \mathbb{P}^m$ and a family f of homogenous polynomials f_i, $1 \leq i \leq r$, in $\mathcal{O}_F[x_0, \ldots, x_m]$ such that $Z = V \cap (f_1 = \cdots = f_r = 0)$. For x in V set $\varphi_{\iota,f}(x) = \min_i(\mathrm{val}(f_i(x)))$. The function $\varphi_{\iota,f} : V \to [0, \infty]$ is clearly F-definable and v+g-continuous and $\varphi_{\iota,f}^{-1}(\infty) = Z$. Any function $V \to [0, \infty]$ of the form $\varphi_{\iota,f}$ for some ι, f will be called a *schematic distance function to Z*.

Chapter Four

Definable compactness

Summary. This chapter is devoted to the study of definable compactness for subsets of $\widehat{V}$. One of the main results is Theorem 4.2.20 which establishes the equivalence between being definably compact and being closed and bounded.

4.1 DEFINITION OF DEFINABLE COMPACTNESS

We will use definable types as a replacement for the curve selection lemma, whose purpose is often to use the definable type associated with a curve at a point. Note that the curve selection lemma itself is not true for Γ_∞, e.g. in $\{(x, y) \in \Gamma_\infty^2 : y > 0, x < \infty\}$ there is no curve approaching $(\infty, 0)$.

As we already observed, one can consider definable types in infinitely many variables, thus the notion of a definable type on a pro-definable set makes sense. Let X be a definable or pro-definable topological space in the sense of 3.3. Let p be a definable type on X.

DEFINITION 4.1.1. A point $a \in X$ is a *limit* of p if for any definable neighborhood U of a (defined with parameters), p concentrates on U.

When X is Hausdorff, it is clear that a limit point is unique if it exists. Recall that by Proposition 3.7.8, $\widehat{V}$ is Hausdorff for any variety V over a valued field (hence also $\widehat{V} \times \Gamma_\infty^n$).

DEFINITION 4.1.2. Let X be a definable or pro-definable topological space. One says X is *definably compact* if any definable type p on X has a limit point in X.

For subspaces of Γ^n with Γ o-minimal, our definition of definable compactness lies between the definition of [32] in terms of curves, and the property of being closed and bounded; so all three are equivalent. This will be treated in more detail later.

4.2 CHARACTERIZATION OF DEFINABLE COMPACTNESS

A subset of VF^n is said to be *bounded* if for some γ in Γ it is contained in $\{(x_1, \dots, x_n) : v(x_i) \geq \gamma, 1 \leq i \leq n\}$. This notion extends to varieties V over a valued field, cf., e.g., [38] p. 81: $X \subset V$ is defined to be bounded if there exists an affine cover $V = \cup_{i=1}^m U_i$, and bounded subsets $X_i \subset U_i$, with $X \subset \cup_{i=1}^m X_i$. Note that the projective space $\mathbb{P}^n$ is bounded within itself, and so any subset of a projective variety V is bounded in V. We shall say a subset of Γ_∞^m is *bounded* if it is contained in $[a, \infty]^m$ for some a. More generally a subset of $V \times \Gamma_\infty^m$ is bounded if its pullback to $V \times \mathrm{VF}^m$ is bounded. We shall say a subset Y of $\widehat{V}$, resp. $\widehat{V \times \Gamma_\infty^m}$, is bounded if there exists a bounded definable subset X of V, resp. $V \times \Gamma_\infty^m$, such that $Y \subset \widehat{X}$.

Let Y be a definable subset of Γ_∞. Let q be a definable type on Y. If Y is bounded there is a unique $\alpha \in \Gamma_\infty$, such that q concentrates on any neighborhood of α. Indeed, consider the $q(x)$-definition of the formula $x > y$; it must have the form $y < \alpha$ or $y \leq \alpha$.

Let V be a definable set and let q be a definable type on $\widehat{V}$. Assume there exists $r \in \widehat{V}$ such that for any continuous pro-definable map $f : \widehat{V} \to \Gamma_\infty$, $\lim f_*(q)$ exists and $f(r) = \lim f_*(q)$. Then $\lim q$ exists and $r = \lim q$.

LEMMA 4.2.1. *Let V be an affine variety over a valued field and let q be a definable type on $\widehat{V}$. We have $\lim q = r$ if and only if for any regular function H on V, setting $h = \mathrm{val} \circ H$,*

$$h_*(r) = \lim h_*(q).$$

Proof. One implication is clear, let us prove the reverse one. Indeed, by hypothesis, for any pro-definable neighborhood W of r, q implies $x \in W$. In particular, if U is a definable neighborhood of $h_*(r)$, q implies $x \in f^{-1}(U)$, hence $h_*(q)$ implies $x \in U$. It follows that $\lim h_*(q) = h_*(r)$. □

LEMMA 4.2.2. *Let X be a bounded definable subset of an algebraic variety V over a valued field and let q be a definable type on $\widehat{X}$. Then $\lim q$ exists in $\widehat{V}$.*

Proof. It is possible to partition V into open affine subsets V_i and X into bounded definable subsets $X_i \subset V_i$. We may thus assume V is affine; and indeed that X is a bounded subset of $\mathbb{A}^n$. For any regular function H on V, setting $h = \mathrm{val} \circ H$, $h(X)$ is a bounded subset of Γ_∞ and $h_*(q)$ is a definable type on $h(X)$, hence has a limit $\lim h_*(q)$.

Now let K be an algebraically closed valued field containing the base of definition of V and q. Fix $\delta \models q|K$ and $d \models p_\delta|K(\delta)$, where p_δ is the type coded by the element $\delta \in \widehat{V}$. Let $B = \Gamma(K)$, $N = K(\delta, d)$ and $B' = \Gamma(N)$. So B is a divisible ordered abelian group. We have $\Gamma(N) = \Gamma(K(\delta))$ by orthogonality to Γ of p_δ. Since q is definable, for any $e \in B'$, $\mathrm{tp}(e/B)$ is definable; in particular the cut of e over B is definable. Set $B'_0 = \{b' \in B' : (\exists b \in B) b < b'\}$. It follows that if $e \in B'_0$ there exists an element $\pi(e) \in B \cup \{\infty\}$ which is nearest e. Note $\pi : B'_0 \to B_\infty$ is an order-preserving retraction and a homomorphism in the obvious sense. The ring $R = \{a \in K(d) : \mathrm{val}(a) \in B'_0\}$ is a valuation ring of $K(d)$, containing K. Also d has its coordinates in R, because of the boundedness assumption on X. Consider the maximal ideal $M = \{a \in K(d) : \mathrm{val}(a) > B\}$ and set $K' = R/M$. We have a canonical homomorphism $R \to K'$; let d' be the image of d. We have a valuation on K' extending the one on K, namely $\mathrm{val}(x+M) = \pi(\mathrm{val}(x))$. So K' is a valued field extension of K, embeddable in some elementary extension. Let $r = \mathrm{tp}(d'/K)$. Then r is definable and stably dominated; the easiest way to see that is to assume K is maximally complete (as we may); in this case stable domination follows from $\Gamma(K(d')) = \Gamma(K)$ by Theorem 2.9.2. The fact that, for any h as above, $r_*(h) = \lim h_*(q)$ is a direct consequence from the definitions. □

REMARK 4.2.3. Let V be a definable set. According to Definition 4.1.2 a pro-definable $X \subset \widehat{V}$ is definably compact if for any definable type q on X we have $\lim q \in X$. Under this definition, any intersection of definably compact sets is definably compact, in particular an interval such as $\cap_n [0, 1/n]$ in Γ. However we mostly have in mind strict pro-definable sets.

LEMMA 4.2.4. *Let V be an algebraic variety over a valued field, Y a closed pro-definable subset of $\widehat{V}$. Let q be a definable type on Y, and suppose* $\lim q$ *exists. Then* $\lim q \in Y$.

Hence if Y is bounded and closed in $\widehat{V}$, then Y is definably compact.

Proof. The fact that $\lim q \in Y$ when Y is closed pro-definable follows from the definition of the topology on $\widehat{V}$. The second statement thus follows from Lemma 4.2.2. □

DEFINITION 4.2.5. Let T be a theory with universal domain $\mathbb{U}$. Let Γ be a stably embedded sort with a $\varnothing$-definable linear ordering. Recall T is said to be *metastable* over Γ if for any small $C \subset \mathbb{U}$, the following condition is satisfied:

(MS) For some small B containing C, for any a belonging to a finite product of sorts, $\mathrm{tp}(a/B, \Gamma(Ba))$ is stably dominated.

Such a B is called a *metastability base.* It follows from Theorem 2.9.2 that ACVF is metastable over Γ and that any maximally complete algebraically closed valued field is a metastability base.

Let T be any theory, X and Y be pro-definable sets, and $f : X \to Y$ a pro-definable map. Then f induces a map $f_* : S_{def,X} \to S_{def,Y}$ from the set of definable types on X to the set of definable types on Y. We remark that if f is injective, then so is f_*. This reduces to the case of definable $f : X \to Y$, where it is clear. We now deal with surjectivity.

LEMMA 4.2.6. *Let $f : X \to Y$ a surjective pro-definable map between pro-definable sets.*

(1) *Assume T is o-minimal. Then $f_* : S_{def,X} \to S_{def,Y}$ is surjective.*

(2) *Assume T is metastable over some o-minimal Γ. Then $f_* : S_{def,X} \to S_{def,Y}$ is surjective.*

(3) *Assume T is metastable over some o-minimal Γ. Then f_* restricts to a surjective map $\widehat{X} \to \widehat{Y}$.*

In (2), if X, Y and f are defined over a metastability base M, then for any M-definable type r on Y there exists an M'-definable type p on X, with M' generated over M by elements of Γ, such that $r = f_*(p)$. More precisely, there exists a set A, possibly infinite, an M'-definable type p' on the pro-definable set $W = \Gamma^A$ and an M-pro-definable map $H : W \to \widehat{X}$ such that $p = \int_{p'} H$ verifies $f_*(p) = r$.

REMARK 4.2.7. (1) In general surjectivity over a given base set does not hold in (2) and (3) (e.g. take X a finite set, Y a point).

(2) It would also be possible to prove the C-minimal case analogously to the o-minimal one, as below.

Proof. Let us prove (1). First note it is enough to consider the case where X consists of real elements. Indeed if X, Y consist of imaginaries, find a set X' of real elements and a surjective map $X' \to X$; then it suffices to show $S_{def,X'} \to S_{def,Y}$ is surjective.

The statement reduces to the case that $X \subset U \times Y$ is a complete type, $f : X \to Y$ is the projection, and U is one of the basic sorts. Indeed, we can first let $U = X$ and replace X by the graph of f. Any given definable type $r(y)$ in Y restricts to some complete type $r_0(y)$, which we can extend to a complete type $r_0(u, y)$ over some model implying X. Thus we can take $X \subset U \times Y$ to be complete. Recall that when $X = \varprojlim_j X_j$, we have $S_{def,X} = \varprojlim_j S_{def,X_j}$ naturally. Thus, by transfinite induction, it is enough to consider the case of 1-variable U.

We can take X, Y to be complete types with $X \subset \Gamma \times Y$, and $f : X \to Y$ the projection. It follows from completeness that for any $b \in Y$, $f^{-1}(b)$ is convex. Let $r(y)$ be a definable type in Y. Let M be a model with r defined over M, let $b \models r|M$, and consider $f^{-1}(b)$.

If for any M, $x \in X \wedge f(x) \models r|M$ is a complete type $p|M$ over M, then $x \in X \cup p(f(x))$ already generates a definable type by Lemma 2.3.1 and we are done. So, let us assume for some M, and $b \models p|M$, $x \in X \wedge f(x) = b$ does not generate a complete type over $M(b)$. Then there exists an $M(b)$-definable set D that splits $f^{-1}(b)$ into two pieces. We can take D to be an interval. Then since $f^{-1}(b)$ is convex, one of the endpoints of D must fall in $f^{-1}(b)$. This endpoint is $M(b)$-definable, and can be written $h(b)$ with h an M-definable function. In this case $\mathrm{tp}(h(b), b/M)$ is M-definable, and has a unique extension to an M-definable type. In either case we found $p \in S_{def,X}$ with $f_*(p) = r$. Note that the proof works when only X is contained in the definable closure of an o-minimal definable set, for any pro-definable Y.

For the proof of (2) consider $r \in S_{def,Y}$. Let M be a metastability base, with f, X, Y, and r defined over M. Let $b \models r|M$, and let $c \in f^{-1}(b)$. Let b_1 enumerate $\Gamma(M(b))$. Then $\mathrm{tp}(b/M(b_1)) = r'|M(b_1)$ with r' stably dominated, and $\mathrm{tp}(b_1/M) = r_1|M$ with r_1 definable. Let c_1 enumerate $\Gamma(M(c))$; then $\mathrm{tp}(cb/M(c_1)) = q'|M(c_1)$ with q' stably dominated. We have $q' = \tau(c_1)$ for some M-definable function into the stable dominated types, and $r' = \sigma(b_1)$ similarly. By (1) (and stable embeddedness of Γ), it is possible to extend $\mathrm{tp}(c_1b_1/M) \cup r_1$ to a definable type $q_1(x_1, y_1)$ over some M', where M' can be taken to be generated over M by elements of Γ. Let $c_1b_1 \models q_1|M'$, and $cb \models q'|M'(c_1b_1)$. So $\mathrm{tp}(b/M') = r|M'$. Now $\mathrm{tp}(bc/M')$ extends to a definable type $p = \int_{q_1} \tau$ by transitivity, and $f_*(p) = \int_{r_1} \sigma = r$.

Note that the proof in [20] 10.7 and 10.8 holds verbatim in the metastable setting, yielding that a definable type p is stably dominated if and only if it is orthogonal to Γ, as in Proposition 2.9.1. Thus, the proof of (3) is similar to the proof of (2); in this case there is no b_1, and q_1 can be chosen so that $c_1 \in M'$. Indeed $\mathrm{tp}(c_1/M)$ implies $\mathrm{tp}(c_1/M(b))$ so it suffices to take M' containing $M(c_1)$. □

REMARK 4.2.8. There should be no difficulty to give an abstract version of Lemma 4.2.6; let us just mention one more case that we will require. Say T has the extension property if f_* is always surjective, in the situation of Lemma 4.2.6. First, let $T = \mathrm{Th}(A)$, where A is a linearly ordered group with a definable convex subgroup B, such that $(*)$ B and A/B are o-minimal. Then T has the extension property. This is proved exactly as in the beginning of the proof of (1) in Lemma 4.2.6, by reduction to 1-types; here we can reduce to B and cosets of B (all o-minimal) and to A/B. Secondly, assume T is metastable with respect to a linearly ordered group with $(*)$; then the proof of (2) shows that T has the extension property.

In particular, the theory $\mathrm{ACV^2F}$ obtained from ACVF by expanding Γ by a predicate for a convex subgroup considered in 9.3 has the extension property.

PROPOSITION 4.2.9. *Let V and V' be algebraic varieties over a valued field. Let W be a definably compact pro-definable subset of $\widehat{V \times \Gamma_\infty^m}$, and let $f : W \to \widehat{V' \times \Gamma_\infty^{m'}}$ be a continuous pro-definable morphism. Then $f(W)$ is definably compact.*

Proof. Let q be a definable type on $f(W)$. By Lemma 4.2.6 there exists a definable type r on W, with $f_*(r) = q$. Since W is definably compact, $\lim r$ exists and belongs to W. But then $\lim q = f(\lim r)$ belongs to $f(W)$ (since this holds after composing with any continuous morphism to Γ_∞). So $f(W)$ is definably compact. □

LEMMA 4.2.10. *Let V be an algebraic variety over a valued field, and let W be a definably compact pro-definable subset of $\widehat{V \times \Gamma_\infty^m}$. Then W is contained in $\widehat{X}$ for some bounded definable v+g-closed subset X of $V \times \Gamma_\infty^m$.*

Proof. By using Proposition 4.2.9 for the projections $\widehat{V \times \Gamma_\infty^m} \to \widehat{V}$ and $\widehat{V \times \Gamma_\infty^m} \to \Gamma_\infty$, one may assume W is a pro-definable subset of Γ_∞ or $\widehat{V}$. The first case is clear. For the second one, one may assume V is affine contained in $\mathbb{A}^n$ with coordinates $(x_1, \dots, x_n)$. Consider the function $\min \mathrm{val}(x_i)$ on V, extended to $\widehat{V}$; it is a continuous function on $\widehat{V}$. The image of W is a definably compact subset of Γ_∞, hence is bounded below, say by α. Let $X = \{(x_1, \dots, x_n) : \mathrm{val}(x_i) \geq \alpha\}$. Then $W \subset \widehat{X}$. □

By a *countably pro-definable* set we mean a pro-definable set isomorphic to one with a countable inverse limit system. Note that $\widehat{V}$ is countably pro-definable.

LEMMA 4.2.11. *Let X be a strict, countably pro-definable set over a model M, Y a relatively definable subset of X over M. If $Y \neq \varnothing$ then $Y(M) \neq \varnothing$.*

Proof. Write $X = \varprojlim_n X_n$ with transition morphisms $\pi_{m,n} : X_m \to X_n$, and X_n and $\pi_{m,n}$ definable. Let $\pi_n : X \to X_n$ denote the projection. Since X is strict pro-definable, the image of X in X_n is definable; replacing X_n with this image, we may assume π_n is surjective. Since Y is relatively definable, it has the form $\pi_n^{-1}(Y_n)$ for some nonempty $Y_n \subset X_n$. We have $Y_n \neq \varnothing$, so there exists $a_n \in Y_n(M)$. Define inductively $a_m \in Y_m(M)$ for $m > n$, choosing $a_m \in Y_m(M)$ with $\pi_{m,m-1}(a_m) = a_{m-1}$. For $m < n$ let $a_m = \pi_{n,m}(a_n)$. Then (a_m) is an element of $Y(M)$. □

Let X be a pro-definable set with a definable topology (in some theory). Given a model M, and an element a of X in some elementary extension of M, we say that $\mathrm{tp}(a/M)$ has a limit b if $b \in X(M)$, and for any M-definable open neighborhood U of b, we have $a \in U$. This extends the notion of a limit of a definable type; if $a \models q|M$ with q an M-definable type, the limits have the same meaning. In the o-minimal setting of Γ_∞, we show however that in fact limits appear only for definable types.

LEMMA 4.2.12. *Let M be an elementary submodel of Γ. Let A be a set and let $a \in \Gamma_\infty^A$. Let $p_0 = \mathrm{tp}(a/M)$ and assume $\lim p_0$ exists. Then there exists a (unique) M-definable type p extending p_0.*

Proof. It is enough to consider the case when A is finite, so we may assume $\Gamma_\infty^A = \Gamma_\infty^n$. In case $n = 1$, $\mathrm{tp}(a/M)$ is determined by a cut in $\Gamma_\infty(M)$. If this cut is irrational then by definition there can be no limit in M. So this case is clear. We have to show that for any formula $\phi(x, y)$ over M, with $x = (x_1, \dots, x_n)$ and $y = (y_1, \dots, y_m)$, the set $\{c \in M : \phi(a, c)\}$ is definable. Any formula is a Boolean combination of unary formulas and of formulas of the form: $\sum \alpha_i x_i + \sum \beta_j y_j + \gamma \diamond 0$, where i, j range over some subset of $\{1, \ldots, n\}$, $\{1, \ldots, m\}$ respectively, $\alpha_i, \beta_j \in \mathbb{Q}, \gamma \in \Gamma(M)$, and $\diamond \in \{=, <\}$. This case follows from the case $n = 1$ already noted, applied to $\mathrm{tp}(\sum \alpha_i a_i/M)$. □

PROPOSITION 4.2.13. *Let X be a pro-definable subset of $\widehat{V} \times \Gamma_\infty^m$ with V an algebraic variety over a valued field. Let a belong to the closure of X. Then there exists a definable type on $\widehat{V} \times \Gamma_\infty^m$ concentrating on X, with limit point a.*

Proof. We may assume V is affine, and by lifting the Γ-coordinates to $\widehat{\mathbb{A}^m}$ and absorbing in the field coordinates, that $m = 0$. Let M be a maximally complete model of ACVF over which the data is defined. It is a metastability base. Let $\mathcal{F} = (f_i)_{i \in I}$ list all functions on $\widehat{V}$ of the form $\mathrm{val}(F)$ for a regular function F on V, defined over M. Since $\mathcal{F}_*(a)$ is a limit point of $\mathcal{F}(X)$, there exists a type q_M on $\mathcal{F}(X)$ over M, with limit $\mathcal{F}_*(a)$. By Lemma 4.2.12, q_M extends to an M-definable type q with limit $\mathcal{F}_*(a)$. By Lemma 4.2.6, there exists a definable type p on X such that $\mathcal{F}_*(p) = q$. Furthermore one can assume p is defined over $M' = M \cup E$ with $E \subset \Gamma$ and $p = \int_{p'} H$, where p' is an M'-definable type on a pro-definable set $W = \Gamma^A$ and H is an M-pro-definable map $W \to \widehat{X}$. Let us prove that $\lim p = a$. Recall the canonical map $\vartheta : \widehat{\widehat{V}} \to \widehat{V}$ from Remark 2.11.4 sending a stably dominated type q on $\widehat{V}$ to $\vartheta(q) = \int_q \mathrm{id}_{\widehat{V}}$. By composing H with ϑ, one obtains an M-pro-definable map $h : W \to \widehat{V}$. To prove that $\lim p = a$, it is enough to check that $\lim h_*(p') = a$. Now assume V is embedded in $\mathbb{A}^m$ and consider the morphisms $J_d : \widehat{\mathbb{A}^m} \to L(H_d)$ defined in 5.1. For every $d \geq 0$, set $h_d = J_d \circ h$. Note that $h_d(W)$ is a Γ-internal subset of $L(H_d)$ defined over M. By Lemma 6.2.2, there exists a finite number of bases of H_d over M such that each semi-lattice in $h_d(W)$ is diagonal for one of these bases. It follows that there exists a common basis B_d, defined over M, that diagonalizes all semi-lattices $h_d(t)$ for $t \models p'$. Since for any basis element $e \in B_d$, the valuative norm of e according to the semi-lattice $h_d(t)$ is given by the functions in $\mathcal{F}$ and $\lim q = \mathcal{F}_*(a)$, it follows that $\lim h_d$ exists for all d. Since by Theorem 5.1.4 the morphism $J : \widehat{\mathbb{A}^m} \longrightarrow \varprojlim_d L(H_d)$ induced by the system (J_d) is injective and induces a homeomorphism between $\widehat{\mathbb{A}^m}$ and its image, it follows that $\lim (J \circ h)_*(p') = J(a)$ and $\lim h_*(p') = a$. □

COROLLARY 4.2.14. *Let X be a pro-definable subset of $\widehat{V} \times \Gamma_\infty^m$, with V an algebraic variety over a valued field. If X is definably compact, then X is closed in $\widehat{V} \times \Gamma_\infty^m$.*

Proof. The fact that X is closed is immediate from Proposition 4.2.13 and the definition of definable compactness. □

REMARK 4.2.15. Let Γ be a Skolemized o-minimal structure, $a \in \Gamma^n$. Let D be a definable subset of Γ^n such that a belongs to the topological closure $cl(D)$ of D. Then there exists a definable type p on D with limit a, in the sense of Definition 4.1.1. Indeed, consider the family F of all rectangles (products of intervals) whose interior contains a. This is a definable family, directed downwards under containment. By Lemma 4.2.18 there exists a definable type q on F concentrating, for each $b \in F$, on $\{b' \in F : b' \subset b\}$. Since $a \in cl(D)$, there exists a definable (Skolem) function g such that for $u \in F$, $g(u) \in u \cap D$. To conclude it is enough to set $p = g_*(q)$. An alternative proof is provided, in our case, by Proposition 4.2.13. It follows that if the limit of any definable type on D exists and lies in D, then D is closed. Conversely, if D is bounded, any definable type on D will have a limit, and if D is closed then this limit is necessarily in D.

Even for Th(Γ), definability of a type tp(ab/M) does not imply that tp($a/M(b)$) is definable. For instance b can approach ∞, while $a \sim \alpha b$ for some irrational real α, i.e. $qb < a < q'b$ if $q, q' \in \mathbb{Q}$, $q < \alpha < q'$. However we do have:

LEMMA 4.2.16. *Let p be a definable type of Γ^n, over M. Then up to a definable change of coordinates, p decomposes as the join of two orthogonal definable types p_f, p_i, such that p_f has a limit in Γ^m, and p_i has limit point ∞^ℓ.*

Proof. If $\alpha \in \mathbb{Q}^n$ and $x \in \Gamma^n$, we write $\alpha \cdot x$ for the scalar product $\sum_i \alpha_i x_i \in \Gamma$. Let $\alpha_1, \dots, \alpha_k$ be a maximal set of linearly independent vectors in $\mathbb{Q}^n$ such that the image of p under $x \mapsto \alpha_i \cdot x$ has a limit point in Γ. Let $\beta_1, \dots, \beta_\ell$ be a maximal set of vectors in $\mathbb{Q}^n$ such that for $x \models p|M$, $\alpha_1 \cdot x, \dots, \alpha_k \cdot x, \beta_1 \cdot x, \dots, \beta_\ell \cdot x$ are linearly independent over M. If $a \models p|M$, let $a' = (\alpha_1 \cdot a, \dots, \alpha_k \cdot a)$, $a'' = (\beta_1 \cdot a, \dots, \beta_\ell \cdot a)$. For $\alpha \in \mathbb{Q}(\alpha_1, \dots, \alpha_k)$ we have that $\alpha \cdot a$ is bounded between elements of M. On the other hand each $\beta \cdot a$, with $\beta \in \mathbb{Q}(\beta_1, \dots, \beta_\ell)$, satisfies $\beta \cdot a > M$ or $\beta \cdot a < M$. For if $m \leq \beta \cdot a \leq m'$ for some $m, m' \in M$, since tp($\beta \cdot a/M$) is definable it must have a finite limit, contradicting the maximality of k. It follows that tp($\alpha \cdot a/M$) $\cup$ tp($\beta \cdot a/M$) extends uniquely to a complete 2-type, namely tp($(\alpha \cdot a, \beta \cdot a)/M$); in particular tp($(\alpha \cdot a) + (\beta \cdot a)/M$) is determined; from this, by quantifier elimination, tp(a'/M)$\cup$tp(a''/M) extends to a unique type in $k + \ell$ variables. So tp(a'/M) and tp(a''/M) are orthogonal. After some sign changes in a'', so that each coordinate is $> M$, the lemma follows. $\square$

REMARK 4.2.17. It follows from Lemma 4.2.16 that to check for definable compactness of X, it suffices to check definable maps from definable types on Γ^k that either have limit 0, or limit ∞. From this an alternative proof of the g- and v-criteria of Chapter 9 for closure in $\widehat{V}$ can be deduced.

For the sake of completeness we shall provide the proof of the following lemma from [23] here.

LEMMA 4.2.18 ([23] Lemma 2.19). *Let P be a definable directed partial ordering in an o-minimal structure Γ. Then there exists a definable type p cofinal in P.*

Proof. We assume P is 0-definable, and work with 0-definable sets; we will find a 0-definable type with this property. Note first that we may replace P with any 0-definable cofinal subset. Also if Q_1, Q_2 are non-cofinal subsets of P, there exist a_1, a_2 such that no element of Q_i lies above a_i; but by directedness there exists $a \geq a_1, a_2$; so no element of $Q_1 \cup Q_2$ lies above a, i.e. $Q_1 \cup Q_2$ is not cofinal. In particular if $P = P' \cup P''$, at least one of P', P'' is cofinal in P (hence also directed).

If $\dim(P) = 0$ then P is finite, so according to the above remarks we may assume it is one point; in which case the lemma is trivial. We use here the fact that in an o-minimal theory, any point of a finite 0-definable set is definable.

If $\dim(P) = n > 0$, we can divide P into finitely many 0-definable sets P_i, each admitting a map $f_i : P_i \to \Gamma$ with fibers of dimension $< n$. We may thus assume that there exists a 0-definable map $f : P \to \Gamma$ with fibers of dimension $< n$. Let $P(\gamma) = f^{-1}(\gamma)$, and $P(a,b) = f^{-1}(a,b)$.

CLAIM 1. *One of the following holds:*

(1) *For any $a \in \Gamma$, $P(a, \infty)$ is cofinal in P.*

(2) *For some 0-definable $a \in \Gamma$, for all $b > a$, $P(a, b)$ is cofinal.*

(3) *For some 0-definable $a \in \Gamma$, $P(a)$ is cofinal.*

(4) *For some 0-definable $a \in \Gamma$, for all $b < a$, $P(b, a)$ is cofinal.*

(5) *For all $a \in \Gamma$, $P(-\infty, a)$ is cofinal.*

Proof of the claim. Suppose (1) and (5) fail. Then $P(a, \infty)$ is not cofinal in P for some a; so $P(-\infty, b)$ must be cofinal, for any $b > a$. Since (5) fails, the set $\{b : P(-\infty, b) \text{ is cofinal}\}$ is a nonempty proper definable subset of Γ, closed upwards, hence of the form $[A, \infty)$ or (A, ∞) for some 0-definable $A \in \Gamma$. In the former case, $P(-\infty, A)$ is cofinal, but $P(-\infty, b)$ is not cofinal for $b < A$, so $P(b, A)$ is cofinal for any $b < A$; thus (4) holds. In the latter case, $(-\infty, b)$ is cofinal for any $b > A$, while $(-\infty, A)$ is not; so $P([A, b))$ is cofinal for any $b > A$. Thus either (2) or (3) hold. □

Let p_1 be a 0-definable type of Γ, concentrating on sets X with $f^{-1}(X)$ cofinal. (For instance in case (1) p_1 concentrates on intervals (a, ∞).)

CLAIM 2. *For any $c \in P$, if $a \models p_1|\{c\}$ then there exists $d \in P(a)$ with $d \geq c$.*

Proof of the claim. Let $Y(c) = \{x : (\exists y \in P(x))(y \geq c)\}$. Then $P^{-1}(\Gamma \setminus Y(c))$ is not cofinal in P, so it cannot be in the definable type p_1. Hence $Y(c) \in p_1|\{c\}$. □

Now let $M \models T$. Let $a \models p_1|M$. By induction, let q_a be an a-definable type, cofinal in $P(a)$, and let $b \models q_a|Ma$. Then $\mathrm{tp}(ab/M)$ is definable. If $c \in M$ then by Claim 2, there exists $d \in P(a)$ with $d \geq c$. So the set $\{y \in P(a) : \neg(y \geq c)\}$ is not cofinal in $P(a)$. Therefore this set is not in q_a. Since $b \models q_a|Ma$, we have $b \geq c$. This shows that $\mathrm{tp}(ab/M)$ is cofinal in P. □

LEMMA 4.2.19. *Let S be a definably compact definable subset of an o-minimal structure. If $\mathcal{D}$ is a uniformly definable family of nonempty closed definable subsets of S, and $\mathcal{D}$ is directed (the intersection of any two elements of $\mathcal{D}$ contains a third one), then $\cap\mathcal{D} \neq \varnothing$.*

Proof. By Lemma 4.2.18 there exists a cofinal definable type $q(y)$ on $\mathcal{D}$ concentrating, for each $U \in \mathcal{D}$, on $\{V \in \mathcal{D} : V \subset U\}$.

Using the lemma on extension of definable types, Lemma 4.2.6, let $r(w, y)$ be a definable type extending q and implying $w \in U_y \cap S$. Let $p(w)$ be the projection of r to the w-variable. By definable compactness $\lim p = a$ exists. Since a is a limit of points in D, we have $a \in D$ for any $D \in \mathcal{D}$. So $a \in \cap\mathcal{D}$. □

Lemma 4.2.19 gives another proof that a definably compact set is closed: let $\mathcal{D} = \{S \setminus U\}$, where U ranges over basic open neighborhoods of a given point a of the closure of S.

THEOREM 4.2.20. *Let V be an algebraic variety over a valued field, and let W be a pro-definable subset of $\widehat{V \times \Gamma_\infty^m}$. Then W is definably compact if and only if it is closed and bounded.*

Proof. If W is definably compact it is closed and bounded by Corollary 4.2.14 and Lemma 4.2.10. If W is closed and bounded, its preimage W' in $\widehat{V \times \mathbb{A}^m}$ under $\mathrm{id} \times \mathrm{val}$ is also closed and bounded, hence definably compact by Lemma 4.2.4. It follows from Proposition 4.2.9 that W is definably compact. □

For Γ^n, Theorem 4.2.20 is a special case of [32], Theorem 2.1.

PROPOSITION 4.2.21. *Let V be a variety over a valued field F, and let W be an F-definable subset of $V \times \Gamma_\infty^m$. Then W is v+g-closed (resp. v+g-open) if and only if $\widehat{W}$ is closed (resp. open) in $\widehat{V}$.*

Proof. A Zariski-locally v-open set is v-open, and similarly for g-open, hence for v+g-open. So we may assume $V = \mathbb{A}^n$ and by pulling back to $V \times \mathbb{A}^m$ that $m = 0$. It is enough to prove the statement about closed subsets.

Let $V_\alpha = (c\mathcal{O})^n$ be the closed polydisc of valuative radius $\alpha = \mathrm{val}(c)$. Let $W_\alpha = W \cap V_\alpha$, so $\widehat{W_\alpha} = \widehat{W} \cap \widehat{V_\alpha}$. Then W is v-closed if and only if W_α is v-closed for each α; by Lemma 9.1.3, the same holds for g-closed; also $\widehat{W}$ is closed if and only if $\widehat{W_\alpha}$ is closed for each α. This reduces the question to the case of bounded W.

By Lemma 3.7.7, if W is v+g-closed then $\widehat{W}$ is closed. In the reverse direction, if $\widehat{W}$ is closed it is definably compact. It follows that W is v-closed. For otherwise there exists an accumulation point w of W, with $w = (w_1, \ldots, w_m) \notin W$. Let $\delta(v) = \min_{i=1}^m \mathrm{val}(v_i - w_i)$. Then $\delta(v) \in \Gamma$ for $v \in W$, i.e. $\delta(v) < \infty$. Hence the induced function $\delta : \widehat{W} \to \Gamma_\infty$ also has image contained in Γ; and $\delta(\widehat{W})$ is definably compact. It follows that $\delta(\widehat{W})$ has a maximal point $\gamma_0 < \infty$. But then the γ_0-neighborhood around w contains no point of W, a contradiction.

To conclude it is enough to show that if $\widehat{W}$ is definably compact, then W must be g-closed. This follows from Corollary 9.1.4. □

COROLLARY 4.2.22. *Let V be an algebraic variety over a valued field, and let W be a definable subset of $V \times \Gamma_\infty^m$. Then W is bounded and v+g-closed* if and only if *$\widehat{W}$ is definably compact.*

Proof. Since W is v+g-closed if and only if $\widehat{W}$ is closed by Proposition 4.2.21, this is a special case of Theorem 4.2.20. □

LEMMA 4.2.23. *Let V be an algebraic variety over a valued field and let Y be a v+g-closed, bounded definable subset of $V \times \Gamma_\infty^m$. Let W be a definable subset of $V' \times \Gamma_\infty^m$, with V' another variety. Let $f : Y \to W$ be a definable map. Assume $\widehat{f} : \widehat{Y} \to \widehat{W}$ is continuous. Then $\widehat{f}$ is a definably closed map in the sense of* Definition 3.9.4.

Proof. We may assume f is surjective, in which case $\widehat{f}$ is also surjective by Lemma 4.2.6. By Proposition 4.2.21 and Theorem 4.2.20, $\widehat{Y}$ is definably compact and any pro-definable closed subset of $\widehat{Y}$ is definably compact, so the result follows from Proposition 4.2.9 and Corollary 4.2.14. □

LEMMA 4.2.24. *Let X and Y be v+g-closed, bounded definable subsets of a product of an algebraic variety over a valued field with some Γ_∞^m. Then, the image of any closed pro-definable subset of $\widehat{X} \times \widehat{Y}$ under the projection $\pi : \widehat{X} \times \widehat{Y} \to \widehat{Y}$ is closed.*

Proof. Let h denote the canonical morphism $\widehat{X \times Y} \to \widehat{X} \times \widehat{Y}$. The map $\pi \circ h : \widehat{X \times Y} \to \widehat{Y}$ is definably closed by Lemma 4.2.23. Let Z be a closed pro-definable subset of $\widehat{X} \times \widehat{Y}$. Its preimage $h^{-1}(Z)$ is a closed pro-definable subset of $\widehat{X \times Y}$, hence $\pi(Z) = (\pi \circ h)(h^{-1}(Z))$ is closed. □

COROLLARY 4.2.25. *Let U and V be v+g-closed, bounded definable subsets of a product of an algebraic variety over a valued field with some Γ_∞^m. If $f : \widehat{U} \to \widehat{V}$ is a pro-definable morphism with closed graph, then f is continuous.*

Proof. By Lemma 4.2.24 and Remark 3.9.5, the projection π_1 from the graph of f to $\widehat{U}$ is a homeomorphism onto its image. The projection π_2 is continuous. Hence $f = \pi_2 \circ \pi_1^{-1}$ is continuous. □

LEMMA 4.2.26. *Let $f : V \to W$ be a proper morphism of algebraic varieties. Then $\widehat{f}$ is a definably closed map. So is $\widehat{f} \times \mathrm{Id} : \widehat{V} \times \Gamma_\infty^m \to \widehat{W} \times \Gamma_\infty^m$.*

Proof. Note that $\widehat{V \times \Gamma_\infty^m}$ can be identified with a subset S of $\widehat{V \times \mathbb{A}^m}$ (projecting on generics of balls around zero in the second coordinate); with this identification, $\widehat{f} \times \mathrm{Id}$ identifies with the restriction of $\widehat{f \times \mathrm{Id}_{\mathbb{A}^m}}$ to S. Thus the second statement, for $V \times \Gamma_\infty^m$, reduces to first for the case of the map $f \times \mathrm{Id} : V \times \mathbb{A}^m \to W \times \mathbb{A}^m$.

To prove the statement on $f : V \to W$, let V', W' be complete varieties containing V, W, and let $\bar{V}$ be the closure of the graph of f in $V' \times W'$. In the Zariski topology, the map $\mathrm{Id} \times f : V' \times V \to V' \times W$ is closed by properness (universal closedness). The image of the diagonal on V, under this map, is the graph of f, a subset of $V \times W$; so it is Zariski closed as a subset of $V' \times W$. Let $g : V \to V \times W$ given by $g(v) = (v, f(v))$; so g is the composition of the isomorphism $v \mapsto (v, f(v))$ of V onto the graph of f, with the inclusion of the graph of f in $V \times W$. Both of these induce definably closed morphisms on $\widehat{\ }$-spaces, so $\widehat{g}$ is definably closed.

Let $\pi : \bar{V} \to W'$ be the projection. Now $\widehat{\pi}$ is a definably closed map by Lemma 4.2.23. So $\widehat{\pi} \circ \widehat{g} = \widehat{\pi \circ g} = \widehat{f}$ is definably closed. (We could also obtain the result directly from Proposition 4.2.13.) □

REMARK 4.2.27. The previous lemmas apply also to ∞-definable sets.

COROLLARY 4.2.28. *Let $f : V \to W$ be a finite radicial surjective morphism of algebraic varieties over a valued field K. Then $\widehat{f} : \widehat{V} \to \widehat{W}$ is a homeomorphism.*

Proof. Since for f to be radicial means that for any field extension K' the induced map $V(K') \to W(K')$ is injective, f is an isomorphism in the category of definable sets. Thus, $\widehat{f} : \widehat{V} \to \widehat{W}$ is a bijection, say by Lemma 4.2.6. On the other hand, f being a universal homeomorphism for the Zariski topology, it is proper, thus $\widehat{f}$ is definably closed by Lemma 4.2.26, hence a homeomorphism, cf. Remark 3.9.5. □

LEMMA 4.2.29. *Let X be a v+g-closed bounded definable subset of an algebraic variety V over a valued field. Let $f : X \to \Gamma_\infty$ be a definable map which is v+g-continuous. Then the maximum of f is attained on X, similarly if X is a closed bounded pro-definable subset of $\widehat{V}$.*

Proof. By Lemma 3.8.4, f extends continuously to $F : \widehat{X} \to \Gamma_\infty$. By Proposition 4.2.21 and Theorem 4.2.20 $\widehat{X}$ is definably compact. It follows from Proposition 4.2.9 that $F(\widehat{X})$ is a definably compact subset of Γ_∞ and hence has a maximal point γ. Take p such that $F(p) = \gamma$, let $c \models p$, then $f(c) = \gamma$. □

PROPOSITION 4.2.30. *Let V be an algebraic variety over a valued field. Then V is complete if and only if $\widehat{V}$ is definably compact.*

Proof. If V is projective, $\widehat{V}$ is definably compact by Theorem 4.2.20. If V is complete, there exists a surjective morphism of algebraic varieties $f : W \to V$ with W projective by Chow's lemma. Hence $\widehat{V}$ is definably compact, since $\widehat{V} = \widehat{f}(\widehat{W})$ by Lemma 4.2.6 and $\widehat{V}$ is Hausdorff by Proposition 3.7.8. Conversely, assume $\widehat{V}$ is definably compact. One may embed V as a Zariski dense open subset of a complete variety W by Nagata's theorem. In particular, $\widehat{V}$ is dense in $\widehat{W}$. On the other hand, $\widehat{V}$ is closed in $\widehat{W}$ by Corollary 4.2.14. Thus $\widehat{V} = \widehat{W}$ and $V = W$. □

Chapter Five

A closer look at the stable completion

Summary. In 5.1 we give a description of $\widehat{\mathbb{A}^n}$ in terms of spaces of semi-lattices which will be used in 6.2. This is provided by constructing a topological embedding of $\widehat{\mathbb{A}^n}$ into the inverse limit of a system of spaces of semi-lattices $L(H_d)$ endowed with the linear topology, where H_d are finite-dimensional vector spaces. The description is extended in 5.2 to the projective setting. In 5.3 we relate the linear topology to the one induced by the finite level morphism $\widehat{\mathbb{A}^n} \to L(H_d)$.

5.1 $\widehat{\mathbb{A}^N}$ AND SPACES OF SEMI-LATTICES

Let K be a model of ACVF and let V be a K-vector space of dimension N. By a *lattice* in V we mean a free $\mathcal{O}$-submodule of rank N. By a *semi-lattice* in V we mean an $\mathcal{O}$-submodule u of V, such that for some K-subspace U_0 of V we have $U_0 \subset u$ and u/U_0 is a lattice in V/U_0. Note that every semi-lattice is uniformly definable with parameters and that the set $L(V)$ of semi-lattices in V is definable. Also, a definable $\mathcal{O}$-submodule u of V is a semi-lattice if and only if there is no $0 \neq v \in V$ such that $Kv \cap u = \{0\}$ or $Kv \cap u = \mathcal{M}v$ where $\mathcal{M}$ is the maximal ideal.

We define a topology on $L(V)$ as follows. The pre-basic open sets are those of the form $\{u : h \notin u\}$ and those of the form $\{u : h \in \mathcal{M}u\}$, where h is any element of V. The finite intersections of these sets clearly form an ind-definable family. We call this family the *linear topology* on $L(V)$.

Any finitely generated $\mathcal{O}$-submodule of K^N is generated by $\leq N$ elements; hence the intersection of any finite number of open sets of the second type is the intersection of N such open sets. However this is not the case for the first kind.

Another description can be given in terms of linear seminorms. By a *linear seminorm* on a finite-dimensional K-vector space V we mean a definable map $w : V \to \Gamma_\infty$ with $w(x_1 + x_2) \geq \min(w(x_1) + w(x_2))$ and $w(cx) = \mathrm{val}(c) + w(x)$.

LEMMA 5.1.1. *If w is a linear seminorm on V, $\Lambda_w = \{x \in V : w(x) \geq 0\}$ is a semi-lattice. Conversely, any semi-lattice $\Lambda \in L(V)$ has the form $\Lambda = \Lambda_w$ for a unique w, namely $w(x) = -\inf\{\mathrm{val}(\lambda) : \lambda x \in \Lambda\}$.* □

We may thus identify $L(V)$ with the set of linear seminorms on V. On the set of linear seminorms there is a natural topology, with basic open sets of the form $\{w : (w(f_1), \dots, w(f_k)) \in O\}$, with $f_1, \dots, f_k \in V$ and O an open subset of Γ_∞^k. The linear topology on $L(V)$ coincides with the linear seminorm topology.

Finally, a description as a quotient by a definable group action: Fix a basis for V, and let Λ_0 be the $\mathcal{O}$-module generated by this basis. Given $M \in \mathrm{End}(V)$, let $\lambda(M) = M^{-1}(\Lambda_0)$. We identify $\mathrm{Aut}(\Lambda_0)$ with the group of automorphisms T of V with $T(\Lambda_0) = \Lambda_0$. So $T \cong \mathrm{Aut}(\Lambda_0) \cong GL_n(\mathcal{O})$. We give $\mathrm{End}(V) = M_n(V)$ the valuation topology, viewing $M_n(V)$ as a copy of K^{n^2}.

LEMMA 5.1.2. *The mapping* $\lambda : M \mapsto \lambda(M) = M^{-1}(\Lambda_0)$ *is surjective and continuous. It induces a bijection between* $\mathrm{Aut}(\Lambda_0)\backslash\mathrm{End}(V)$ *and* $L(V)$.

Proof. It is clear that $M \mapsto \lambda(M)$ is a surjective map from $\mathrm{End}(V)$ to $L(V)$, and also that $\Lambda(N) = \Lambda(TN)$ if $T \in \mathrm{Aut}(\Lambda_0)$. Conversely suppose $\Lambda(M) = \Lambda(N)$. Then M and N have the same kernel $E = \{a : Ka \subset M^{-1}(\Lambda_0)\}$. So NM^{-1} is a well-defined homomorphism $MV \to NV$. Moreover, $MV \cap \Lambda_0$ is a free $\mathcal{O}$-submodule of V, and $(NM^{-1})(MV \cap \Lambda_0) = (NV \cap \Lambda_0)$. Let C (resp. C') be a free $\mathcal{O}$-submodule of Λ_0 complementary to $MV \cap \Lambda_0$ (resp. $NV \cap \Lambda_0$), and let $T_2 : C \to C'$ be an isomorphism. Let $T = (NM^{-1})|(MV \cap \Lambda_0) \oplus T_2$. Then $T \in \mathrm{Aut}(\Lambda_0)$, and $NM^{-1}\Lambda_0 = T^{-1}\Lambda_0$, so (using $\ker M = \ker N$) we have $M^{-1}\Lambda_0 = N^{-1}\Lambda_0$. This shows the bijectivity of the induced map $\mathrm{Aut}(\Lambda_0)\backslash\mathrm{End}(V) \to L(V)$.

Continuity is clear: the inverse image of $\{u : h \notin u\}$ is $\{M : Mh \notin \mathcal{O}^n\}$, while the inverse image of $\{u : h \in \mathcal{M}u\}$ is $\{M : Mh \in \mathcal{M}^n\}$. These are in fact v+g-closed. □

The mapping λ is far from being closed or open, with respect to the v-topology on $\mathrm{End}(V)$. In that topology, $\mathrm{Aut}(\Lambda_0)$ is open, so $\mathcal{O}^n$ is an isolated point in the pushforward topology.

We say a subset of $L(V)$ is *bounded* if its pullback with respect to the map above to $\mathrm{End}(V)$ is bounded. Note that if $X \subset M_n(K)$ is bounded then so is $\mathrm{GL}_n(\mathcal{O})X$ (even $M_n(\mathcal{O})X$); so the image of a bounded set is bounded. Thus a bounded subset of $L(V)$ is a set of semi-lattices admitting bases in a common bounded ball in V. In terms of linear seminorms, if Λ_w ranges over a bounded set, then for any $h \in V$, $w(h)$ lies in a bounded subset of Γ_∞, i.e. bounded on the left.

LEMMA 5.1.3. *The space* $L(V)$ *with the linear topology is Hausdorff. Moreover, any definable type on a bounded subset of* $L(V)$ *has a (unique) limit point in* $L(V)$.

Proof. Let $u' \neq u'' \in L(V)$. One, say u', is not a subset of the other. Let $a \in u', a \notin u''$. Let $I = \{c \in K : ca \in u''\}$. Then $I = \mathcal{O}c_0$ for some c_0 with $\mathrm{val}(c_0) > 0$. Let c_1 be such that $0 < \mathrm{val}(c_1) < \mathrm{val}(c_0)$ and let $a' = c_1 a$.

Then $a' \in \mathcal{M}u'$ but $a' \notin u''$. This shows that u' and u'' are separated by the disjoint open sets $\{u : a' \notin u\}$ and $\{u : a' \in \mathcal{M}u\}$.

For the second statement, let Z be a bounded set of linear seminorms. Let p be a definable type on Z. Let $w(h) = \lim_p w_x(h)$, where w_x is the norm corresponding to $x \models p$. This limit is not $-\infty$ since Z is bounded. It is easy to see that w is a linear seminorm. Moreover any pre-basic open set containing Λ_w must also contain a generic point of p. □

Let $H_{m;d}$ be the space of polynomials of degree $\leq d$ in m variables. For the rest of this section m will be fixed; we will hence suppress the index and write H_d. For p in $\widehat{\mathbb{A}^m}$, consider the definable $\mathcal{O}$-submodule of H_d

$$J_d(p) = \{h \in H_d : p_*(\mathrm{val}(h)) \geq 0\}.$$

Since $h \mapsto p_*(\mathrm{val}(h))$ is a linear seminorm, $J_d(p)$ belongs to $L(H_d)$. Hence we have a mapping $J_d = J_{d,m} : \widehat{\mathbb{A}^m} \to L(H_d)$ given by $p \mapsto J_d(p)$. It is clearly a continuous map, when $L(H_d)$ is given the linear topology: $f \notin J_d(p)$ if and only if $p_*(\mathrm{val}(f)) < 0$, and $f \in \mathcal{M}J_d(p)$ if and only if $p_*(\mathrm{val}(f)) > 0$.

THEOREM 5.1.4. *The system $(J_d)_{d=1,2,\ldots}$ induces a continuous morphism of pro-definable sets*

$$J : \widehat{\mathbb{A}^m} \longrightarrow \varprojlim_d L(H_d),$$

the transition maps $L(H_{d+1}) \to L(H_d)$ being the natural maps induced by the inclusions $H_d \subset H_{d+1}$. The morphism J is injective and induces a homeomorphism between $\widehat{\mathbb{A}^m}$ and its image.

Proof. Let $f : \mathbb{A}^m \times H_d \to \Gamma_\infty$ given by $(x, h) \mapsto \mathrm{val}(h(x))$. Since J_d factors through $Y_{H_d,f}$, J is a morphism of pro-definable sets (here $Y_{H_d,f}$ is defined as in the proof of Theorem 3.1.1).

For injectivity, recall that types on $\mathbb{A}^n$ correspond to equivalence classes of K-algebra morphisms $\varphi : K[x_1, \ldots, x_n] \to F$ with F a valued field, with φ and φ' equivalent if they are restrictions of a same φ''. In particular, if φ_1 and φ_2 correspond to different types, one should have

$$\{f \in K[x_1, \ldots, x_m] : \mathrm{val}(\varphi_1(f)) \geq 0\} \neq \{f \in K[x_1, \ldots, x_m] : \mathrm{val}(\varphi_2(f)) \geq 0\},$$

whence the result.

We noted already continuity. Let us prove that J is an open map onto its image. The topology on $\widehat{\mathbb{A}^n}$ is generated by sets S of the form $\{p : p_*(\mathrm{val}(f)) > \gamma\}$ or $\{p : p_*(\mathrm{val}(f)) < \gamma\}$, where $f \in H_d$ for some d. For such an S, $J(S) = \pi_d^{-1}(J_d(S))$, with $\pi_d : \varprojlim L(H'_d) \to L(H_d)$ the natural map. Thus, it is enough to check that $J_d(S)$ is open. Replacing f by cf for appropriate c, it suffices to consider S of the form $\{p : p_*(\mathrm{val}(f)) > 0\}$ or $\{p : p_*(\mathrm{val}(f)) < 0\}$. Now the image of these sets is precisely the intersection with the image of J of the open sets $p_*(\mathrm{val}(f)) \notin \Lambda$ or $p_*(\mathrm{val}(f)) \in \mathcal{M}\Lambda$. □

REMARK 5.1.5. Note that the image of J consists of all sequences $(\Lambda_d \in L(H_d))_{d=1,2,\ldots}$, with corresponding linear seminorms w_d on $L(H_d)$, such that for any $f_i \in H_{d_1}, f_2 \in H_{d_2}$ we have, $w_{d_1+d_2}(f_1 f_2) = w_{d_1}(f_1) + w_{d_2}(f_2)$.

REMARK 5.1.6. Theorem 5.1.4 describes the $\widehat{V}$-topology in terms of the linear topology when one takes all "jets" into account. It remains interesting to describe the topology induced on the S_n by the individual maps J_d. The image of J_d is described in section 7 of [25]; it may hint at the induced topology as well.

5.2 A REPRESENTATION OF $\widehat{\mathbb{P}^N}$

Let us define the tropical projective space $\operatorname{Trop}\mathbb{P}^n$, for $n \geq 0$, as the quotient $\Gamma_\infty^{n+1} \smallsetminus \{\infty\}^{n+1}/\Gamma$ where Γ acts diagonally by translation. This space may be topologically embedded in Γ_∞^{n+1} since it can be identified with

$$\{(a_0, \ldots, a_n) \in \Gamma_\infty^{n+1} : \min a_i = 0\}.$$

Over a valued field L, we have a canonical definable map $\tau : \mathbb{P}^n \to \operatorname{Trop}\mathbb{P}^n$, sending $[x_0 : \ldots : x_n]$ to

$$[v(x_0) : \ldots : v(x_n)] = ((v(x_0) - \min_i v(x_i), \ldots, v(x_n) - \min_i v(x_i))).$$

Let us denote by $H_{n+1;d,0}$ the set of homogeneous polynomials in $n+1$ variables of degree d with coefficients in the valued field sort. Again we view n as fixed and omit it from the notation, letting $H_{d,0} = H_{n+1;d,0}$. Denote by $H_{d,m}$ the definable subset of $H_{d,0}^{m+1}$ consisting of $m+1$-tuples of homogeneous polynomials with no common zeroes other than the trivial zero. Hence, one can consider the image $PH_{d,m}$ of $H_{d,m}$ in the projectivization $P(H_{d,0}^{m+1})$. We have a morphism $c : \mathbb{P}^n \times H_{d,m} \to \mathbb{P}^m$, given by $c([x_0 : \ldots : x_n], (h_0, \ldots, h_m)) = [h_0(x) : \ldots : h_m(x)]$. Since $c(x, h)$ depends only on the image of h in $PH_{d,m}$, we obtain a morphism $c : \mathbb{P}^n \times PH_{d,m} \to \mathbb{P}^m$. Composing c with the map $\tau : \mathbb{P}^m \to \operatorname{Trop}\mathbb{P}^m$, we obtain $\tau : \mathbb{P}^n \times PH_{d,m} \to \operatorname{Trop}\mathbb{P}^m$. For h in $PH_{d,m}$ (or in $H_{d,m}$), we denote by τ_h the map $x \mapsto \tau(x, h)$. Thus τ_h extends to a map $\widehat{\tau_h} : \widehat{\mathbb{P}^n} \to \operatorname{Trop}\mathbb{P}^m$.

Let $T_{d,m}$ denote the set of functions $PH_{d,m} \to \operatorname{Trop}\mathbb{P}^n$ of the form $h \mapsto \widehat{\tau_h}(x)$ for some $x \in \widehat{\mathbb{P}^n}$. Note that $T_{d,m}$ is definable.

PROPOSITION 5.2.1. *The space $\widehat{\mathbb{P}^n}$ may be identified via the canonical mappings $\widehat{\mathbb{P}^n} \to T_{d,m}$ with the projective limit of the spaces $T_{d,m}$. If one endows $T_{d,m}$ with the topology induced from the Tychonoff topology, this identification is a homeomorphism.* □

The proof of the proposition is a straightforward reduction to the affine case, by using standard affine charts, that we omit.

REMARK 5.2.2. By composing with the embedding $\mathrm{Trop}\,\mathbb{P}^m \to \Gamma_\infty^{m+1}$, one gets a definable map $\widehat{\mathbb{P}^n} \to \Gamma_\infty^{m+1}$. The topology on $\widehat{\mathbb{P}^n}$ can be defined directly using the above maps into Γ_∞, without an affine chart.

5.3 RELATIVE COMPACTNESS

Let H be a finite-dimensional K-vector space. In this section we take $L(H)$ to have the linear topology.

We say that an ∞-definable subset X of $L(H)$ is *relatively compact for the linear topology* if for any definable type q on X, if q has a limit point a in $L(H)$, then $a \in X$. The closed sets of the linear topology are clearly relatively compact.

LEMMA 5.3.1. *The image of a closed pro-definable set by the morphism* $J_d : \widehat{\mathbb{A}^m} \to L(H_d)$ *is relatively compact.*

Proof. Let Y be a closed subset of $\widehat{\mathbb{A}^m}$. Let q be a definable type on $J_d(Y)$, and let b be a limit point of q for the linear topology. We have to show that $b \in J_d(Y)$. The case $d = 0$ is easy as J_0 is a constant map, so assume $d \geq 1$. We have in H_d the monomials x_i. For some nonzero $c_i' \in K$ we have $c_i' x_i \in b$, since b generates H_d as a vector space. Choose a nonzero c_i such that $c_i x_i \in \mathcal{M}b$. Let $U = \{b' : c_i x_i \in \mathcal{M}b', i = 1, \ldots, m\}$. Then U is a pre-basic open neighborhood of b; as b is a limit point of q, it follows that q concentrates on U. Note that $J_d^{-1}(U)$ is contained in $\widehat{B}$ where B is the polydisc $\mathrm{val}(x_i) \geq -\mathrm{val}(c_i), i = 1, \ldots, m$. Thus $J_d^{-1}(U)$ is bounded. By Lemma 4.2.6 we may lift q to a definable type p on $Y \cap \widehat{B}$. Then as $Y \cap \widehat{B}$ is closed and bounded, p has a limit point a on $Y \cap \widehat{B}$. By continuity we have $J_d(a) = b$, hence $b \in J_d(Y)$. □

It follows, writing $X = J_d(J_d^{-1}(X))$, that if a definable set in $L(H_d)$ is an intersection of relatively compact sets, then it is itself relatively compact. Thus the relatively compact sets are the closed sets of a certain topology.

For $b \in L(H)$, we denote by v_b the linear seminorm associated with b.

We consider definable metrics in a different sense than in 3.10. Namely a *definable g-metric* on a definable set X is a map $d : X^2 \to \Gamma_{\geq 0}$, satisfying symmetry, the triangle law $d(x, z) \leq d(x, y) + d(y, z)$, and $d(x, y) = 0$ iff $x = y$. It induces a topology in the obvious way (from the g-topology on Γ).

REMARK 5.3.2. (1) Let $L^*(H)$ be the set of lattices on H. This is easily seen to be a dense subset of $L(H)$ for the linear topology.

(2) On $L^*(H)$, we have a definable g-metric defined as follows. Each lattice corresponds to an actual linear norm on H, i.e. a linear seminorm such that $w(h) = \infty$ iff $h = 0$. We obtain a definable g-metric between norms:

$$d(w, w') = \sup\{|w(h) - w'(h)| : h \in H \setminus (0)\}$$

(3) This g-metric induces a definable topology on $L^*(H)$ (in the sense of Ziegler), finer than the linear topology.

(4) The space $L(H)$ fibers over the (Grassmannian) space of linear subspaces of H, and each fiber admits a similar metric.

(5) $L^*(H)$ is not linearly open in $L(H)$ when H is of dimension > 1. Fix a lattice Λ in H. Given a finite number of vectors $h_1, \dots, h_k$ and $h'_1, \dots, h'_l$ with $h_i \notin \Lambda$, $h'_i \in \mathcal{M}\Lambda$, let $f : H \to K$ be a linear map so that $\ker(f)$ does not pass through any of the vectors h_i or h'_i; renormalize it so such that $f(\Lambda) = \mathcal{O}$. Then $\mathrm{val}(f(h_i)) < 0$ and $\mathrm{val}(f(h'_j)) > 0$. So $h_i \notin f^{-1}(\mathcal{O})$, $h'_j \in f^{-1}(\mathcal{M}) = \mathcal{M}f^{-1}(\mathcal{O})$. Hence $f^{-1}(\mathcal{O})$ belongs to a prescribed neighborhood of Λ in $L(H)$, but it is not a lattice as soon as H is of dimension > 1.

(6) Let $-1 \in \Gamma$ be negative, let $m \geq 1$, and let Y be the set of lattices in $L(K^m)$ of volume -1: $Y = \{M\mathcal{O}^n : \mathrm{val}(\det(M)) = -1\}$. Then Y is relatively definably compact, $\mathcal{O}^n \notin Y$, but $\mathcal{O}^n \in cl(Y)$ in the linear topology. To see this last point consider the lattice $M\mathcal{O}^n$, where M is a lower-triangular matrix with rows $(a, 0), (c, d)$, where $\mathrm{val}(a) = \mathrm{val}(c) < 0$, $\mathrm{val}(d) < 0$ and $\mathrm{val}(a) + \mathrm{val}(d) = -1$.

Chapter Six

Γ-internal spaces

Summary. This chapter is devoted to the topological structure of Γ-internal spaces. The main results about the topological structure of Γ-internal spaces are proved in 6.2. In 6.1 several related issues are discussed. The rather technical results in 6.3 are used in 6.4 which deals with the study of the topology of relatively Γ-internal spaces.

6.1 PRELIMINARY REMARKS

Let V be an algebraic variety over a valued field. Recall an iso-definable subset X of $\widehat{V}$ is said to be Γ-internal if it is in pro-definable bijection with a definable set which is Γ-internal. A number of delicate issues arise here. Let us say a pro-definable subset X of $\widehat{V}$ is *Γ-parameterized* if there exists a definable subset Y of Γ^n, for some n, and a pro-definable map $g : Y \to \widehat{V}$ with image X. By the following example, there exists Γ-parameterized subsets of $\widehat{V}$ which are not iso-definable, whence not Γ-internal.

EXAMPLE 6.1.1. Let A be a base structure consisting of a trivially valued field F and a value group containing $\mathbb{Z}$. Let $\varphi = \sum_{i=0}^{\infty} a_i x^i$ be a formal series with coefficients $a_i \in F$. Assume φ is not algebraic. For any nonnegative integer n, set $\varphi_n(x) = \sum_{i\leq n} a_i x^i$. For any $\gamma \in \Gamma_\infty$, consider the complete type p_γ, in the variables x and y, generated over $A(\gamma)$ by the generic type of the closed ball $\mathrm{val}(x) \geq 1$ together with the formulas

$$\mathrm{val}(y - \varphi_n(x)) \geq \min(n + 1, \gamma).$$

If $\gamma \leq n_0$, for some $n_0 \in \mathbb{N}$, this is the image under $(x, z) \mapsto (x, z + \varphi_{n_0}(x))$ of the generic type of the polydisc $\mathrm{val}(x) \geq 1, \mathrm{val}(z) \geq \gamma$. If $\gamma > \mathbb{N}$, it is the type described in Example 3.2.2. Consider the continuous pro-definable map $g : \Gamma_\infty \to \widehat{\mathbb{A}^2}$ sending γ to p_γ. With the notation of 8.1, we have $p_\gamma \in \mathbb{A}^{2\#}$ iff $\gamma \leq n_0$ for some $n_0 \in \mathbb{N}$. The composition of g with the projection to the space of lattices on polynomials of degree $\leq n$ is constant for $\gamma \geq n$. The image of g is Γ-parameterized, but is not iso-definable and hence not Γ-internal.

However, the image of a complete type is iso-∞-definable, as the following lemma shows.

LEMMA 6.1.2. *Let $P \subset \Gamma^n$ be the solution set of a complete type over some base structure A. Let Y be a pro-definable set and $f : P \to Y$ be a pro-definable map. Then the kernel of f is a definable equivalence relation E on P.*

In case $Y = \widehat{V}$, with V an algebraic variety over a valued field, we have $\dim(P/E) \leq \dim(V)$, where the former dimension is the o-minimal dimension and the latter, the dimension of the algebraic variety V.

Proof. Write $Y = \varprojlim_i Y_i$, where $(Y_i)_{i \in I}$ is a directed system of definable sets, and denote by $\pi_i : Y \to Y_i$ the natural projection. Let E_i be the kernel of $f_i = \pi_i \circ f$. Since $\dim(P/E_i)$ is non-decreasing, there exists some element 0 of I such that, for $i \geq 0$, E_i splits each E_0-class into finitely many classes. Using elimination of imaginaries for Γ, there exists an A-definable map $\phi_i : P \to \Gamma^n$ such that xE_iy iff $\phi_i(x) = \phi_i(y)$. So the image under ϕ_i of each E_0-class is finite. In particular for each E_0-class X, some element $c \in X$ has smallest possible image $\phi_i(c)$, under the lexicographic ordering on Γ^n. But all elements $c \in X$ have the same type: if $c, c' \in X$, then $\mathrm{tp}(c/A) = \mathrm{tp}(c'/A)$ since $c, c' \in P$; let $d = \phi_0(c) = \phi_0(c')$, so $\mathrm{tp}(c/Ad) = \mathrm{tp}(c'/Ad)$. Thus all elements $c \in X$ have smallest possible image under ϕ_i, i.e. they have the same image under ϕ_i, so X is a single E_i-class. This shows that $E_i = E_0$ for all $i \geq 0$. It follows that the kernel of f is E_0, and $f(P)$ is iso-∞-definable.

Now assume $Y = \widehat{V}$. By Corollary 6.2.5 and Remark 6.2.6 there exist finitely many polynomials $h_1, \dots, h_r$ such that $h = (\mathrm{val}(h_1), \dots, \mathrm{val}(h_r))$ induces an injective map on $f(P)$. The image of h in Γ^r has dimension $\leq \dim(V)$, proving the dimension inequality. □

The above discussion referred to the pro-definable category; we will now move to topological questions. When concerned with the definable category alone, there is no point mentioning Γ_∞, since ∞ has the same role as any other element. But from the point of view of the definable topology, the point ∞ does not have the same properties as any points of Γ, nor of the point 0 of $[0, \infty]$; Γ_∞ does not (even locally) embed into Γ^n, and the point ∞ must be taken into account.

DEFINITION 6.1.3. Let V be an algebraic variety over a valued field and let X be an iso-definable Γ internal subset of $\widehat{V}$ (or of $\widehat{V} \times \Gamma_\infty^s$, for some s). We say X is *topologically* Γ*-internal* if X is pro-definably homeomorphic to a definable subset of Γ_∞^r, for some r.

REMARK 6.1.4. In 6.2, we shall prove that, when V is quasi-projective, for any iso-definable Γ-internal subset X of $\widehat{V}$ there exists a pro-definable continuous injection $f : X \hookrightarrow \Gamma_\infty^r$, for some r. In particular, if X is definably compact, f is a homeomorphism and X is topologically Γ-internal. Indeed, in this case the image of a closed pro-definable subset of X is a closed subset of Γ_∞^r. Thus f is definably closed and being injective it is closed, cf.

Remark 3.9.5. In general, we do not know whether every iso-definable Γ-internal subset of $\widehat{V}$ is topologically Γ-internal. The ones that will occur in our constructions will always be contained in some definably compact iso-definable Γ-internal set, thus will be topologically Γ-internal.

We now discuss briefly the role of parameters. We fix a valued field F. The term "definable" refers to ACVF_F. Varieties are assumed defined over F. At the level of definable sets and maps, Γ has elimination of imaginaries. Let us say that Γ admits *topological elimination of imaginaries* if whenever $X \subset \Gamma_\infty^n$ and E is a closed definable equivalence relation on X, there exists a definable map $f : X \to \Gamma_\infty^n$ inducing a homeomorphism between the topological quotient X/E, and $f(X)$ with the topology induced from Γ_∞^n. It seems that any o-minimal expansion of RCF admits elimination of imaginaries in the topological sense.

In another direction, the pair $(\mathbf{k}, \Gamma)$ also eliminates imaginaries (where $\mathbf{k}$ is the residue field, with induced structure), and so does (RES, Γ), where RES denotes the generalized residue structure of [26]. However, $(\mathbf{k}, \Gamma)$ or (RES, Γ) do not eliminate imaginaries topologically. One reason for this, due to Eleftheriou (cf. Remark 14.3.3, [15]) and valid already for Γ, is that the theory DOAG of divisible ordered abelian groups is not sufficiently flexible to identify simplices of different sizes. A more essential reason for us is the existence of spaces with nontrivial Galois action on cohomology. For instance take $\pm\sqrt{-1} \times [0, 1]$ with $\pm\sqrt{-1} \times \{0\}$ and $\pm\sqrt{-1} \times \{1\}$ each collapsed to a point. However for connected spaces topologically embedded in Γ_∞^n, the Galois action on cohomology is trivial. Hence there is no embedding of the above circle in Γ_∞^n compatible with the Galois action. The best we can hope for is that it may be embedded in a twisted form Γ_∞^w, for some finite set w; after base change to w, this becomes isomorphic to Γ_∞^n. It will follow from Theorem 6.2.9 that such an embedding in fact exists for topologically Γ-internal sets.

It would be interesting to study more generally the definable spaces occurring as closed iso-definable subsets of $\widehat{V}$ parameterized by a subset of $\mathrm{VF}^n \times \Gamma^m$. In the case of VF^n alone, a key example should be the set of generic points of subvarieties of V lying in some constructible subset of the Hilbert scheme. This includes the variety V embedded with the valuation topology via the simple points functor (Lemma 3.6.1); possibly other components of the Hilbert scheme obtain the valuation topology too, but the different components (of distinct dimensions) are not topologically disjoint.

6.2 TOPOLOGICAL STRUCTURE OF Γ-INTERNAL SUBSETS

LEMMA 6.2.1. *Let V be a quasi-projective variety over an infinite valued field F, and let $f : \Gamma^n \to \widehat{V}$ be F-definable. There exists an affine open $V' \subset V$*

with $f(\Gamma^n) \subset \widehat{V'}$. *If* $V = \mathbb{P}^n$, *there exists a linear hyperplane* H *such that* $f(\Gamma^n) \cap \widehat{H} = \varnothing$.

Proof. Since V embeds into $\mathbb{P}^n$, we can view f as a map into $\widehat{\mathbb{P}^n}$; so we may assume $V = \mathbb{P}^n$. For $\gamma \in \Gamma^n$, let $V(\gamma)$ be the linear Zariski closure of $f(\gamma)$; i.e. the intersection of all hyperplanes H such that $f(\gamma)$ concentrates on H. The intersection of $V(\gamma)$ with any $\mathbb{A}^n$ is the zero set of all linear polynomials g on $\mathbb{A}^n$ such that $f(\gamma)_*(g) = 0$. So $V(\gamma)$ is definable uniformly in γ. Now $V(\gamma)$ is an ACF_F-definable set, with canonical parameter $e(\gamma)$; by elimination of imaginaries in ACF_F, we can take $e(\gamma)$ to be a tuple of field elements. But functions $\Gamma^n \to \mathrm{VF}$ have finitely many values (every infinite definable subset of VF contains an open subset, and admits a definable map onto $\mathbf{k}$). So there are finitely many sets $V(\gamma)$. Let H be any hyperplane containing none of these. Then no $f(\gamma)$ can concentrate on H. □

Let K be a model of ACVF and let H be a K-vector space of dimension n. We shall make use of the space $L(H)$ of semi-lattices considered in 5.1. Given a basis $v_1, \dots, v_n$ of H, we say that a semi-lattice is diagonal if it is a direct sum $\sum_{i=1}^{n} I_i v_i$, with I_i an ideal of K or $I_i = K$.

LEMMA 6.2.2. *Let* Y *be a* Γ*-internal subset of* $L(H)$. *Then there exists a finite number of bases* $b^1, \dots, b^\ell$ *of* H *such that each* $y \in Y$ *is diagonal for some* b^i. *If* Y *is defined over a valued field* F, *these bases can be found over* F^{alg}.

Proof. For $y \in Y$, let $U_y = \{h \in H : Kh \subset y\}$. Then U_y is a subspace of H, definable from Y. The Grassmannian of subspaces of H is an algebraic variety, and has no infinite Γ-internal definable subsets. Hence there are only finitely many values of U_y. Partitioning Y into finitely many sets we may assume $U_y = U$ for all $y \in Y$. Replacing H by H/U, and Y by $\{y/U : y \in Y\}$, we may assume $U = (0)$. Thus Y is a set of lattices.

Now the lemma follows from Theorem 2.4.13 (iii) of [19], except that in this theorem one considers f defined on Γ (or a finite cover of Γ) whereas Y is the image of Γ^n under some definable function f. In fact the proof of 2.4.13 works for functions from Γ^n; however we will indicate how to deduce the n-dimensional case from the statement there, beginning with 2.4.11. We first formulate a relative version of 2.4.11. Let $U = G_i$ be one of the unipotent groups considered in 2.4.11 (we only need the case of $U = U_n$, the full strictly upper triangular group). Let X be a definable set, and let g be a definable map on $X \times \Gamma$, with $g(x, \gamma)$ a subgroup of U, for any (x, γ) in the domain of g. Let f be another definable map on $X \times \Gamma$, with $f(x, \gamma) \in U/g(x, \gamma)$. Then there exist finitely many definable functions $p_j : X \to \Gamma$, with $p_j \le p_{j+1}$, definable functions b_j on X, such that letting $g_j^*(x) = \cap_{p_j(x) < \gamma < p_{j+1}(x)} g(x, \gamma)$ we have $b_j(x) \in U/g_j^*(x)$, and

$$(*) \quad f(x, \gamma) = b_j(x) g(x, \gamma)$$

whenever $p_j(x) < \gamma < p_{j+1}(x)$. This relative version follows immediately from 2.4.11 using compactness, and noting that $(*)$ determines $b_j(x)$ uniquely as an element of $U/g_j^*(x)$.

By induction, we obtain the multidimensional version of 2.4.11:

Let g be a definable map on a definable subset I of Γ^n, with $g(\gamma)$ a subgroup of U for each $\gamma \in I$. Suppose f is also a definable map on I, with $f(\gamma) \in U/g(\gamma)$. Then there is a partition of I into finitely many definable subsets I' such that for each I' there is $b \in U$ with $f(\gamma) = bg(\gamma)$ for all $\gamma \in I'$.

To prove this for $\Gamma^{n+1} = \Gamma^n \times \Gamma$, apply the case Γ^n to the functions b_j, g_j as well as $f, g(x, p_j(x))$ (at the endpoints of the open intervals).

Now the lemma for the multidimensional case follows as in [19] 2.4.13. Namely, each lattice Λ has a triangular $\mathcal{O}$-basis; viewed as a matrix, it is an element of the triangular group B_n. So there exists an element $A \in U_n$ such that Λ is diagonal for A, i.e. Λ has a basis DA with $D \in T_n$ a diagonal matrix. If $D'A'$ is another basis for Λ of the same form, we have $DA = ED'A'$ for some $E \in B_n(\mathcal{O})$. Factoring out the unipotent part, we find that $D^{-1}D' \in T_n(\mathcal{O})$. So D is well-defined modulo $T_n(\mathcal{O})$, the group $D^{-1}B_n(\mathcal{O})D$ is well-defined, we have $D^{-1}ED' \in D^{-1}B_n(\mathcal{O})D \cap U_n$, and the matrix A is well-defined up to translation by an element of $g(\Lambda) = D^{-1}B_n(\mathcal{O})D \cap U_n$. By the multidimensional 2.4.11, since Y is Γ-internal, it admits a finite partition into definable subsets Y_i, such that for each i, there exists a basis A diagonalizing each $y \in Y_i$.

Moreover, A is uniquely defined up to $\bigcap_{y \in Y_i} g(y)$. The rationality statement now follows from Lemma 6.2.3. □

LEMMA 6.2.3. *Let F be a valued field, let h be an F-definable subgroup of the unipotent group U_n, and let c be an F-definable coset of h. Then c has a point in F^{alg}. If F has residue characteristic 0, or if F is trivially valued and perfect, c has a point in F.*

Proof. As in [19], 2.4.11, the lemma can be proved for all unipotent algebraic groups by induction on dimension, so we are reduced to the case of the one-dimensional unipotent group G_a. In the nontrivially valued case the statement is clear for F^{alg}, since F^{alg} is a model. If F is nontrivially valued and has equal characteristic 0, any definable ball has a definable point, obtained by averaging a definable finite set of points.

There remains the case of trivially valued, perfect F. In this case the subgroup must be $G_a, (0), \mathcal{O}$ or $\mathcal{M}$. The group $\mathcal{O}$ has no other F-definable cosets. As for $\mathcal{M}$ the definable cosets correspond to definable elements of the residue field; as the residue field (isomorphic to F) is perfect, the definable closure is just the residue field itself; but each element of the residue field of F is the residue of a (unique) point of F. □

REMARK 6.2.4. Is the rationality statement in Lemma 6.2.3 valid in positive characteristic, for the groups encountered in Lemma 6.2.2, i.e. intersections of conjugates of $B_n(\mathcal{O})$ with U_n? This is not important for our purposes since the partition of Y may require going to the algebraic closure at all events.

COROLLARY 6.2.5. *Let $X \subset \widehat{\mathbb{A}^N}$ be iso-definable over an algebraically closed valued field F and Γ-internal. Then for some d, and finitely many polynomials h_i of degree $\leq d$, the map $p \mapsto (p_*(\mathrm{val}(h_i)))_i$ is injective on X.*

Proof. By Theorem 5.1.4, the maps

$$p \mapsto J_d(p) = \{h \in H_d : p_*(\mathrm{val}(h)) \geq 0\}$$

separate points on $\widehat{\mathbb{A}^N}$ and hence on X. So for each $x \neq x' \in X$, for some d, $J_d(x) \neq J_d(x')$. Since X is iso-definable, for some fixed d, J_d is injective on X. Let F be a finite set of bases as in Lemma 6.2.2, and let $\{h_i\}$ be the set of elements of these bases. Pick x and x' in X; if $x_*(\mathrm{val}(h_i)) = x'_*(\mathrm{val}(h_i))$ for all i, we have to show that $x = x'$, or equivalently that $J_d(x) = J_d(x')$; by symmetry it suffices to show that $J_d(x) \subset J_d(x')$. Choose a basis, say $b = (b^1, \dots, b^m)$, such that $J_d(x)$ is diagonal with respect to b; the b^i are among the h_i, so $x_*(b^i) = x'_*(b^i)$ for each i. It follows that $J_d(x) \cap Kb^i = J_d(x') \cap Kb^i$. But since $J_d(x)$ is diagonal for b, it is generated by $\cup_i(J_d(x) \cap Kb^i)$; so $J_d(x) \subset J_d(x')$ as required. □

REMARK 6.2.6. Let us observe that the proof goes through for iso-∞-definable sets X, definably parameterized by an ∞-definable subset of Γ. In quoting Lemma 6.2.2, note that an ∞-definable subset of a definable set such as $L(H)$ is always contained in a definable set, and in the present case in a Γ-internal one.

PROPOSITION 6.2.7. *Let $V \subset \mathbb{P}^N$ be a quasi-projective variety over a valued field F. Let $X \subset \widehat{V}$ be F-iso-definable and Γ-internal. Then there exist m, d and $h \in H_{d,m}(F^{\mathrm{alg}})$ such that, with the notations of 5.2, the restriction $\widehat{\tau_h} : \widehat{\mathbb{P}^n} \to \mathrm{Trop}\,\mathbb{P}^m$ to X is injective. If V is projective and X is closed, $\widehat{\tau_h}$ restricts to a homeomorphism between X and its image.*

Proof. We may take $V = \mathbb{P}^N$. Note that if $\widehat{\tau_h}$ is injective, and $g \in \mathrm{Aut}(\mathbb{P}^n) = \mathrm{PGL}(N+1)$, it is clear that $\widehat{\tau_{h \circ g}}$ is injective too. By Lemma 6.2.1, there exists a linear hyperplane H with $\widehat{H}$ disjoint from X. We may assume H is the hyperplane $x_0 = 0$. Let $X_1 = \{(x_1, \dots, x_N) : [1 : x_1 : \dots : x_N] \in X\}$. By Corollary 6.2.5, there exist finitely many polynomials $h_1, \dots, h_r$ such that $p \mapsto (p_*(h_i))_i$ is injective on X_1. Say h_i has degree $\leq d$. Let $H_i(x_0, \dots, x_d) = x_0^d h_i(x_1/x_0, \dots, x_d/x_0)$, and let $h = (x_0^d, \dots, x_N^d, H_1, \dots, H_r)$, $m = N + r$. Then $h \in H_{d,m}$, and it is clear that $\widehat{\tau_h}$ is injective on X. □

THEOREM 6.2.8. *Let V be a quasi-projective variety over a valued field F. Let $X \subset \widehat{V}$ be F-iso-definable and Γ-internal. Then there exists an F-definable map $\beta : V \to [0,\infty]^w$, for some finite set w definable over F, such that $\widehat{\beta} : \widehat{V} \to [0,\infty]^w$ is continuous and restricts to an injective F-definable continuous map $\alpha : X \to [0,\infty]^w$.*

Proof. By Proposition 6.2.7, such map β_a exists over a finite Galois extension $F(a)$ over F with values in $[0,\infty]^n$. Let w_0 be the set of Galois conjugates of a over F and set $w = w_0 \times \{1,\dots,n\}$. Define $\beta : V \to [0,\infty]^w$ by taking all the conjugates of the function β_a. Then the statement is clear. □

Theorem 6.2.8 applies only when the base structure is a valued field; it may not have elements of Γ other than $\mathbb{Q}$-multiples of valuations of field elements. We now extend the result to the case when the base structure may contain additional elements of Γ.

THEOREM 6.2.9. *Let A be a base structure consisting of a field F, and a set S of elements of Γ. Let V be a projective variety over F, and let X be an A-iso-definable and Γ-internal subset of $\widehat{V}$. Then there exists an A-definable continuous injective map $\phi : X \to [0,\infty]^w$ for some finite A-definable set w. If furthermore X is closed, then ϕ is a topological embedding.*

Proof. We have $\mathrm{acl}(A) = \mathrm{dcl}(A \cup F^{\mathrm{alg}}) = F^{\mathrm{alg}}(S)$ by Lemma 2.7.6. It suffices to show that a continuous, injective $\phi : X \to [0,\infty]^n$ is definable over $\mathrm{acl}(A)$, for then the descent to A can be done as in Theorem 6.2.8. So we may assume $F = F^{\mathrm{alg}}$, hence $A = \mathrm{acl}(A)$. We may also assume S is finite, since the data is defined over a finite subset. Say $S = \{\gamma_1,\dots,\gamma_n\}$. Let q be the generic type of field elements $(x_1,\dots,x_n)$ with $\mathrm{val}(x_i) = \gamma_i$. Then q is stably dominated. If $c \models q$, then by Proposition 6.2.7 there exists an $A(b)$-definable continuous injective map $f_b : X \to \Gamma^n$ for some n and some $b \in F(c)^{\mathrm{alg}}$. Since q is stably dominated, and $A = \mathrm{acl}(A)$, $\mathrm{tp}(b/A)$ extends to a stably dominated A-definable type p. If $(a,b) \models p^2|A$ then $f_a f_b^{-1} : X \to X$; but $\mathrm{tp}(ab/A)$ is orthogonal to Γ while X is Γ-internal, so the canonical parameter of $f_a f_b^{-1}$ is defined over $A \cup \Gamma$ and also over $A(a,b)$, hence over A. Thus $f_a f_b^{-1} = g$. If $(a,b,c) \models p^3$ we have $f_b f_c^{-1} = f_a f_c^{-1} = g$ so $g^2 = g$ and hence $g = \mathrm{Id}_X$. So $f_a = f_b$, and f_a is A-definable, as required. If X is closed in $\widehat{V}$, it is definably compact, hence ϕ is a topological embedding by Remark 6.1.4. □

REMARK 6.2.10. With the notation in Theorem 6.2.9, if X is topologically Γ-internal, the morphism $\phi : X \to [0,\infty]^w$ induces a homeomorphism between X and its image Y. Indeed, X is definably homeomorphic to a definable subset Y' of Γ_∞^s for some s, and any definable continuous map $Y' \to Y$ is a homeomorphism.

6.3 GUESSING DEFINABLE MAPS BY REGULAR ALGEBRAIC MAPS

LEMMA 6.3.1. *Let V be a normal, irreducible, complete variety, Y an irreducible variety, X a closed subvariety of V, $g : Y \to X \subset V$ a dominant constructible map (i.e.* ACF*-definable) with finite fibers, all defined over a field F. Then there exists a pseudo-Galois covering $f : \widetilde{V} \to V$ such that each component U of $f^{-1}(X)$ dominates Y rationally, i.e. there exists a dominant rational map $g : U \to Y$ over X.*

Proof. First an algebraic version. Let K be a field, R an integrally closed subring, $G : R \to k$ a ring homomorphism onto a field k. Let k' be a finite field extension. Then there exists a finite normal field extension K' and a homomorphism $G' : R' \to k''$ onto a field, where R' is the integral closure of R in K', such that k'' contains k'.

Indeed we may reach k' as a finite tower of 1-generated field extensions, so we may assume $k' = k(a)$ is generated by a single element. Lift the monic minimal polynomial of a over k to a monic polynomial P over R. Then since R is integrally closed, P is irreducible. Let K' be the splitting field of P. The kernel of G extends to a maximal ideal M' of the integral closure R' of R in K', and R'/M' is clearly a field containing k'.

To apply the algebraic version let $K = F(V)$ be the function field of V. Let R be the local ring of X, i.e. the ring of regular functions on some Zariski open set not disjoint from X, and let $G : R \to k$ be the evaluation homomorphism to the function field $k = F(X)$ of X. Let $k' = F(Y)$ the function field of Y, and K', R', G', M' and k'' be as above. Let $f : \widetilde{V} \to V$ be the normalization of V in K'. Then k'' is the function field of a component X' of $f^{-1}(X)$, mapping dominantly to X. Since k' is contained in k'' as extensions of k there exists a dominant rational map $g : X' \to Y$ over X. But $\mathrm{Aut}(K'/K)$ acts transitively on the components of $f^{-1}(X)$, proving the lemma. □

LEMMA 6.3.2. *Let V be an algebraic variety over a field F, X_i a finite number of locally closed subvarieties, $g_i : Y_i \to X_i$ a surjective constructible map with finite fibers. Then there exists a surjective finite morphism of varieties $f : \widetilde{V} \to V$ and a finite number of locally closed subsets U_{ij} of $f^{-1}(X_i)$ and morphisms $g_{ij} : U_{ij} \to Y_i$ such that, for every i, and every $a \in X_i$, $b \in Y_i$, $c \in \widetilde{V}$ with $g_i(b) = a$ and $f(c) = a$, we have $c \in U_{ij}$ and $b = g_{ij}(c)$ for some j. Furthermore, if V is normal, we may take $f : \widetilde{V} \to V$ to be a pseudo-Galois covering.*

Proof. If the lemma holds for each irreducible component V_j of V, with $X_{j,i} = V_j \cap X_i$ and $Y_{j,i} = g_i^{-1}(X_{j,i})$, then it holds for V with X_i, Y_i: assuming $f_j : \widetilde{V_j} \to V_j$ is as in the conclusion of the lemma, let f be the disjoint union of the f_j. In this way we may assume that V is irreducible. Clearly we may assume V is complete. Finally, we may assume V is normal, by lifting the X_i to the normalization V_n of V, and replacing Y_i by $Y_i \times_{g_i} V_n$. We thus assume V is irreducible, normal and complete. We may also assume the varieties Y_i and X_i to be irreducible.

Let $X_1, \ldots, X_\ell$ be the varieties of maximal dimension d among the locally closed subvarieties $X_1, \ldots, X_n$. We use induction on d. By Lemma 6.3.1 there exist pseudo-Galois coverings $f_i : \widetilde{V}_i \to V$ such that each component of $f_i^{-1}(X_i)$ of dimension d dominates Y_i rationally. Let V^* be an irreducible subvariety of the fiber product $\Pi_V \widetilde{V}_i$ with dominant (hence surjective) projection to each $\widetilde{V}_i$. (The function field of V^* is an amalgam of the function fields of the $\widetilde{V}_i$, finite extensions of the function field of V.) Define $f : V^* \to V$, $f(x) = f_1(x_1) = \ldots = f_n(x_n)$ for $x = (x_1, \ldots, x_n) \in V^*$. Take $a \in X_i$ and $b \in Y_i$, with $g_i(b) = a$. Let F' be a field extension such that $a \in X_i(F')$ (hence $b \in Y_i((F')^{\mathrm{alg}})$). If a is sufficiently generic in X_i, then there exists $c \in V^*((F')^{\mathrm{alg}})$, $c = (c_1, \ldots, c_n)$ with $f(c) = a$. Since f_i is a pseudo-Galois covering, and $\widetilde{V}_i$ dominates Y_i, we have $b \in F'(c_i)$. So there exists a dense open subset $W_i \subset X_i$ such that for any a, b, c and F' as above, with $a \in W_i$, $f(c) = a$, $g_i(b) = a$, we have $b \in F'(c)$.

We may apply the above to the generic point a of X_i, with $F' = F(X_i)$. For any point $c \in V^*$ with $f(c) = a$, any $b \in Y_i$ with $g_i(b) = a$ may be expressed as a rational function of c with coefficients in F'. Each of these rational functions extends to a rational morphism g_{ij} defined on some dense affine Zariski open subset U_{ij} of $f_i^{-1}(W_i)$. After shrinking W_i, we may assume that g_{ij} is in fact a regular morphism $g_{ij} : U_{ij} \to Y_i$ such that, for any $a \in W_i$, $b \in Y_i$ and $c \in \widetilde{V}$ with $g_i(b) = a$ and $f(c) = a$, we have $c \in U_{ij}$ and $b = g_{ij}(c)$ for some j.

Let C_i be the complement of W_i in X_i; so $\dim(C_i) < d$. We now consider the family $\{X'_\nu\}$ of subvarieties of V^* consisting of components of the preimages of the X_i for $i > \ell$ and of the C_i for $i \leq \ell$, and the $\{Y'_\nu\}$ consisting of the pullback of Y_i to X_i for $i > \ell$ and to C_i for $i \leq \ell$. By induction, there exists a finite morphism $f' : \widetilde{V}' \to V^*$ dominating the Y'_ν in the sense of the lemma. Let $\widetilde{V}$ be the normalization of $\widetilde{V}'$ in the normal hull over $F(V)$ of the function field $F(V^*)$. To insure that $\widetilde{V}$ is pseudo-Galois, one may proceed as follows. One replaces V^* by its normalization, and one chooses $\widetilde{V}'$ to be pseudo-Galois over V^*, which is possible by induction. Then $\widetilde{V} \to V$ is pseudo-Galois, and clearly satisfies the conditions of the lemma. □

Note that since finite morphisms are projective (cf. [18] 6.1.11), if V is projective then so is $\widetilde{V}$.

LEMMA 6.3.3. *Let V be a normal projective variety and L an ample line bundle on V. Let H be a finite-dimensional vector space, and let $h : V \to H$ be a rational map. Then for any sufficiently large integer m there exists sections $s_1, \ldots, s_k$ of $\mathfrak{L} = L^{\otimes m}$ such that there is no common zero of the s_i*

inside the domain of definition of h, and such that for each i, $s_i \otimes h$ extends to a morphism $V \to \mathfrak{L} \otimes H$.

Proof. Say $H = \mathbb{A}^n$. We have $h = (h_1, \ldots, h_n)$. Let D_i be the polar divisor of h_i and $D = \sum_{i=1}^n D_i$. Let L_D be the associated line bundle. Then $h \otimes 1$ extends to a section of $H \otimes L_D$. Since L is ample, for some m, $L^{\otimes m} \otimes L_D^{-1}$ is generated by global sections $\sigma_1, \ldots, \sigma_k$. Since 1 is a global section of L_D, $s_i = 1 \otimes \sigma_i$ is a section of $L_D \otimes (L^{\otimes m} \otimes L_D^{-1}) \cong L^{\otimes m}$. Since away from the support of the divisor D, the common zeroes of the s_i are also common zeroes of the σ_i, they have no common zeroes there. Now $h \otimes s_i = (h \otimes 1) \otimes (1 \otimes s_i)$ extends to a section of $(H \otimes L_D) \otimes (L_D^{-1} \otimes L^{\otimes m}) \cong H \otimes L^{\otimes m}$. □

A theory of fields is called an *algebraically bounded theory*, cf. [41] or [10], if for any subfield F of a model M, $F^{\mathrm{alg}} \cap M$ is model-theoretically algebraically closed in M. By Proposition 2.7.1 (4), ACVF is algebraically bounded. The following lemma is valid for any algebraically bounded theory. We work over a base field $F = \mathrm{dcl}(F)$.

LEMMA 6.3.4. *Let F be a valued field. Let V be an irreducible normal F-variety and let H be a finite-dimensional F-vector space. Let ϕ be an ACVF_F-definable subset of $V \times H$ whose projection to V has finite fibers. Then there exists a pseudo-Galois covering $\pi : \widetilde{V} \to V$, a finite family of Zariski open subsets $U_i \subset V$, $\widetilde{U_i} = \pi^{-1}(U_i)$, and morphisms $\psi_i : \widetilde{U_i} \to H$ such that for any $\tilde{v} \in \widetilde{V}$, if $(\pi(\tilde{v}), h) \in \phi$ then $\tilde{v} \in \widetilde{U_i}$ and $h = \psi_i(\tilde{v})$ for some i.*

Proof. For a in V write $\phi(a) = \{b : (a, b) \in \phi\}$; this is a finite subset of H. Let p be an ACVF-type over F (located on V) and $a \models p$. By the algebraic boundedness of ACVF, $\phi(a)$ is contained in a finite normal field extension $F(a')$ of $F(a)$. Let $q = \mathrm{tp}_{\mathrm{ACF}}(a'/F)$, and let $h_p : q \to V$ be a rational map with $h_p(a') = a$.

We can also write each element c of $\phi(a)$ as $c = \psi(a')$ for some rational function ψ over F. This gives a finite family $\Psi = \Psi(p)$ of rational functions ψ; enlarging it, we may take it to be Galois invariant. For any $c' \models q$ with $h_p(c') = a$, we have $\phi(a) \subset \Psi(c') := \{\psi(c') : \psi \in \Psi\}$.

The type q can be viewed as a type of elements of an algebraic variety W, and after shrinking W we can take h_p to be a quasi-finite morphism on W, and assume each $\psi \in \Psi : W \to H$ is defined on W; moreover we can find W such that:

(∗) for any $c' \in W$ with $h_p(c') = a \models p$, we have $\phi(a) \subset \Psi(c')$.

By compactness, there exist finitely many triples (W_i, Ψ_i, h_i) such that for any p, some triple has (∗) for p. We may now use Lemma 6.3.2 to conclude. Indeed, let $Y \subset V \times H$ be the set of points (x, y) such that for some $w \in W_i$, $x = h_i(w)$ and $y = \psi(w)$ for some $\psi \in \Psi_i$. Let $X_i \subset V$ be the image of Y_i under the projection to V. We may assume X_i and Y_i are locally closed subvarieties and we denote by $g_i : Y_i \to X_i$ the morphism induced by the projection to V. Applying Lemma 6.3.2, we obtain a pseudo-Galois covering $f : \widetilde{V} \to V$, a finite number of locally closed subsets U_{ij} of $f^{-1}(X_i)$ and morphisms $g_{ij} : U_{ij} \to Y_i$ satisfying the conditions of Lemma 6.3.2. There is no harm in assuming that each U_{ij} is closed in some affine nonempty open $\widetilde{\Omega}_{ij} = \pi^{-1}(\Omega_{ij}) \subset \widetilde{V}$, with Ω_{ij} Zariski open in V. Let $\phi_{ij} : U_{ij} \to H$ be the morphism obtained by composing g_{ij} with the projection to H. We may extend ϕ_{ij} to a morphism $\psi_{ij} : \widetilde{\Omega}_{ij} \to H$. Now the pseudo-Galois covering $\widetilde{V} \to V$ together with the family of open subsets Ω_{ij} and morphisms ψ_{ij} does the job. □

If H is a vector space, or a vector bundle over V, let H^n be the n-th direct power of H, and let $P(H^n)$ denote the projectivization of H^n. Let $h \mapsto : h :$ denote the natural map $H \smallsetminus \{0\} \to PH$. Let $r_k : P(H^n) \to PH$ be the natural rational map, $r_k(h_1 : \ldots : h_n) = (: h_k :)$. For any vector bundle L over V, there is a canonical isomorphism $L \otimes H^n \cong (L \otimes H)^n$. When L is a line bundle, we have $P(L \otimes E) \cong P(E)$ canonically for any vector bundle E. Composing, we obtain an identification of $P((L \otimes H)^n)$ with $P(H^n)$.

LEMMA 6.3.5. *Let F be a valued field. Let V be a normal irreducible quasi-projective F-variety, H a vector space with a basis of F-definable points, and ϕ an* ACVF$_F$*-definable subset of $V \times (H \smallsetminus (0))$ whose projection to V has finite fibers. Then there exists a pseudo-Galois covering $\pi : \widetilde{V} \to V$, a regular morphism $\theta : \widetilde{V} \to P(H^m)$ for some m, such that for any $\tilde{v} \in \widetilde{V}$, if $(\pi(\tilde{v}), h) \in \phi$ then for some k, $r_k(\theta(\tilde{v}))$ is defined and equals $: h :$.*

Proof. Replacing V by the normalization of the closure of V in some projective embedding, we may assume V is projective and normal. Let ψ_i be as in Lemma 6.3.4. Let L, s_{ij} be as in Lemma 6.3.3, applied to $\widetilde{V}$, ψ_i; choose m that works for all ψ_i. Let θ_{ij} be the extension to $\widetilde{V}$ of $s_{ij} \otimes \psi_i$. Define $\theta = (\ldots : \theta_{ij} : \ldots)$, using the identification above the lemma. □

6.4 RELATIVELY Γ-INTERNAL SUBSETS

We proceed towards a relative version of Proposition 6.2.7. First let us clarify some relations of V with $\widehat{\widehat{V}}$, where V is any pro-definable set. We have an embedding $s_V : V \to \widehat{V}$ of V in $\widehat{V}$ as simple points. We can thus form two natural embeddings $\widehat{V} \to \widehat{\widehat{V}}$, namely $\widehat{s_V}$ and $s_{\widehat{V}}$. If $a \in \widehat{\widehat{V}}$, $b \models a|A(a)$, and $c \models b|A(a,b)$, then a lies in the image of $\widehat{s_V}$ iff $c \in A(a,b)$, while it is in the image of $s_{\widehat{V}}$ iff $b \in A(a)$. In other words, the image of $s_{\widehat{V}}$ consists of the types on $\widehat{V}$ that concentrate on a point of $\widehat{V}$, while the image of $\widehat{s_V}$ consists of the types on $\widehat{V}$ concentrating on the set of simple points of $\widehat{V}$. Thus the intersection of the two images is equal to the image of V in $\widehat{\widehat{V}}$, where v is mapped to the type concentrating on the type concentrating on the single point v. So, away from degenerate cases, when $V = \widehat{V}$ already, the two images are distinct and neither contains the other. It is $\widehat{s_V}$ that will concern us below.

Let $\pi : V \to U$ be a morphism of algebraic varieties over a valued field F. We denote by $\widehat{V/U}$ the subset of $\widehat{V}$ consisting of types $p \in \widehat{V}$ such that $\widehat{\pi}(p)$ is a simple point of $\widehat{U}$. Note that it follows from Lemma 3.6.1 (1) that $\widehat{V/U}$ is a relatively definable subset of $\widehat{V}$.

We say $X \subset \widehat{V/U}$ is *relatively* Γ-*internal* over U, if X is a relatively definable subset of $\widehat{V}$, and the fibers X_u of $X \to U$ are iso-definable and Γ-internal, uniformly in $u \in U$.

LEMMA 6.4.1. *Let $\pi : V \to U$ be a morphism of algebraic varieties over a valued field F, and let $X \subset \widehat{V/U}$ be relatively* Γ*-internal over U. Then there exists a natural embedding $\theta : \widehat{X} \to \widehat{V}$ over $\widehat{U}$, determined by: $\widehat{s_V} \circ \theta = \widehat{j}$, where j is the inclusion map $X \to \widehat{V}$. Over a simple point $u \in \widehat{U}$, θ restricts to the identification of $\widehat{X_u}$ with X_u.*

Proof. Let $\pi_X : X \to U$ be the natural map. Let $p \in \widehat{X}$; let $A = \mathrm{acl}(A)$ be such that p is A-definable; and let $c \models p|A$, $u = \pi_X(c)$. Since $\mathrm{tp}(c/A(u))$ is contained in an $\mathrm{acl}(A(u))$-iso-definable Γ-internal set, by Lemma 2.8.1 (5) there exists an $\mathrm{acl}(A(u))$-definable injective map j with $j(c) \in \Gamma^m$. But $\mathrm{acl}(A(c)) \cap \Gamma = \Gamma(A)$. So $j(c) = \alpha \in \Gamma(A)$, and $c = j^{-1}(\alpha) \in \mathrm{acl}(A(u))$. Let $v \models c|\,\mathrm{acl}(A(u))$, and let $\theta(p)$ be the unique stably dominated, A-definable type extending $\mathrm{tp}(v/A)$. So $\theta(p) \in \widehat{V}$, and $\widehat{\pi}_X(p) = \pi_*\theta(p)$. □

Assume now that $X \subset \widehat{V/U}$ is iso-definable and relatively Γ-internal. By Lemma 6.4.1 we may identify $\widehat{X}$ with a pro-definable subset of $\widehat{V}$; namely the set $\int_U X$ of $p \in \widehat{V}$ such that if p is A-definable and $c \models p|A$, then $\mathrm{tp}(c/A(\pi(c))) = q|A(\pi(c))$ for some $q \in X$. It is really this set that we have in mind when speaking of $\widehat{X}$ below. In particular, it inherits a topology from $\widehat{V}$.

THEOREM 6.4.2. *Let $V \to U$ be a projective morphism of quasi-projective varieties over a valued field F. Let $X \subset \widehat{V/U}$ be F-iso-definable and relatively* Γ*-internal. Then there exists a finite pseudo-Galois covering $U' \to U$, such that letting $X' = U' \times_U X$ and $V' = U' \times_U V$, there exists an F-definable morphism $g : V' \to U' \times \Gamma_\infty^N$ over U', such that the induced map $g : \widehat{V'} \to \widehat{U'} \times \Gamma_\infty^N$ is continuous, and such that the restriction of g to $\widehat{X}'$ is injective. In fact Zariski locally each coordinate of g is obtained as a composition of regular maps and the valuation map.*

Proof. After pulling back the data to some $\mathbb{P}^n$ we may assume U is irreducible and normal. By Proposition 6.2.7, for each $u \in U$, there exists $h \in H_{d,m}(F(u)^{\mathrm{alg}})$ such that τ_h is injective on the fiber X_u above u. By compactness, a finite number of pairs (m, d) will work for all u; by taking a large enough (m,d), we may take it to be fixed. Again by compactness, there exists an F-definable $\phi \subset U \times H_{d,m}$ whose projection to U has finite fibers, such that if $(u,h) \in \phi$ then τ_h is injective on X_u. By Lemma 6.3.5, there is a finite pseudo-Galois covering $\pi : U' \to U$, and a regular morphism $\theta : U' \to P(H'^{M}_{d,m})$ for some M, with $H'_{d,m}$ the vector space generated by $H_{d,m}$, such that for any $u' \in U'$, if $(\pi(u'), h) \in \phi$ then, for some k, $r_k(\theta(u'))$ is defined and equals $: h :$. Note that since $h \in H_{d,m}$, it follows that $\theta(u') \in PH_{Mm,d}$. Let $g(u',v) = (u', \tau_{\theta(u')})(v)$. Then it is clear that g is continuous and that its restriction to X' is injective. It follows that its restriction to $\widehat{X}'$ is injective. □

Note that the proposition has content even when the fibers of X/U are finite. Under certain conditions, the continuous injection of Theorem 6.4.2 can be seen to be a homeomorphism. This is clear when X is definably compact, but we will need it in somewhat greater generality.

Let X be a pro-definable subset of $\widehat{V} \times \Gamma_\infty^\ell$, for V an algebraic variety. If $\rho : X \to \Gamma_\infty$ is a definable continuous function, we shall say X is *compact at* $\rho = \infty$ if any definable type q on X with $\rho_* q$ unbounded has a limit point in X. Compactness at $\rho = \infty$ implies that $\rho^{-1}(\infty)$ is definably compact. If X is a subspace of a definably compact space Y, ρ extends to a continuous definable function ρ_Y on Y, and $\rho_Y^{-1}(\infty) \subset X$, then X is compact at $\rho = \infty$. In the applications, this will be the case. We say X is *σ-compact via* a continuous definable function $\xi : X \to \Gamma$, if for any $\gamma \in \Gamma$, $\{x \in X : \xi(x) \leq \gamma\}$ is definably compact.

More generally, let $\rho, \xi : X \to \Gamma_\infty$ be definable continuous functions. We say that X is *σ-compact via* (ρ, ξ) if $\xi^{-1}(\infty) \subset \rho^{-1}(\infty)$, X is compact at $\rho = \infty$, and $X \smallsetminus \xi^{-1}(\infty)$ is σ-compact via ξ.

Assume $f : V \to U$ is a morphism of algebraic varieties, $\rho : V \to \Gamma_\infty$ and $\xi : U \to \Gamma_\infty$ are definable v+g-continuous functions. We say that a pro-definable subset X of $\widehat{V}$ is σ-compact over U via (ρ, ξ) if X is σ-compact via $(\rho, \xi \circ f)$, where we omit the $\widehat{\ }$ on morphisms.

Lemma 6.4.3. *In* Theorem 6.4.2, *assume* $\widehat{X}$ *is σ-compact over* U *via* (ρ, ξ), *where* $\rho : V \to \Gamma_\infty$ *and* $\xi : U \to \Gamma_\infty$ *are definable and v+g-continuous. Then one can find* g *as in* Theorem 6.4.2 *inducing a homeomorphism of* $\widehat{X'}$ *with its image in* $\widehat{U'} \times \Gamma_\infty^N$.

Proof. Let $f : V' \to V$ denote the projection and $f : \widehat{V'} \to \widehat{V}$ its extension. After replacing $g : V' \to U' \times \Gamma_\infty^N$ in the construction of Theorem 6.4.2 by $V' \to U' \times \Gamma_\infty^{N+1}$ sending x to $(g(x), \rho \circ f)$, one may assume that $\rho \circ f = \rho' \circ g$ with ρ' the projection on the last factor; and $\xi \circ \pi \circ f = \xi' \circ g$, with ξ' the penultimate projection, and $\pi : V \to U$. As in Theorem 6.4.2 we still denote by g its extension $\widehat{V'} \to \widehat{U'} \times \Gamma_\infty^N$. The restriction $g_{|\widehat{X'}}$ of g to $\widehat{X'}$ is injective and continuous. We have to show that its inverse $g^{-1}_{|\widehat{X'}}$ is continuous too, or equivalently that $g^{-1}_{|\widehat{X'}} \circ \phi$ is continuous for any continuous definable $\phi : \widehat{X'} \to \Gamma_\infty$. It suffices thus to show that if W is a closed relatively definable subset of $\widehat{X'}$, then $g(W)$ is closed. By Proposition 4.2.13, it suffices to show this: if p is a definable type on W, and $g(w)$ is a limit of g_*p in $\widehat{U'} \times \Gamma^N_\infty$ for $w \in W$, then w is the limit of p in $\widehat{X'}$. As g is injective and continuous on $\widehat{X'}$, it suffices to show that p has a limit in $\widehat{X'}$.

Let us first show that if $f_*(p)$ has a limit point in $\widehat{X}$, then p has a limit point in $\widehat{X'}$. Since $V' \to V$ is a finite morphism, it is proper, so the morphism $\widehat{V'} \to \widehat{V}$ is definably closed by Lemma 4.2.26. It follows that the morphism $f' : \widehat{X'} \to \widehat{X}$ induced by f is definably closed. Furthermore it is surjective since $X' \to X$ is surjective, by Lemma 4.2.6. Let α be the limit of $f_*(p)$. Its fiber $f'^{-1}(\alpha)$ is finite and nonempty, say equal to $\{\beta_1, \ldots, \beta_n\}$. If p has a limit in $\widehat{X'}$, by continuity of f', it should be one of the β_i. Hence, if p does not have a limit in $\widehat{X'}$, there exists open relatively definable subsets O_i of $\widehat{X'}$ containing β_i, such that $O_i \cap O_j = \varnothing$ if $i \neq j$, and such that p is on $Z = \widehat{X'} \smallsetminus \cup_{1 \leq i \leq n} O_i$. Since Z is closed, its image $f'(Z)$ is closed, hence $\Omega = \widehat{X} \smallsetminus f'(Z)$ is open and contains α. Thus $f_*(p)$ is on Ω. But $f'^{-1}(\Omega) \subset \cup_{1 \leq i \leq n} O_i$, which contradicts the fact that p is on Z. Hence it suffices to show that $f_*(p)$ has a limit point in $\widehat{X}$.

Assume first $\rho_*(f_*(p))$ is not bounded. Then $f_*(p)$ has a limit point in $\widehat{X}$ by compactness at $\rho = \infty$.

Otherwise, ρ' is bounded on g_*p, hence as ρ' is continuous, $\rho'(g(w)) < \infty$. So $\rho(f(w)) \in \Gamma$. It follows that $\xi'(g(w)) = \xi(\pi(f(w)) \in \Gamma$ also. Since $g(w)$ is a limit of g_*p, the type $(\xi' \circ g)_*p$ concentrates on a bounded subset of Γ. Hence the type $f_*(p)$ includes a formula $\xi \circ \pi \leq \alpha$ for some $\alpha \in \Gamma$. Thus, by σ-compactness, f_*p concentrates on a definably compact relatively definable subset of $\widehat{X'}$, containing $f(w)$; so f_*p has a limit in this set, hence in $\widehat{X}$. □

The following lemma shows that o-minimal covers may be replaced by finite covers carrying the same information, at least as far as homotopy lifting goes.

Given a morphism $g : U' \to U$ and homotopies $h : I \times U \to \widehat{U}$ and $h' : I \times U' \to \widehat{U'}$, we say h and h' are *compatible* or that h' *lifts* h if $\widehat{g}(h'(t, u')) = h(t, g(u'))$ for all $t \in I$ and $u' \in U'$. Here, I refers to any closed generalized interval, with final point e_I. Let H be the canonical homotopy $I \times \widehat{U} \to \widehat{U}$ extending h, cf. Lemma 3.8.5. Note that if $h(e_I, U)$ is iso-definable and Γ-internal, then $h(e_I, U) = H(e_I, \widehat{U})$.

THEOREM 6.4.4. *Let $\phi : V \to U$ be a projective morphism of algebraic varieties with U normal and quasi-projective, over a valued field F. Let $X \subset \widehat{V/U}$ be iso-definable over F and relatively Γ-internal over U. Assume $\widehat{X}$ is σ-compact over U via (ρ, ξ), where $\rho : V \to \Gamma_\infty$ and $\xi : U \to \Gamma_\infty$ are definable and v+g-continuous. Then there exists a pseudo-Galois covering U'*

of U, and an F-definable function $j : U' \times_U X \to U' \times \Gamma_\infty^m$ over U', inducing a homeomorphism between $\widehat{U' \times_U X}$ and its image in $\widehat{U'} \times \Gamma_\infty^m$. Moreover:

(1) *There exists a finite family of F-definable functions $\xi_i' : U \to \Gamma_\infty$, such that, for any compatible pair of F-definable homotopies $h : I \times U \to \widehat{U}$ and $h' : I \times U' \to \widehat{U'}$, if h respects the functions ξ_i', then h lifts to an F-definable homotopy $H_X : I \times \widehat{X} \to \widehat{X}$. Furthermore, if h' is a deformation retraction with iso-definable Γ-internal image Σ', and h is a deformation retraction with iso-definable Γ-internal image Σ, then one may impose that H_X is also a deformation retraction with iso-definable Γ-internal image $\Upsilon = \widehat{\phi}^{-1}(\Sigma) \cap \widehat{X}$.*

(2) *Given a finite number of F-definable functions $\tilde{\xi}_j : X \to \Gamma_\infty$ on X, and a finite group action on X over U, one can choose the functions $\xi_i' : U \to \Gamma_\infty$ such that the lift $I \times \widehat{X} \to \widehat{X}$ respects the given functions $\tilde{\xi}_j$ and the group action.*

(3) *If h' satisfies condition $(*)$ of* Definition 3.9.3, *one may also impose that H_X satisfies $(*)$.*

Proof. We take U' and j as given by Theorem 6.4.2 and Lemma 6.4.3 (that is, j is the restriction of g). First consider the case when $X \subset U \times \Gamma_\infty^N$. There exists a finite number of Γ_∞-valued F-definable functions ξ_i'' on U such that the set of values $\xi_i''(u)$ determine the fiber $X_u = \{x : (u, x) \in X\}$, as well as the functions $\tilde{\xi}_j | X_u$ (with $\tilde{\xi}_j$ as in (2)), and the group action on X_u. In other words if $\xi_i''(u) = \xi_i''(u')$ for simple points u, u' then $X_u = X_{u'}$, $\tilde{\xi}_j(u, x) = \xi(u', x)$ for $x \in X_u$, and $g(u, x) = (u, x')$ iff $g(u', x) = (u', x')$ for g a group element from the group acting in (2). Clearly any homotopy $h : I \times U \to \widehat{U}$ respecting the functions ξ_i'' lifts to a homotopy $H_X : I \times \widehat{X} \to \widehat{X} \subset \widehat{U} \times \Gamma_\infty^N$ given by $(t, (u, \gamma)) \mapsto (H(t, u), \gamma)$, where H is the canonical homotopy $I \times \widehat{U} \to \widehat{U}$ lifting h provided by Lemma 3.8.5. Moreover H_X respects the functions of (2) and the group action.

This applies to $X' = U' \times_U X$, via the homeomorphism induced by j; so for any pair (h, h') as in (1), if h' respects the functions ξ_i'', then h' lifts to a definable homotopy $H' : I \times \widehat{X'} \to \widehat{X'}$, respecting the data of (2), in particular the Galois action on X'. As already noted in the proof of Lemma 6.4.3, the morphism $\widehat{X'} = \int_{U'} X' \to \int_U X = \widehat{X}$ is definably closed and surjective. Moreover H' respects the fibers of $\widehat{X'} \to \widehat{X}$ in the sense of Lemma 3.9.6. Hence by this lemma, H' descends to a homotopy $H_X : I \times \widehat{X} \to \widehat{X}$.

By Corollary 9.7.5, the condition that h' respects the ξ'' can be replaced with the condition that h respects certain other definable $\xi' : U \to \Gamma_\infty$.

Since X is iso-definable uniformly over U, Corollary 2.8.4 applies to the image of H'; so this image is iso-definable and Γ-internal. The image of H_X is obtained by factoring out the action of the Galois group of U'/U; by Lemma 2.2.5, the image of H_X is also iso-definable, and hence Γ-internal.

The statement regarding condition $(*)$ is verified by construction, using density of simple points and continuity. □

EXAMPLE 6.4.5. In dimension > 1 there exist definable topologies on definable subsets of Γ^n, induced from function space topologies, for which Theorem 6.2.8 fails. For instance, let $X = \{(s,t) : 0 \le s \le t\}$. For $(s,t) \in X$ consider the continuous function $f_{s,t}$ on $[0,1]$ supported on $[s,t]$, with slope 1 on $(s, s+\frac{s+t}{2})$, and slope -1 on $(s+\frac{s+t}{2}, t)$. The topology induced on X from the Tychonoff topology on the space of functions $[0,1] \to \Gamma$ is a definable topology, and definably compact. Any neighborhood of the function 0 (even if defined with nonstandard parameters) is a finite union of bounded subsets of Γ^2, but contains a "line" of functions $f_{s,s+\varepsilon}$ whose length is at least $1/n$ for some standard n, so this topology is not induced from any definable embedding of X in Γ_∞^m. By Theorem 6.2.8, such topologies do not occur within $\widehat{V}$ for an algebraic variety V.

Chapter Seven

Curves

Summary. In 7.1 we prove the iso-definability of $\widehat{C}$ when C is a curve. This is done using Riemann-Roch. In 7.2 we explain how definable types on C correspond to germs of paths on $\widehat{C}$. The remainder of the chapter is devoted to the construction of the retraction on skeleta for curves. A key result is the finiteness of forward-branching points proved in Proposition 7.4.5.

7.1 DEFINABILITY OF $\widehat{C}$ FOR A CURVE C

Recall that a pro-definable set is called iso-definable if it is isomorphic, as a pro-definable set, to a definable set.

THEOREM 7.1.1. *Let C be an algebraic curve defined over a valued field F. Then $\widehat{C}$ is an iso-definable set. The topology on $\widehat{C}$ is definably generated, that is, generated by a definable family of (iso)-definable subsets. In other words, there is a definable family giving a pre-basis of the topology.*

Proof. One may assume C is a projective curve. There exists a finite purely inseparable extension F' of F such that the normalization of $C \otimes F'$ is smooth over F'. Since this does not change the notion of definability over F, we may assume $F' = F$. Hence we may assume C is projective and smooth over F, and that it is irreducible. Let g be its genus. Let $L= F(C)$ be the function field of C and let Y be the set of elements $f \in L$ with at most $g+1$ poles (counted with multiplicities).

CLAIM. *Any element of $L^{\times}$ is a product of finitely many elements of Y.*

Proof of the claim. We use induction on the number of poles of $f \in L^{\times}$. If this number is $\leq g+1$, then $f \in Y$. Otherwise, let $a_1, \ldots, a_H$ be poles of f, not necessarily distinct, and let b be a zero of f. By Riemann-Roch, any divisor of degree $\geq g$ has a nontrivial global section, which provides one a function f_1 with poles at most at $a_1, \ldots, a_{g+1}$, and a zero at b. Then $f_1 \in Y$, and f/f_1 has fewer poles than f (say f_1 has m poles; they are all among the poles of f; and f_1 has at most $m-1$ zeroes other than b). The statement follows by induction. □

Choose an embedding of $i : C \to \mathbb{P}^m$ in some projective space. Thus, for every positive integer N, the line bundle $i^*\mathcal{O}(N)$ has degree Nd with d

the degree of the embedding. By Riemann-Roch, if N is large enough, for every line bundle $\mathcal{L}$ on C of degree $\leq g+1$, $i^*\mathcal{O}(N) \otimes \mathcal{L}^{-1}$ is generated by its global sections. Also, for N large enough, the restriction mapping $H^0(\mathbb{P}^m, \mathcal{O}(N)) \to H^0(C, i^*\mathcal{O}(N))$ is surjective. It follows that, for N large enough, any function on C with at most $g+1$ poles is the quotient of two homogeneous polynomials of degree N.

Fix such an N. Let W be the set of pairs of homogeneous polynomials of degree N. We consider the morphism $f : C \times W \to \Gamma_\infty$ mapping (x, φ, ψ) to $v(\varphi(x)) - v(\psi(x))$ or to 0 if x is a zero of both φ and ψ.

With notations from the proof of Theorem 3.1.1, f induces a mapping $\widehat{C} \to Y_{W,f}$ with $Y_{W,f}$ definable. Now, let us remark that any type p on C induces a valuation on L in the following way: let $c \models p$ send g in L to $v(g(c))$ (or say to the symbol $-\infty$ if c is a pole of g), and that different types give rise to different valuations. It follows that the map $\widehat{C} \to Y_{W,f}$ is injective, since if two valuations agree on Y they agree on $L^\times$. This shows that $\widehat{C}$ is an iso-∞-definable set. Since $\widehat{C}$ is strict pro-definable by Theorem 3.1.1 it follows it is iso-definable. The statement on the topology is clear. □

Let $h : C \to V$ be a relative curve over an algebraic variety V, that is, h is flat with fibers of dimension one. Let $\widehat{C/V}$ be the set of $p \in \widehat{C}$ such that $\widehat{h}(p)$ is a simple point of $\widehat{V}$. Then we have the following relative version of Theorem 7.1.1:

THEOREM 7.1.2. *Let $h : C \to V$ be a relative curve over an algebraic variety V. Then $\widehat{C/V}$ is iso-definable.*

Proof. The proof is the obvious relativization of the proof of Theorem 7.1.1. Indeed, after replacing V by a dense open subset we may assume that h is projective, and that there exists a finite purely inseparable morphism $V' \to V$ such that the normalization $h' : C' \to V'$ of the pullback of C to V' is a smooth morphism. Thus, one may assume $h : C \to V$ is projective and smooth. Furthermore, by Stein factorization, h factors as the composition of a morphism $g : C \to U$ with connected fibers and a finite surjective morphism $U \to V$. Since $\widehat{C/U}$ may be canonically identified with $\widehat{C/V}$, one may assume each fiber C_a of h to be connected. We embed C in $\mathbb{P}^m_V$ and note that for N large enough, for any $a \in V$, any function on C_a with $\leq g+1$ poles is the quotient of two homogeneous polynomials of degree N. Let W_1 be the set of pairs of homogeneous polynomials of degree N, W_2 be the set of characteristic functions of points of V, and set $W = W_1 \cup W_2$. Let $f : C \times W \to \Gamma_\infty$ mapping (x, φ, ψ) to $v(\varphi(x)) - v(\psi(x))$ or to 0 if x is a zero of both φ, ψ, for (φ, ψ) in W_1 and mapping (x, φ) to $v(\varphi(h(x)))$ for φ in W_2. The map $\widehat{C} \to Y_{W,f}$ is injective, and we may proceed as in Theorem 7.1.1. □

REMARK 7.1.3. The statement of Theorem 7.1.1 is specific to dimension one. Indeed, assume we work over a base valued field of equicharacteristic zero. By Example 3.2.2, $\widehat{\mathcal{O}^2}(\mathbb{Q}(t))$ is uncountable, when $\mathbb{Q}(t)$ is endowed with the t-adic discrete valuation, thus $\widehat{\mathcal{O}^2}$ cannot be iso-definable. By rescaling, it follows that for any nontrivial closed ball b, $\widehat{b^2}$ is not iso-definable and thus also $\widehat{D}$ for D a definable subset of $\mathbb{A}^2$ of dimension two. By projecting to $\mathbb{A}^2$ and using Lemma 4.2.6, it follows that for any definable set X in the VF-sort of dimension two, $\widehat{X}$ is not iso-definable. Clearly the same holds in any dimension ≥ 2, over any nontrivially valued field of any residue characteristic (by a similar argument involving, e.g., the construction in Example 13.1 in [20] instead of the one in Example 3.2.2).

QUESTION 7.1.4. If $f : U \to V$ is a finite morphism of algebraic varieties, is the inverse image of an iso-definable subset of $\widehat{V}$ iso-definable?

When the answer is positive, the definability of $\widehat{C}$ follows from that of $\widehat{\mathbb{P}^1}$ which is clear by Example 3.2.1.

7.2 DEFINABLE TYPES ON CURVES

Let V be an algebraic variety and $a, b \in \Gamma_\infty$. Two pro-definable functions $f, g : [a, b) \to \widehat{V}$ are said to have the same germ at b if $f|[a', b) = g|[a', b)$ for some a'.

PROPOSITION 7.2.1. *Let C be a curve, defined over A. There is a canonical bijection between:*

(1) *A-definable types on C.*

(2) *A-definable germs at b of paths $[a, b) \to \widehat{C}$, up to reparameterization.*

Under this bijection, the stably dominated types on C correspond to the germs of constant paths on $\widehat{C}$.

Proof. A constant path, up to reparameterization, is just a point of $\widehat{C}$. In this way the stably dominated types correspond to germs of constant paths into $\widehat{C}$. Let p be a definable type on C, which is not stably dominated. Then, by Lemma 2.11.2, for some definable $\delta : C \to \Gamma$, $\delta_*(p)$ is a nonconstant definable type on Γ. Changing sign if necessary, either $\delta_*(p)$ is the type of very large elements of Γ, or else for some b, $\delta_*(p)$ concentrates on elements in some interval $[a, b]$; in the latter case there is a smallest b such that p concentrates on $[a, b)$, so that it is the type of elements just $< b$, or else dually. Thus we may assume $\delta_*(p)$ is the generic at b of an interval $[a, b)$ (where possibly $b = \infty$).

By Theorem 2.11.5 there exists a $\delta_*(p)$-germ f of definable function to $\widehat{C}$ whose integral is p. It is the germ of a definable function $f = f_{p,\delta}$:

$[a_0, b) \to \widehat{C}$; since $\widehat{C}$ is definable and the topology is definably generated by Theorem 7.1.1, for some (not necessarily definable) a, the restriction $f = f_{p,\delta} : [a, b) \to \widehat{C}$ is continuous. The germ of this function f is well-defined. A change in the choice of δ corresponds to reparameterization. Conversely, given $f : [a, b) \to \widehat{C}$, we obtain a definable type p_f on C; namely $p_f|E = \mathrm{tp}(e/E)$ if t is generic over E in $[a, b)$, and $e \models f(t)|E(t)$. It is clear that p_f depends only on the germ of f. Furthermore, with the above notation, $p = p_{f_{p,\delta}}$. On the other hand, for any δ as above, f and $f_{p_f,\delta}$ have the same germ, up to reparameterization. Finally, if the germ of f is A-definable, then each ϕ-definition $d_{p_f}\phi$ is A-definable, and so p_f is A-definable. □

REMARK 7.2.2. (1) Over a general base set A, the germ may not have an M-definable representative. For instance assume A is the canonical code for an open ball of valuative radius γ (e.g. $A = \mathrm{dcl}(\beta)$ with β a transcendental element of the residue field, and $b = \mathrm{res}^{-1}(\beta)$; in this case $\gamma = 0$). The path in question takes $t \in (\gamma, \infty)$ to the generic type of a closed sub-ball of M, of valuative radius t, containing a given point p_0. The germ at b does not depend on p_0, but there is no definable representative over A.

(2) Assume C is M-definable, and p an M-definable type on C. If $M = \mathrm{dcl}(F)$ for a field F, the germ in Proposition 7.2.1 (2) is represented by an M-definable path.

(3) The same proof gives a correspondence between invariant types on C, and germs at b of paths to $\widehat{C}$, up to reparameterization, where now b is a Dedekind cut in Γ. The analogue of (2) remains true if M is a maximally complete model.

7.3 LIFTING PATHS

Let us start by an easy consequence of Hensel's lemma, valid in all dimensions, but applicable only near simple points.

LEMMA 7.3.1. *Let $f : X \to Y$ be a finite morphism between smooth varieties, and let $x \in X$ be a closed point. Assume f is étale at $x \in X$. Then there exists neighborhoods N_x of x in $\widehat{X}$ and N_y of y in $\widehat{Y}$ such that $\widehat{f} : \widehat{X} \to \widehat{Y}$ induces a homeomorphism $N_x \to N_y$.*

Proof. By Hensel's lemma, there exist valuative neighborhoods V_x of x and V_y of y such that f restricts to a bijection $V_x \to V_y$. We take V_x and V_y to be defined by weak inequalities; let U_x and U_y be defined by the corresponding strict inequalities. Then f induces a continuous bijection $\widehat{V_x} \to \widehat{V_y}$ which is a homeomorphism by definable compactness. In particular, f induces a homeomorphism $N_x \to N_y$, where $N_x = \widehat{U_x}$ and $N_y = \widehat{U_y}$. □

In fact this gives a notion of a small closed ball on a curve, in the following sense:

LEMMA 7.3.2. *Let F be a valued field, C be a smooth curve over F, and let $a \in C(F)$ be a point. Then there exists an ACVF_F-definable decreasing family $b(\gamma)$ of g-closed, v-clopen definable subsets of C, with intersection $\{a\}$. Any two such families agree eventually up to reparameterization, in the sense that if b' is another such family then for some $\gamma_0, \gamma_1 \in \Gamma$ and $\alpha \in \mathbb{Q}_{>0}$, for all $\gamma \geq \gamma_1$ we have $b(\gamma) = b'(\alpha\gamma + \gamma_0)$.*

Proof. Choose $f : C \to \mathbb{P}^1$, étale at a. Then f is injective on some v-neighborhood U of a. We may assume $f(a) = 0$. Let b_γ be the closed ball of radius γ on $\mathbb{A}^1$ centered at 0. For some γ_1, for $\gamma \geq \gamma_1$ we have $b_\gamma \subset f(U)$ since $f(U)$ is v-open. Let $b(\gamma) = f^{-1}(b_\gamma) \cap U$. Note that $A = \{(x, y) \in C \times b_\gamma : f(x) = y\}$ is a v+g-closed and bounded subset of $C \times \mathbb{P}^1$. It follows from Proposition 4.2.21, Theorem 4.2.20 and Lemma 4.2.23 that $b(\gamma)$ is g-closed. Since f is a local v-homeomorphism it is v-clopen.

Now suppose $b'(\gamma)$ is another such family. Let $b'_\gamma = f(b'(\gamma))$. Then by the same reasoning b'_γ is a v-clopen, g-closed definable subset of $\mathbb{A}^1$, with $\cap_{\gamma \geq \gamma_2} b'_\gamma = \{0\}$. Each b'_γ (for large γ) is a finite union $\cup_{i=1}^m c_i(\gamma) \smallsetminus d_i(\gamma)$, where $c_i(\gamma)$ is a closed ball and $d_i(\gamma)$ is a finite union of open sub-balls of $c_i(\gamma)$, whose number is uniformly bounded, cf. Holly Theorem, Theorem 2.1.2 of [19]. From [19] it is known that there exists an F-definable finite set S, meeting each $c_i(\gamma)$ (for large γ) in one point a_i. The valuative radius of $c_i(\gamma)$ must approach ∞, otherwise it has some fixed radius γ_i for large γ, forcing the balls in $d_i(\gamma)$ to have eventually fixed radius and contradicting $\cap_\gamma b'_\gamma = \{0\}$. So, for every i and large γ, $c_i(\gamma)$ are disjoint closed balls centered at a_i. It follows that $c_i(\gamma') \smallsetminus d_i(\gamma') \subset c_i(\gamma) \smallsetminus d_i(\gamma)$ for $\gamma \ll \gamma'$. We have $a_i \notin d_i(\gamma)$, or else for large γ' we would have $c_i(\gamma') \subset d_i(\gamma)$. Thus $a_i \in \cap_\gamma c_i(\gamma) \smallsetminus d_i(\gamma)$ and $a_i = 0$, hence $m = 1$.

Now the balls of $d_1(\gamma)$ must also be centered in a point of S' for some finite set S', and for large γ we have $c_1(\gamma)$ disjoint from these balls; so $b'_\gamma = c_1(\gamma)$ is a closed ball around 0. For large γ it must have valuative radius $\alpha\gamma + \gamma_0$, for some $\alpha \in \mathbb{Q}_{>0}, \gamma_0 \in \Gamma$. □

DEFINITION 7.3.3. A continuous map $f : X \to Y$ between topological spaces with finite fibers is *topologically étale* if the diagonal Δ_X is open in $X \times_Y X$.

LEMMA 7.3.4. *Let $f : X \to Y$ be a finite morphism between varieties over a valued field. Let $c : I \to \widehat{Y}$ be a path, and $x_0 \in \widehat{X}$. If $\widehat{f} : \widehat{X} \to \widehat{Y}$ is topologically étale above $c(I)$, then c has at most one lift to a path $c' : I \to \widehat{X}$, with $c'(i_I) = x_0$.*

Proof. Let c' and c'' be two such lifts. The set $\{t : c'(t) = c''(t)\}$ is definable, it contains the initial point, and is closed by continuity. So it suffices to show that if $c'(a) = c''(a)$ then $c'(a+t) = c''(a+t)$, for sufficiently small t, which is clear by openness of the diagonal. $\square$

EXAMPLE 7.3.5. In characteristic $p > 0$, let $f : \mathbb{A}^1 \to \mathbb{A}^1$, $f(x) = x^p - x$. Let $a \in \mathbb{A}^1$ be a closed point, and consider the standard path $c_a : (-\infty, \infty] \to \widehat{\mathbb{A}^1}$, with $c_a(t)$ the generic of the closed ball of valuative radius t around a. Then $\widehat{f}^{-1}(c_a(t))$ consists of p distinct points for $t > 0$, but of a single point for $t \leq 0$. In this sense $c_a(t)$ may be said to be backwards-branching. The set of backwards-branching points is the set of balls of valuative radius 0, which is not a Γ-internal set. The complement of the diagonal within $\widehat{\mathbb{A}^1} \times_f \widehat{\mathbb{A}^1}$ is the union over $0 \neq \alpha \in \mathbb{F}_p$ of the sets $U_\alpha = \{(c_a(t), c_b(t)) : a - b = \alpha, t > 0\}$. The closure (at $t = 0$) intersects the diagonal in the backwards-branching points.

Because of Example 7.3.5, we will rely on the classical notion of étale only near initial simple points.

LEMMA 7.3.6. *Let C be an algebraic curve defined over a valued field F and let a be a closed point of C.*

(1) *There exists a path $c : [0, \infty] \to \widehat{C}$ with $c(\infty) = a$, but $c(t) \neq a$ for $t < \infty$.*

(2) *If a is a smooth point, and c and c' are two such paths, then they eventually agree, up to definable reparameterization.*

(3) *If a is in the valuative closure of an F-definable subset W and $a \notin W$, then for large $t \neq \infty$ one has $c(t) \in \widehat{W}$.*

Proof. One first reduces to the case where C is smooth. As in the proof of Theorem 7.1.1, there exists a finite purely inseparable extension F' of F such that the normalization of $C \otimes F'$ is smooth over F'. Since this does not change the notion of definability over F, we may assume $F' = F$. Let $n : \tilde{C} \to C$ be the normalization, and let $\tilde{a} \in \tilde{C}$ be a point such that, if a W is given as above, then $\tilde{a}$ is a limit point of $n^{-1}(W)$. Then the lemma for $\tilde{C}$ and $\tilde{a}$ implies the same for C and a. So, we may assume C is normal. For $\mathbb{P}^1$ the lemma is clear by inspection. In general, find a morphism $p : C \to \mathbb{P}^1$, with $p(c) = 0$ which is unramified above 0. By Lemma 7.3.1 and its proof, there exists a definable homeomorphism for the valuation topology between a definable neighborhood Y of c and a definable neighborhood W' of 0 in $\mathbb{P}^1$ which extends to a homeomorphism between $\widehat{Y}$ and $\widehat{W'}$. If c and c' are two paths to a then eventually they fall into $\widehat{W'}$. This reduces to the case of $\mathbb{P}^1$. For (3) it is enough to notice that one can assume $p(W) \cup \{0\} = W'$. (2) comes from Lemma 7.3.2. $\square$

REMARK 7.3.7. More generally let $p \in \widehat{C}$, where C is a curve. If $c \models p$, let $\mathrm{res}(F)(\bar{c})$ be the set of points of St_F definable over $F(c)$. This is the function field of a curve $\bar{C}$ in St_F. One has a definable family of paths in $\widehat{C}$ with initial point p, parameterized by $\bar{C}$. And any such path eventually agrees with some member of the family, up to definable reparameterization.

7.4 BRANCHING POINTS

Let C be a (noncomplete) curve over F together with a finite morphism of algebraic varieties $f : C \to \mathbb{A}^1$ defined over F. Given a closed ball $b \subset \mathbb{A}^1$, let $p_b \in \widehat{\mathbb{A}^1}$ be the generic type of b.

By an *outward path* on $\mathbb{A}^1$ we mean a path $c : I \to \widehat{\mathbb{A}^1}$ with I an interval in Γ_∞ such that $c(t) = p_{b(t)}$, with $b(t)$ a ball around some point c_0 of valuative radius t.

Let X be a definable subset of C. By an *outward path* on (X, f) with initial point p we mean a germ of path $c : (a, d] \to \widehat{X}$, with $a < d$, such that $f_* \circ c$ is an outward path on $\mathbb{A}^1$ and $c(d) = p$. We first consider the case $X = C$.

In the next lemma, we do not worry about the field of definition of the path; this will be considered later.

LEMMA 7.4.1. *Let $p \in \widehat{C}$. Then p is the initial point of at least one outward path on (C, f).*

Proof. The case of simple p was covered in Lemma 7.3.6, so assume p is not simple. The point $\widehat{f}(p)$ is a non-simple element of $\widehat{\mathbb{A}^1}$, i.e. the generic of a closed ball b_p, of valuative radius $\alpha \neq \infty$. Fix a model F of ACVF over which C, p and f are defined, $b_p(F) \neq \varnothing$, and $\alpha = \mathrm{val}(a_0)$ for some $a_0 \in F$. We will show the existence of an F-definable outward path with initial point p. For this purpose we may renormalize, and assume b is the unit ball $\mathcal{O}$.

Let $c \models p|F$. Then $f(c)$ is generic in $\mathcal{O}$. Since C is a curve, $\mathbf{k}(F(c))$ is a function field over $\mathbf{k}(F)$ of transcendence degree 1. Let $z : \mathbf{k}(F(c)) \to \mathbf{k}(F)$ be a place, mapping the image of $f(c)$ in $\mathbf{k}(F(c))$ to ∞. We also have a place $Z : F(c) \to \mathbf{k}(F(c))$ corresponding to the structural valuation on $F(c)$. The composition $z \circ Z$ gives a place $F(c) \to \mathbf{k}(F)$, yielding a valuation v' on $F(c)$. Since $z \circ Z$ agrees with Z on F, we can take v' to agree with val on F. We have $0 < -v'(f(c)) < \mathrm{val}(y)$ for any $y \in F$ with $\mathrm{val}(y) > 0$.

Let $q = \mathrm{tp}(c/F; (F(c), v'))$ be the quantifier-free type of c over F in the valued field $(F(c), v')$. In other words, find an embedding of valued fields $\iota : (F(c), v') \to \mathbb{U}$ over F, and let $q = \mathrm{tp}(\iota(c)/F)$. Similarly, set $r = \mathrm{tp}(f(c)/F; (F(c), v')) := \mathrm{tp}(\iota(f(c))/F)$. Clearly r is definable, thus, by Lemma 2.3.4 it follows that q is a definable type over F, so we can extend it to a global F-definable type. Note that q comes equipped with a definable map $\delta \to \Gamma$ with $\delta_*(q)$ nonconstant, namely $\mathrm{val}(f(c))$. According to Proposition 7.2.1, q corresponds to a germ at 0 of a path $c : (-\infty, 0) \to C$. Since for any rational function $g \in F(C)$ regular on p, we have $v'(g(c)) =$

$\operatorname{val}(g(c)) \bmod \mathbb{Z}v'(f(c))$, one may extend c by continuity to $(-\infty, 0]$ by $c(0) = p$. It is easy to check that (the germ of) c is an outward path, since $f_* \circ c$ is a standard outward path on $\mathbb{A}^1$. $\square$

We note immediately that the number of germs at a of paths as given in the lemma is finite. Let $p \in \widehat{C}$. Fix an outward path $c_0 : (-\infty, d] \to \widehat{\mathbb{A}^1}$, with $c_0(d) = f_*(p)$. Let $\mathrm{OP}(p)$ be the set of germs of paths $c : (a, d] \to \widehat{C}$ with $c(d) = p$ and $f_* \circ c = c_0$ on $(a, d]$ for some $a < d$. If $c_1, \ldots, c_N \in \mathrm{OP}(p)$ have distinct germs at d, then for $d' < d$ sufficiently close to d the points $c_i(d')$ are distinct; in particular $N \leq \deg(f)$.

Definition 7.4.2. A point $p \in \widehat{C}$ is called *forward-branching* for f if there exists more than one germ of outward paths $c : (a, d] \to \widehat{C}$ with $a < d$ and $c(d) = p$, above a given outward path on $\mathbb{A}^1$. We will also say in this case that $f_*(p)$ is forward-branching for f, and even that b is forward-branching for f where $f_*(p)$ is the generic type of b.

Let b be a closed ball in $\mathbb{A}^1$, p_b the generic type of b. Let $M \models \mathrm{ACVF}$, with $F \leq M$ and b defined over M, and let $a \models p_b|M$. Define $n(f, b)$ to be the number of types $\{\mathrm{tp}(c/M(a)) : f(c) = a\}$. This is also the number of types $\{\mathrm{tp}(c/\operatorname{acl}(F(b))(a)) : f(c) = a\}$ (where M is not mentioned), using the stationarity lemma Proposition 3.4.13 of [19]. Equivalently it is the number of types $q(y, x)$ over M extending $p_b(x)|M \cup \{f(y) = x\}$; or again:

$$n(f, b) = |\{\mathrm{tp}(c/M) : c \in C, f(c) = a\}|.$$

In other words $n(f, b)$ is the cardinal of the fiber of $\widehat{f}^{-1}(p_b)$, with $\widehat{f} : \widehat{C} \to \widehat{\mathbb{A}^1}$. In particular, the function $b \mapsto n(f, b)$ is definable.

If b is a closed ball of valuative radius α, and $\lambda > \alpha$, both defined over F, we define a *generic closed sub-ball of b of valuative radius λ* (over F) to be a ball of valuative radius λ around c, where c is generic in b over F. Equivalently, c is contained in no proper $\operatorname{acl}(F)$-definable sub-ball of b.

Lemma 7.4.3. *Assume b and λ are in $\operatorname{dcl}(F)$, and let b' be a generic closed sub-ball of b of valuative radius λ, over F. Then $n(f, b') \geq n(f, b)$.*

Proof. Let $F(b) \leq M \models \mathrm{ACVF}$, and $M(b') \leq M' \models \mathrm{ACVF}$. Take a generic in b' over M'. Then a is also a generic point of b over F. Now $n(f, b)$ is the number of types $\{\mathrm{tp}(c/M) : f(c) = a\}$, while $n(f, b')$ is the number of types $\{\mathrm{tp}(c/M') : f(c) = a\}$. As the restriction map sending types over M' to types over M is well-defined and surjective, we get $n(f, b) \leq n(f, b')$. $\square$

Lemma 7.4.4. *The set FB' of closed balls b such that, for some closed $b' \supsetneqq b$, for all closed b'' with $b \subsetneqq b'' \subsetneqq b'$, we have $n(f, b) < n(f, b'')$ is a finite definable set, uniformly with respect to the parameters.*

Proof. The statements about definability of FB' are clear since $b \mapsto n(f,b)$ is definable. Let us prove that for $\alpha \in \Gamma$, the set FB'_α of balls in FB' of valuative radius α is finite. Otherwise, by the Swiss cheese description of 1-torsors in Lemma 2.3.3 of [19], FB' would contain a closed ball b^* of valuative radius $\alpha' < \alpha$ such that every sub-ball of b^* of valuative radius α is in FB'. For each such sub-ball b', for some λ with $\alpha' \leq \lambda < \alpha$, we have $n(f,b') < n(f,b'')$ for any ball b'' of valuative radius γ with $\lambda < \gamma < \alpha$ containing b'. Let $\lambda(b')$ be the infimum of such λ's. Now λ is a definable function into Γ, so it is constant generically on b^*. Replacing b^* with a slightly smaller ball, we may assume λ is actually constant; so we find b of valuative radius λ such that for any sub-ball b' of b of valuative radius α, we have $n(f,b') < n(f,b)$. But this contradicts Lemma 7.4.3.

Hence FB' has only finitely many balls of each valuative radius, so it can be viewed as a function from a finite cover of Γ into the set of closed balls. Suppose FB' is infinite. Then it must contain all closed balls of valuative radius γ containing a certain point $c_0 \in K$, for γ in some proper interval $\alpha < \gamma < \alpha'$ (again by Lemma 2.3.3 of [19]). But then by definition of FB' we find $b_1 \subset b_2 \subset \ldots$ with $n(f,b_1) < n(f,b_2) < \ldots$, a contradiction. □

PROPOSITION 7.4.5. *The set of forward-branching points for f is finite.*

Proof. By Lemma 7.4.4 it is enough to prove that if p_b is forward-branching, then $b \in \mathrm{FB}'$. Let $n = n(f,b) = |\widehat{f}^{-1}(p_b)|$. Let c be an outward path on $\widehat{\mathbb{A}^1}$ beginning at p_b. For each $q \in \widehat{f}^{-1}(p_b)$ there exists at least one path starting at q and lifting c by Lemma 7.4.1, and for some such q, there exist more than one germ of such path. So in all there are $> n$ distinct germs of paths c_i lifting c. For b'' along c sufficiently close to b, the $c_i(b'')$ are distinct; so $n(f,b'') > n$. □

PROPOSITION 7.4.6. *Let $f : C \to \mathbb{A}^1$ be a finite morphism of curves over a valued field F. Let $x_0 \in C$ be a closed point where f is unramified, $y_0 = f(x_0)$, and let c be an outward path on $\widehat{\mathbb{A}^1}$, with $c(\infty) = y_0$. Let t_0 be maximal such that $c(t_0)$ is a forward-branching point of f, or $t_0 = -\infty$ if there is no such point. Then there exists a unique F-definable path $c' : [t_0, \infty] \to \widehat{C}$ with $\widehat{f} \circ c' = c$, and $c'(\infty) = x_0$.*

Proof. Let us first prove uniqueness. Suppose c' and c'' are two such paths. By Lemma 7.3.1 and Lemma 7.3.4, $c'(t) = c''(t)$ for sufficiently large t. By continuity, $\{t : c'(t) = c''(t)\}$ is closed. Let t_1 be the smallest t such that $c'(t) = c''(t)$. Then we have two germs of paths lifting c beginning with $c'(t)$, namely the continuations of c', c''. So $c'(t)$ is a forward-branching point, and hence $t \leq t_0$. This proves uniqueness on $[t_0, \infty)$.

Now let us prove existence. Since we are aiming to show existence of a unique and definable object, we may increase the base field; so we may assume the base field F is a maximally complete model of ACVF.

Claim 1. *Let $P \subset (t_0, \infty]$ be a complete type over F, with $n(f, a) = n$ for $a \in c(P)$. Then there exist continuous definable $c_1, \dots, c_n : P \to \widehat{C}$ with $\widehat{f} \circ c_i = c$, such that $c_i(\alpha) \neq c_j(\alpha)$ for $\alpha \in P$ and $i \neq j \leq n$.*

Proof of the claim. The proof is similar to that of Proposition 7.2.1, but we repeat it. Let $\alpha \in P$. We consider the distinct preimages $\beta_1, \dots, \beta_n$ of $c(\alpha)$ on $\widehat{C}$, and for each β_i we chose a realization b_i of the corresponding type. The morphism f is finite, so $\Gamma(F(f(b_i)))$ has finite index in $\Gamma(F(b_i))$. Since $\Gamma(F(f(b_i)))$ is generated by $\Gamma(F)$ and α, it follows from Theorem 2.9.2 that $\mathrm{tp}(b_i/\operatorname{acl}(F(\alpha)))$ is stably dominated. By [19], Corollary 3.4.3 and Theorem 3.4.4, $\operatorname{acl}(F(\alpha)) = \operatorname{dcl}(F(\alpha))$. Thus $\mathrm{tp}(b_i/F(\alpha)) \in \widehat{C}$ is α-definable over F, and we can write $\mathrm{tp}(b_i/F(\alpha)) = c_i(\alpha)$. □

Claim 2. *For each complete type $P \subset (t_0, \infty]$ over F, there exists a half-open interval $(\alpha_P, \beta_P]$, $\alpha_P, \beta_P \in \Gamma_\infty(F)$, with $P \subset (\alpha_P, \beta_P]$, and for each $y \in \widehat{f}^{-1}(c(\beta_P))$, a (unique) $F(y)$-definable path $c' : (\alpha_P, \beta_P] \to \widehat{C}$ with $\widehat{f} \circ c' = c$ and $c'(\beta_P) = y$.*

Proof of the claim. For $P = \{\infty\}$ this again follows from Lemma 7.3.1. When P is a realized type different from ∞, the statement for P follows from the one for the F-type P^- of elements infinitely close to P and smaller than P. Thus it remains to consider the case when P is not realized. Then P is an intersection of open intervals defined over F. Say $n(f, a) = n$ for $a \in c(P)$. By Claim 1 there exist disjoint $c_1, \dots, c_n$ on P with $\widehat{f} \circ c_i = c$. By definability of the space $\widehat{C}$, and compactness, they may be extended to an open interval (α, β) around P defined over F, such that moreover $n(f, c(a)) = n$ for $a \in I$, and the $c_i(a)$ are distinct. So $\{c_i(a) : i = 1, \dots, n\} = \widehat{f}^{-1}(c(a))$. Since $\beta > t_0$ it is not forward-branching, so we have $n(f, c(\beta)) = n$ also, and the paths c_i remain distinct at $c(\beta)$. The claim follows. □

Now by compactness of the space of types over F, $(t_0, \infty]$ is covered by a finite union of open intervals where the conclusion of Claim 2 holds. It is now easy to produce c', beginning at ∞ and gluing along these intervals. □

Remark 7.4.7. Here we continue the path till the first time t such that some point of C above $c(t)$ is forward-branching. It is possible to continue the path c' a little further, to the first point such that $c'(t)$ itself is forward-branching. However in practice, with the continuity with respect to nearby starting points in mind, we will stop short even of t_0, reaching only the first t such that $c(t)$ *contains* a forward-branching ball.

7.5 CONSTRUCTION OF A DEFORMATION RETRACTION

Let $\mathbb{P}^1$ endowed with the standard metric of Lemma 3.10.1, dependent on a choice of open embedding $\mathbb{A}^1 \to \mathbb{P}^1$. Define $\psi : [0, \infty] \times \mathbb{P}^1 \to \widehat{\mathbb{P}^1}$ by letting

$\psi(t,a)$ be the generic of the closed ball around a of valuative radius t, for this metric. By definition of the metric, the homotopy preserves $\widehat{\mathcal{O}}$ (in either of the standard copies of $\mathbb{A}^1$). We will refer to ψ as the *standard homotopy* of $\mathbb{P}^1$.

Note that $\widehat{\mathbb{P}^1}$ has a natural tree structure. Given two points x and y in $\widehat{\mathbb{P}^1}$ there exists a unique iso-definable subset $[x,y]$ definably isomorphic to a closed generalized interval with endpoints x and y. If D is a subset of $\widehat{\mathbb{P}^1}$, one defines the convex hull of D as the union of all the sets $[x,y]$, for $x,y \in D$.

Given a Zariski closed subset $D \subset \mathbb{P}^1$, let $\rho(a,D) = \max\{\rho(a,d) : d \in D\}$. Define $\psi_D : [0,\infty] \times \mathbb{P}^1 \to \widehat{\mathbb{P}^1}$ by $\psi_D(t,a) = \psi(\max(t,\rho(a,D)),a)$. We call ψ_D the *standard homotopy with stopping time defined by D*. In case $D = \mathbb{P}^1$ this is the identity homotopy, $\psi_D(t,a) = a$; but we will be mostly interested in the case of finite D. In this case ψ_D has a Γ-internal image, namely the convex hull of D. (Note: it is important to use the metric minimum distance, and not schematic distance. For instance if one uses the latter for the subscheme on $\mathbb{A}^1$ having a double point at 0, the image would not be Γ-internal.)

Let C be a projective curve over F together with a finite morphism $f : C \to \mathbb{P}^1$ defined over F. Working in the two standard affine charts A_1 and A_2 of $\mathbb{P}^1$, one may extend the definition of forward-branching points of f to the present setting. The set of forward-branching points of f is contained in a finite definable set, uniformly with respect to the parameters. Factor f as $C \xrightarrow{h} C' \xrightarrow{f'} \mathbb{P}^1$ with h finite radicial and f' generically étale. By Corollary 4.2.28, $\widehat{h} : \widehat{C} \to \widehat{C'}$ is a homeomorphism. Note that h induces a bijection between the set of forward-branching points of f and of f'.

THEOREM 7.5.1. *Fix a finite F-definable subset G_0 of $\widehat{C'}$, including all forward-branching points of f', all singular points of C' and all ramification points of f'. Set $G = \widehat{f'}(G_0)$ and fix a nonempty divisor D in $\mathbb{P}^1$ having a nonempty intersection with all balls in G (i.e. all balls of either affine line in $\mathbb{P}^1$, whose generic point lies in G). In other words, the convex hull of D contains all the aforementioned points. Then $\psi_D : [0,\infty] \times \mathbb{P}^1 \to \widehat{\mathbb{P}^1}$ lifts uniquely to a v+g-continuous F-definable function $[0,\infty] \times C \to \widehat{C}$ extending to a deformation retraction $H : [0,\infty] \times \widehat{C} \to \widehat{C}$ onto an iso-definable Γ-internal subset of $\widehat{C}$.*

Proof. Since $\widehat{h} : \widehat{C} \to \widehat{C'}$ is a homeomorphism we may assume $C = C'$ and $f = f'$. Fix $y \in \mathbb{P}^1$. The function $c'_y : [0,\infty] \to \widehat{\mathbb{P}^1}$ sending t to $\psi_D(t,y)$ is v+g-continuous. By Proposition 7.4.6, for every x in C there exists a unique path $c_x : [0,\infty] \to \widehat{C}$ lifting $c'_{f(x)}$. This path remains within the preimage of either copy of $\mathbb{A}^1$. By Lemma 10.1.1 with $X = \mathbb{P}^1$, it follows that the function $h : [0,\infty] \times C \to \widehat{C}$ defined by $(t,x) \mapsto c_x(t)$ is v+g-continuous. By Lemma 3.8.5, h extends to a deformation retraction $H : [0,\infty] \times \widehat{C} \to \widehat{C}$.

To show that $H(0,C)$ is Γ-internal, it is enough to check that $\widehat{f}(H(0,C))$ is Γ-internal, which is clear. Uniqueness is clear by Proposition 7.4.6. □

EXAMPLE 7.5.2. Assume the residual characteristic of the valued field F is not 2. Fix $\lambda \in F$, $\lambda \neq 0$, with $\mathrm{val}(\lambda) > 0$. Let C_λ be the projective model of the Legendre curve $y^2 = x(x-1)(x-\lambda)$ and let $f : C_\lambda \to \mathbb{P}^1$ be the projection to the x coordinate. With the notation of Theorem 7.5.1, we may take D to be the divisor consisting of the four points 0, 1, λ and ∞. For $x \in F$ with $\mathrm{val}(x) \geq 0$, denote by η_x the generic point of the smallest closed ball containing 0 and x. Thus, the final image of $\mathbb{P}^1$ under ψ_D is the finite graph K that consists of the union of five segments connecting respectively 0 to η_λ, λ to η_λ, 1 to η_1, η_λ to η_1 and ∞ to η_1. The final image of H is the preimage K' of K under $\widehat{f}$ which may be described as follows: over each point of the interior of the segment connecting η_λ to η_1 there are exactly two points in K' and over all other points of K there is exactly one (note that $\widehat{f}^{-1}(\eta_\lambda)$ is a forward-branching point). Thus K' retracts on the preimage of the segment connecting η_λ to η_1 which is combinatorially a circle (see Example 14.2.2 for the translation of this example in the Berkovich setting).

EXAMPLE 7.5.3. Let C be the union of the three lines $x = 0$, $y = 0$ and $x + y = 1$ in $\mathbb{A}^2_F$ or its closure in $\mathbb{P}^2_F$. On each line L consider ψ_D with D the divisor consisting of the intersection points with the two other lines. They paste together to produce a retraction of $\widehat{C}$ to an iso-definable Γ-internal subset definably homeomorphic to the subset Σ of Γ^3_∞ defined as follows. Let $Y = \{(\infty, t, 0); 0 \leq t \leq \infty\}$ be the segment connecting $(\infty, \infty, 0)$ to $(\infty, 0, 0)$ and let the symmetric group S_3 act on Γ^3_∞ by permuting the coordinates. Then Σ is the hexagon $\cup_{\sigma \in S_3} \sigma(Y)$. One may check, similarly as in the example of Remark 13.2.3, that Σ is not homotopically equivalent to a definable subset of some Γ^n (or Γ^w with finite definable w). In particular, there is no way to retract definably $\widehat{C}$ onto an iso-definable Γ-internal subset definably homeomorphic to a subset of some Γ^n or Γ^w. Note that this phenomenon detects the singularities of C; for instance, a similar statement would hold when C is a nodal cubic ($\widehat{C}$ would retract to a "circle" containing the singular point and such a circle is not definably homotopy equivalent to a definable subgraph of some Γ^n).

Chapter Eight

Strongly stably dominated points

Summary. In 8.1 we study further the properties of strongly stably dominated types over valued fields bases. In this setting, strong stability corresponds to a strong form of the Abhyankar property for valuations: the transcendence degrees of the extension and of the residue field extension coincide. In 8.2 we prove a Bertini type result and also that the strongly stable points form a strict ind-definable subset $V^\#$ of $\widehat{V}$. In 8.3 we prove a rigidity statement for iso-definable Γ-internal subsets of maximal o-minimal dimension of $\widehat{V}$, namely that they cannot be deformed by any homotopy leaving appropriate functions invariant. This result will be used in 11.6. In 8.4, we study the closure of iso-definable Γ-internal sets in $V^\#$ and we prove that $V^\#$ is exactly the union of all skeleta (using Theorem 11.1.1).

8.1 STRONGLY STABLY DOMINATED POINTS

Recall the notion of being strongly stably dominated from Definition 2.6.9. This definition makes sense for types of arbitrary imaginaries, but we will be interested here in the case of types on an algebraic variety.

Let q be a definable type on a variety V over a valued field. Write $\dim(q)$ for the dimension of the Zariski closure of q, i.e. of the smallest subvariety of V on which q concentrates.

We call a definable type *sequentially stably dominated* if for all $A = \mathrm{acl}(A)$ with q based on A and $q|A = \mathrm{tp}(c/A)$, there exist $c_1, \ldots, c_n \in A(c)$ with $\mathrm{tp}(c_i/A(c_1, \ldots, c_{i-1}))$ stably dominated, and $c \in \mathrm{acl}(A(c_1, \ldots, c_n))$. Here each c_i is a singleton from the field sort. We will see in Proposition 8.1.2 that this is the same notion, on a variety, as being strongly stably dominated; and that it suffices to check the property for *some* $A = \mathrm{acl}(A)$ with q based on A.

We call a type $\mathrm{tp}(c/A)$ over A strongly stably dominated, respectively sequentially stably dominated, if it extends to a definable type over $\mathrm{acl}(A)$, with the corresponding property. In this case, the definable type is uniquely determined by $\mathrm{tp}(c/\mathrm{acl}(A))$.

LEMMA 8.1.1. *Assume $A = \mathrm{acl}(A)$ is generated by $\mathrm{VF}(A) \cup \Gamma(A)$. Let V be an algebraic variety defined over $\mathrm{VF}(A)$. Then the set of sequentially stably dominated types on V over A is dense in the space of types on V over A. If in addition $\Gamma(A) \neq (0)$, this remains true if one restricts to Zariski dense types on V.*

Proof. Let P be the class of sequentially stably dominated types (respectively sequentially stably dominated Zariski dense types) over A. To show $\mathrm{tp}(cd/A)$ is approximated by types of the given class P, we may use transitivity. Consider a formula $\phi(x, y) \in \mathrm{tp}(cd/A)$. If we know the density for 1-types, we can find d' with $\mathrm{tp}(d'/A)$ in P, and such that $(\exists x)\phi(x, d')$. Then we can find c' with $\phi(c', c')$ and $\mathrm{tp}(c'/\operatorname{acl}(A(d'))) \in P$, and by transitivity (Proposition 2.6.12 (3)) we have $\mathrm{tp}(c'd'/A) \in P$.

Let $(c_1, \ldots, c_n)$ be affine coordinates of c in an appropriate affine embedding. It suffices to approximate $\mathrm{tp}(c_i/\operatorname{acl}(A(c_1, \ldots, c_{i-1})))$ for each i; so we may assume $c \in \mathbb{A}^1$.

Let D be a nonempty A-definable subset of $\mathbb{A}^1$. By C-minimality, D contains either a subset B which is an A-definable closed ball of finite radius in Γ possibly with finitely many proper A-definable sub-balls removed or an A-definable point. Moreover, if $\Gamma(A) \neq (0)$ and D is Zariski dense, D always contains such a B. Note that such a definable set B has a canonical definable type, namely the type of elements in this diminished ball avoiding any proper sub-ball and that this type yields a (sequentially) stably dominated type over A within D. □

PROPOSITION 8.1.2. *Let q be an A-definable type on a variety V over a valued field. Let F be a valued field with $A \leq \operatorname{dcl}(F)$. The following conditions are equivalent:*

(1) *q is strongly stably dominated;*

(2) *over F there exists a locally closed subvariety W of V with $q \in \widehat{W}$ and q Zariski dense in W, and a quasi-finite morphism $f : W \to \mathbb{A}^n$ of varieties, such that $f_*q = p_{\mathcal{O}}^n$ where $p_{\mathcal{O}}$ is the generic type of $\mathcal{O}$;*

(3) $\dim(q) = \dim(g_*q)$ *for some F-definable map g into a variety over the residue field;*

(4) $\dim(q) = \dim(h_*q)$ *for some A-definable map h into a stable sort; here* $\dim(h_*q)$ *refers to Morley dimension;*

(5) *there exist singletons $c_1, \ldots, c_n \in A(c)$ with* $\mathrm{tp}(c_i/A(c_1, \ldots, c_{i-1}))$ *stably dominated, and* $c \in \operatorname{acl}(A(c_1, \ldots, c_n))$.

(6) *$q|A$ is sequentially stably dominated over A.*

Proof. (1) implies (2): Assume first that F is not trivially valued. Let $c \models q|F$. Then $\mathrm{tp}(c/\mathrm{St}_F(c))$ is isolated. Now $\mathrm{St}_F = \operatorname{acl}(F \cup \mathbf{k})$, where $\mathbf{k}$ is the residue field (over the model $\operatorname{acl}(F)$, a $\mathbf{k}$-internal set is contained in $\operatorname{dcl}(\mathbf{k})$). So $\mathrm{St}_F(c) = \operatorname{dcl}(F(c)) \cap \operatorname{acl}(F \cup \mathbf{k})$. Thus $\mathrm{tp}(c/F(d_1, \ldots, d_n))$ is isolated for some $d_1, \ldots, d_n \in \mathbf{k}(\operatorname{acl}(F(c))$; by taking conjugates over $F(c)$ we may assume $d_1, \ldots, d_n \in \mathbf{k}(F(c))$. Let n be minimal, thus $d_1, \ldots, d_n$ are algebraically independent over F. We may write $d_i = \operatorname{res} f_i(c)$ where f_i is an F-definable function. In fact, upon replacing d_i with $d_i^{p^m}$ for high enough m, if the residue characteristic is $p > 0$, we can take f_i to be a rational function. So $\mathrm{tp}(c/F(f_1(c), \ldots, f_n(c)))$ is isolated. But $F(f_1(c), \ldots, f_n(c))^{\mathrm{alg}} \models$ ACVF; so $\mathrm{tp}(c/F(f_1(c), \ldots, f_n(c)))$ is realized in $F(f_1(c), \ldots, f_n(c))^{\mathrm{alg}}$, i.e. $c \in F(f_1(c), \ldots, f_n(c))^{\mathrm{alg}}$. It follows that $n = \dim(q)$. Now one may easily find W such that $f|W$ is quasi-finite.

If F is trivially valued then so is $F(c)$, since $\mathrm{tp}(c/F)$ is orthogonal to Γ; this case is proved similarly to the above but more easily and is left to the reader.

(2) implies (3) is clear; we may take $\mathbb{A}^n$ over the residue field, and $g = \mathrm{res} \circ f$.

(3) implies (4) and (1): It follows from (3) that q is stably dominated via a function defined over F. Indeed, the image under a map into the residue field of a definable type q on an n-dimensional variety is never more than n, and if it equals n then the image of any definable map into Γ must be constant. As definable types orthogonal to Γ are stably dominated, q must be stably dominated, and any dominating function would be algebraic over the given one over F, so q is already dominated by that function. It follows from the Descent Theorem 4.9 in [20] that q is stably dominated via some A-definable function h into a stable sort. Thus $q|F$ is stably dominated via h, and hence g_*q is dominated by h_*q. It follows that $\dim h_*q \geq \dim g_*q = \dim q \geq \dim h_*q$, so equality holds. This yields (4). To prove (1), we may assume g_*q is a Zariski dense type of $\mathbb{A}^n$ over the residue field; then $g = \mathrm{res} \circ f$ for some f as in (2). As $\dim(q) = n$, if $c \models q|F$, then $c \in \mathrm{acl}_M(f(c))$. In particular, as $\mathrm{tp}(f(c)/g(c))$ is isolated and implies a type over $M(g(c))$, $\mathrm{tp}(c/M, g(c))$ is isolated; so $\mathrm{tp}(c, h(c)/M, g(c))$ is isolated, hence also $\mathrm{tp}(c/M, g(c), h(c))$. But $g(c) \in \mathrm{acl}(M, h(c))$. So $\mathrm{tp}(c/M, h(c))$ is isolated, proving (1).

(4) implies (5): We may assume V is affine and use affine coordinates. Take $c = (c_1, \ldots, c_m)$ such that $q|A = \mathrm{tp}(c/A)$. Reordering the coordinates we may assume $c_1, \ldots, c_n$ are algebraically independent over A, while $c \in \mathrm{acl}(A(c_1, \ldots, c_n))$. So $\dim(q) = n$. Let $C_i = \mathrm{St}_C(A(c_1, \ldots, c_i))$ and let d_i be the Morley transcendence degree of C_i over C_{i-1}, i.e. the supremum of the Morley rank of $\mathrm{tp}(e/C_{i-1})$, with $e \in C_i$. Then $\sum_{i=1}^{n} d_i = n$. It follows that $d_i = 1$ for each i. Hence (this was seen in the proof of (3) implies (1), as a special case) $\mathrm{tp}(c_i/A(c_1, \ldots, c_{i-1}))$ is stably dominated.

(5) implies (6) is clear, since (5) holds for every base A.

(6) implies (1): By transitivity of strong stable domination, Proposition 2.6.12 (3), this reduces to the case $\dim(q) = 1$. In this case, taking a maximally complete model M containing A, it is clear that (3) holds over M. The implication (3) to (1) was seen above. □

EXAMPLE 8.1.3. If some closed ball b is A-definable and q is the generic type of b, then q is strongly stably dominated. Extending A by a realization of q

may not add any residue field points, but it does add a point of a torsor of the residue field, corresponding to b.

If V is a definable set, we denote the set of strongly stably dominated types on V by $V^{\#}$.

LEMMA 8.1.4. *Let U, V and W be varieties over a valued field, $f : V \to U$ be a definable map.*

1. *If $\dim(V) = 1$, then $\widehat{V} = V^{\#}$.*
2. *Let $q \in V^{\#}$. Then $f_*q \in U^{\#}$.*
3. *If f has finite fibers, $(f_*)^{-1}(U^{\#}) = V^{\#}$.*
4. *If f is surjective, then $f_*(V^{\#}) = U^{\#}$.*
5. *Let $g : V \to W^{\#}$ be a pro-definable morphism, $p \in V^{\#}$. Then $\int_p g \in W^{\#}$.*

Proof. (1) By (3) we may assume $V = \mathbb{P}^1$ in which case it is clear.

(2) follows from Proposition 2.6.12 (2).

(3) Clear from the characterization of being strongly stably dominated in Proposition 8.1.2 in terms of dimensions.

(4) Let $p \in U^{\#}$ based on a model M and write $p|M = \mathrm{tp}(c/M)$. By the density statement in Lemma 8.1.1, there exists $d \in f^{-1}(c)$ such that $\mathrm{tp}(d/\operatorname{acl}(M(c)))$ is sequentially stably dominated, hence strongly stably generated by Proposition 8.1.2. Thus, by the transitivity property (Proposition 2.6.12 (3)), $\mathrm{tp}(d/M)$ is also strongly stably generated. This yields a definable type $q \in V^{\#}$ such that $f_*q = p$.

(5) follows from Proposition 2.6.12 (3). □

REMARK 8.1.5. It follows from Example 13.1 in [20], already mentioned in Example 3.2.3, that $(\mathbb{A}^2)^{\#} \neq \widehat{\mathbb{A}^2}$. Thus, for any $n \geq 2$, $(\mathbb{A}^n)^{\#} \neq \widehat{\mathbb{A}^n}$. By rescaling, it follows that for any nontrivial closed ball b, $(b^n)^{\#} \neq \widehat{b^n}$. Thus, the same holds for any definable subset of $\mathbb{A}^n$ of dimension n, hence, by projecting to $\mathbb{A}^n$ and using Lemma 8.1.4 (2), for any definable set X in the VF-sort of dimension n.

8.2 A BERTINI THEOREM

Let F_0 be a valued field with infinite residue field and set $F = \operatorname{acl}(F_0)$

Let $p_{\mathcal{O}}$ denote the generic type of $\mathcal{O}$. We will view the tensor power $p_{\mathcal{O}}^{mk}$ as the generic type of the matrices $M_{m,k}(\mathcal{O})$; thus a *generic* matrix over F is one realizing $p_{\mathcal{O}}^{mk}$.

Since F_0 is a field with infinite residue field, $p_{\mathcal{O}}$ and hence also the generic type of $M_{m,k}(\mathcal{O})$ are finitely satisfiable in F_0. Thus a definable property that holds for a generic matrix also holds for many matrices with entries in F_0.

Recall that $F(e)$ denotes $\mathrm{dcl}(F \cup \{e\})$; this is generally bigger than the field generated by F and e.

PROPOSITION 8.2.1. *Let V be an algebraic variety over F_0. Let $c \in V$ such that* $\mathrm{tp}(c/F_0)$ *is stationary and strongly stably dominated. Assume that* $\mathrm{trdeg}_F F(c) = m$. *Then, for some locally closed subvariety W of V defined over F_0 and containing c, and some F_0-morphism $g : W \to \mathbb{A}^{m-1}$, with $\mathbf{c} = g(c)$, $\mathbf{c} \models p_{\mathcal{O}}^{m-1}$, and* $\mathrm{tp}(c/F(\mathbf{c}))$ *is stationary and strongly stably dominated.*

Proof. Let $f : W \to \mathbb{A}^m$ be as in Proposition 8.1.2 (2). We will take g of the form $L \circ f$, with $L : \mathcal{O}^m \to \mathcal{O}^{m-1}$ an $\mathcal{O}$-linear function. In fact, we will show that a generic such L will work. By *generic*, we mean a realization of the generic type of $M_{m,k}(\mathcal{O})$. Since $g^{-1}(\mathbf{c})$ is a curve, $\widehat{g^{-1}(\mathbf{c})}$ is uniformly iso-definable. The stationarity statement is equivalent to the existence of a *unique* element of $\widehat{g^{-1}(\mathbf{c})}$ extending $\mathrm{tp}(c/F(\mathbf{c}))$; it follows easily that the required property holds not only for realizations of $p_{\mathcal{O}}$ but for all sufficiently close approximations.

Thus it suffices to prove the claim below for $k = m - 1$. For the simple existence statement of L, the claim for any k follows inductively from the case $k = 1$; but we prefer to exhibit the genericity. $\square$

CLAIM. *Let $k < m$. For a generic $L : \mathcal{O}^m \to \mathcal{O}^k$, with $\mathbf{c} = L(f(c))$, $\mathbf{c} \models p_{\mathcal{O}}^k | F$ and* $\mathrm{tp}(c/F(\mathbf{c}))$ *is stationary.*

Proof of the claim. For $k = 0$ there is nothing to prove. Assume the claim holds for $k - 1$. Consider a generic realization L of the generic type of $M_{m,k}(\mathcal{O})$, over F. Let $\mathbf{c} = L(f(c))$. It is clear that $\mathbf{c} \models p_{\mathcal{O}}^k | F(L)$ and in particular $\mathbf{c} \models p_{\mathcal{O}}^k | F$. Let us prove that, moreover, $F(f(c)) \cap \mathrm{acl}(F(\mathbf{c})) = F(\mathbf{c})$. Indeed, in appropriate coordinates, over $F(L)$, $\mathbf{c}$ is the first k coordinates of a tuple $f(c)$ realizing $p_{\mathcal{O}}^m$; so $F(f(c)) \cap \mathrm{acl}(F(\mathbf{c})) \subset F(L, \mathbf{c})$; but $L, \mathbf{c}$ are independent over F, and L/F is stationary, so $F(L, \mathbf{c}) \cap \mathrm{acl}(F(\mathbf{c})) \subset F(\mathbf{c})$. Now suppose $\mathrm{tp}(c/F(\mathbf{c}))$ is not stationary, so

$$F(c) \cap \mathrm{acl}(F(\mathbf{c})) \neq F(\mathbf{c}).$$

Let $G = \mathrm{Aut}(F(c)^{\mathrm{alg}}/F(f(c)))$ be the (profinite) Galois group. We have a canonical isomorphism $\phi : G \to \mathrm{Aut}(F(c)^{\mathrm{alg}}/(F(f(c), \mathbf{c})))$ which is the inverse of the restriction map. The displayed inequality above implies that $\mathrm{Aut}(F(c)^{\mathrm{alg}}/\mathrm{acl}(F(\mathbf{c})))$ is a proper subgroup of $\mathrm{Aut}(F(c)^{\mathrm{alg}}/F(f(c), \mathbf{c}))$. Let H be the pullback under ϕ of this subgroup. Let $\hat{J} = \mathrm{Fix}(H)$, so $\hat{J}$ is a proper algebraic extension of $F(f(c))$, and $\hat{J} \subset \mathrm{dcl}(F(\mathbf{c})^{\mathrm{alg}}(f(c)))$. In fact by Galois theory, there exists an algebraic extension E of $F(\mathbf{c})$ such that

$$\hat{J}(\mathbf{c}) = H(f(c)).$$

Now let L and L' be mutually generic realizations of the generic type of $M_{m,k}(\mathcal{O})$. Let $\mathbf{c}' = L'(f(c))$. If $k \leq m/2$ then $F(\mathbf{c})$ and $F(\mathbf{c}')$ are linearly disjoint over $F(L, L')$ and hence over F. If $m > k > m/2$, they are linearly independent over their intersection, which is generated over F by a realization of $p_{\mathcal{O}}^{2k-m}$. (To see this, it is convenient to express $L = L_1 \oplus L_2, L' = L'_1 \oplus L'_2$ where $L_1 = L'_1 \circ L_3$ for some invertible L_3 so that they have the same image, and L_1, L_2, L'_2, L_3 are generic.) At any rate,

$$\hat{J}(\mathbf{c}, \mathbf{c}') = H(f(c), \mathbf{c}') = H'(f(c), \mathbf{c}).$$

Now $\mathrm{tp}(f(c)/F(\mathbf{c}, \mathbf{c}'))$ is strongly stably dominated and stationary. It follows that there exist finite extensions J of $F(\mathbf{c})$ and J' of $F(\mathbf{c}')$, with $F(\mathbf{c}, \mathbf{c}', J) = F(\mathbf{c}, \mathbf{c}'J')$. This contradicts the inductive hypothesis. □

REMARK 8.2.2.

(1) From the fact that $\mathrm{tp}(c/A(\mathbf{c}))$ extends to an $A(\mathbf{c})$-definable type, it follows that

$$\mathrm{acl}(A(\mathbf{c})) \cap \mathrm{dcl}(A(c)) = \mathrm{dcl}(A(\mathbf{c})).$$

(2) The same argument within ACF shows that for almost all L (outside of a proper Zariski closed subset of $M_{n,k}$), we have $\mathrm{acl}(A(\mathbf{c})) \cap \mathrm{dcl}(A(c)) = \mathrm{dcl}(A(\mathbf{c}))$ *in the sense of* ACF. Hence this can be required at the same time, i.e. we can require $W_{\mathbf{c}}$ is an irreducible curve.

We briefly digress to mention a geometric picture for Proposition 8.2.1, that should be developed elsewhere. Let F be a valued field, algebraically closed for simplicity. Consider a subset of affine space of the form $A = \{x : \mathrm{val}(f_i(x)) \geq 0, i \in I\}$, where $(f_i)_{i \in I}$ is a set of polynomials over F. These are ∞-definable sets in ACVF_F that we will call *polynomially convex*. If W is the Zariski closure of A, we prefer to write $A = \{x \in W : \mathrm{val}(f_i(x)) \geq 0, i \in I\}$. Any $p \in \widehat{\mathbb{A}^n}$ has an associated polynomially convex set $A(p)$, where f_i is the set of polynomials over F such that $p_*(\mathrm{val}(f_i)) \geq 0$; call polynomially convex sets arising in this way *irreducible*.

The generically stable type can be recovered from $A(p)$, via $p_*(\mathrm{val}(f)) = \inf_{a \in A(p)}\{\mathrm{val}(f(a))\}$. If p is strongly stably dominated, call $A(p)$ a *strictly algebraic irreducible affinoid*. Note that $(f_i)_{i \in I}$ may be taken to have finitely many polynomials of any given degree (generators of the appropriate lattice).

It probably follows from results in [23] that if one can take I to be finite, then A is a strictly algebraic irreducible affinoid. The (close) relation between these two notions should be clarified.

In this language, the proof of Proposition 8.2.1 can be adapted to show:

PROPOSITION 8.2.3. *A strictly algebraic irreducible affinoid of dimension* > 2 *admits strictly algebraic irreducible hyperplane sections.*

REMARK 8.2.4. It may be possible to approximate any affinoid (possibly including analytic affinoids in the Berkovich setting) by a strictly algebraic one, leading to a more general Bertini theorem. Strict irreducibility is roughly the same as having a Shilov boundary consisting of a single element.

Proposition 8.2.1 will allow us to think of a strongly stably dominated type of dimension n as the integral over $p_{\mathcal{O}}^{n-1}$ of a definable function into $\widehat{V/\mathbb{A}^{n-1}}$, where $\dim(V) = n$.

PROPOSITION 8.2.5. *Let V be an algebraic variety over a valued field and let $q \in V^{\#}$ such that $\dim(q) = m$. Then there exists a Zariski open subvariety W of the Zariski closure of q, a morphism $W \to \mathbb{A}^{m-1}$ making W a relative curve over an open subset of $\mathbb{A}^{m-1}$, and a definable map $j : \mathcal{O}^{m-1} \to \widehat{W/\mathbb{A}^{m-1}}$, such that $q = \int_{p_{\mathcal{O}}{}^{m-1}} j$. Conversely for any such W and j, $\int_{p_{\mathcal{O}}{}^{m-1}} j$ lies in $V^{\#}$.*

Proof. Let A a base for q, $c \models q|A$, and let notation $(W, m, g, \mathbf{c}, q_{\mathbf{c}})$ be as in Proposition 8.2.1. By Remark 8.2.2 the generic fiber of g can be taken to be an irreducible curve. Restricting to a Zariski open subset of W, we can arrange that $g : W \to U \subset \mathbb{A}^{m-1}$ is a relative curve. We view $q_{\mathbf{c}}$ as an element of the iso-definable set $\widehat{W_{\mathbf{c}}}$ (cf. Theorem 7.1.1). As $q_{\mathbf{c}} \in \operatorname{dcl}(A, \mathbf{c})$, and $\mathbf{c} \models p_{\mathcal{O}}^{m-1}$, there exists an A-definable $j : \mathcal{O}^{m-1} \to \widehat{W/\mathbb{A}^{m-1}}$ such that $j(\mathbf{c}) = q_{\mathbf{c}}$. Now $c \models q_{\mathbf{c}}|A(\mathbf{c})$; by definition, $\int_{p_{\mathcal{O}}{}^{m-1}} j$ is the unique stably dominated type based on A and extending $\operatorname{tp}(c/A)$; but q has these properties, so $\int_{p_{\mathcal{O}}{}^{m-1}} j = q$.

The converse statement is a special case of Lemma 8.1.4 (5). It holds for any definable $j : \mathcal{O}^{m-1} \to \widehat{W/\mathbb{A}^{m-1}}$, though the natural case is when j is a section of $\widehat{W/\mathbb{A}^{m-1}} \to \mathbb{A}^{m-1}$. □

For a binary map $R(x, v)$, we write R_x for the unary map defined by $R_x(v) = R(x, v)$.

DEFINITION 8.2.6. A *uniform parameterization* is a definable set X with a pro-definable map $p : X \to \widehat{V}$, along with a definable map R on $X \times V$ such that for any $x \in X$, R_x is a definable map $V \to \mathrm{St}_x$, and $p(x)$ is stably dominated via R_x.

If in addition there exist formulas ϕ_ν, $1 \leq \nu \leq n$, and a definable partition $X = \cup_{\nu=1}^n X_i$, such that $\dim(p(x))$ is constant on X_ν, and for any $1 \leq \nu \leq n$ and $x \in X_\nu$, $p(x)$ is strongly stably dominated via ϕ_ν and R_x, we say that p is a *strong uniform parameterization.*

A *uniform ind-parameterization*, resp. a *strong uniform ind-parameterization*, is a morphism $p : X \to \widehat{V}$ with X an ind-definable set, along with an ind-definable R on $X \times V$, such that the restriction to any definable $X' \subset X$ is a uniform parameterization, resp. a strong uniform parameterization.

We say a subset W of $\widehat{V}$ is *uniformly stably dominated* (resp. *strongly uniformly stably dominated, ind-uniformly stably dominated, strongly ind-uniformly stably dominated*) if there exists a uniform parameterization (resp. a strong uniform parameterization, a uniform ind-parameterization, a strong uniform ind-parameterization) $p : X \to \widehat{V}$ with $p(X) = W$.

For $p \in V^{\#}$, note that p is stably dominated via r iff for any base model M for p and any $c \models p|M$, $r(c)$ algebraically generates the residue field of $M(c)$ over $\mathrm{res}(M)$; while if p is strongly stably dominated via ϕ and r, then $r(c)$ generates the residue field of $M(c)$ as a field over $\mathrm{res}(M)$.

LEMMA 8.2.7. *Let $\pi : W \to V$ be a relative curve, X a definable set, and let $j : X \to \widehat{W/V}$ be a definable map. Then j is a strong uniform parameterization.*

Proof. First suppose $j : X \to \widehat{\mathbb{P}^1}$; then it is easy to see explicitly that j is a strong uniform parameterization. In the general case, for $x \in X$, let $W_x = \pi^{-1}(\pi(j(x)))$. After partitioning X into definable pieces, we may assume that for some morphism $h : W \to \mathbb{P}^1$, and some fixed k, for any $x \in X$, $\widehat{h}(j(x))$ has exactly k preimages in $\widehat{W_x}$. The lemma follows by a standard compactness argument. □

We denote by $V^{\#}_m$ the set of elements $p \in V^{\#}$ of dimension $\dim(p) = m$.

LEMMA 8.2.8. *Let $p : X \to \widehat{V}$ be a uniform parameterization with image contained in $V^{\#}_m$. Then p is a strong uniform parameterization.*

Proof. By compactness it suffices to show that for each $x \in X$, $p(x)$ has a definable neighborhood where the parameterization is strong. Fix $x \in X$. By Proposition 8.2.5 there exists a Zariski open subvariety W_x of the Zariski closure of $p(x)$, a morphism $f_x : W_x \to \mathbb{A}^{m-1}$ making W_x a relative curve over an open subset of $\mathbb{A}^{m-1}$, and a definable map $j_x : \mathcal{O}^{m-1} \to \widehat{W_x/\mathbb{A}^{m-1}}$, such that $p(x) = \int_{p_{\mathcal{O}^{m-1}}} j_x$. The fact that $(f_x)_*(p(x)) = p_{\mathcal{O}}{}^{m-1}$ is equivalent to $(f_x)_*(p(x)) \subset \mathcal{O}^{m-1}$ along with $(\mathrm{res} \circ f)_* p(x)$ having transcendence degree $m-1$; the latter is equivalent to R_x having transcendence degree ≤ 1 over $(\mathrm{res} \circ f)_* p(x)$; so it can be witnessed in a definable neighborhood of x. On the other hand, by Lemma 8.2.7, j_x is a strong uniform parameterization over $\mathcal{O}^{m-1}$. Now isolation is transitive, in a uniform way: if $\mathrm{tp}(c/Eb)$ is isolated via $\phi(y, b, e)$, and $\mathrm{tp}(b/E)$ is isolated via $\psi(x, e')$, then $\mathrm{tp}(bc/E)$ is isolated via $\psi(x, e') \wedge \phi(y, x, e)$, and $\mathrm{tp}(c/E)$ is isolated via $(\exists x)(\psi(x, e') \wedge \phi(y, x, e))$, so that the form of the isolating formula is fixed. Putting this together, using transitivity of isolation, we see that p is a strong uniform parameterization as well. □

LEMMA 8.2.9. *Let $p : X \to \widehat{V}$ and $q : Y \to \widehat{W}$ be strong uniform parameterizations. Let $H : V \to W$ be a definable map. Then the set*

$$\{(x, y) \in X \times Y : H_*(p(x)) = q(y)\}$$

is definable.

Proof. Say the data is defined over C_0. As $\{(x,y) \in X \times Y : H_*(p(x)) = q(y)\}$ is clearly ∞-definable, it suffices to show that it is also ind-definable. We may again work in a definable neighborhood of a given type over the base set; in particular we may assume $p(X) \subset V^{\#}_m$ and $q(Y) \subset W^{\#}_{m'}$. As p is a strong uniform parameterization, there exists a definable map $R(x, v)$ such that $R_x(v)$ generates $\mathrm{St}_{C_0(x)}(v)$ over $C_0(x)$, whenever $v \models p(x)|C_0(x)$. Let R' and ϕ witness that q is a strong uniform parameterization (partition again so that one ϕ works). Find a formula $\theta(y, z)$ such that for any $y \in Y$, $\theta(y, w)$ is a formula of Morley dimension m' and multiplicity 1 in the stable definable type $(R'_y)_*q(y)$. (Note that Morley dimension and multiplicity vary definably in definable families of formulas of St; this reduces to the case of ACF.) Then $H_*(p(x)) = q(y)$ iff for some C_0-definable h, $h(x, R_x(v)) = R'(y, H(v))$, and $H(v) \models y|C_0(y)$ whenever $v \models x|C_0(x)$. The latter condition reduces to the following three conditions:

(1) $\phi(H(v), h(x, R_x(v)))$;

(2) $\theta(y, h(x, R_x(v)))$;

(3) $R_x(v)$ has Morley rank $\geq m'$ over y.

The first two conditions are clearly definable, and the third can be ascertained ind-definably using a formula that shows $R_x(v)$ to have Morley rank $\leq m - m'$ over $h(x, R_x(v))$. □

REMARK 8.2.10. Applying Lemma 8.2.9 in the case $V = W, H = \mathrm{Id}$, we see that an ind-uniformly strongly stably dominated set $X \subset V^{\#}$ admits a strict ind-definable structure. Moreover by the same lemma, the strict ind-definable structure induced from any other ind-uniformly strong parameterization is the same.

PROPOSITION 8.2.11. *Let V be an algebraic variety over a valued field. Then $V^{\#}_m$ admits a unique strict ind-definable structure, so that it becomes ind-uniformly stably dominated. With this structure, it is in fact ind-uniformly strongly stably dominated.*

Proof. The set $\mathbb{S}_1$ of subvarieties of V is a strict ind-definable set, already in the theory ACF. The same is true of the set $\mathbb{S}_2$ of pairs (W, f) where W is a locally closed subvariety of V of dimension m, and $f : W \to U \subset \mathbb{A}^{m-1}$ is a morphism to an open subset of $\mathbb{A}^{m-1}$, whose fibers are absolutely irreducible curves. Let $\mathbb{S}_3$ be the set of triples (W, f, g), where $(W, f) \in \mathbb{S}_2$, $U = f(W)$, and $g : U \to \widehat{W/U}$ is a definable section of f (in ACVF now). It is clear that $\mathbb{S}_3$ is an ind-definable set (recall that $\widehat{W/U}$ is iso-definable by Theorem 7.1.2; this is uniform in $(W, f) \in \mathbb{S}_2$). Define a map $h : \mathbb{S}_3 \to \widehat{V}$ by $h(W, f, g) = \int_{p_{\mathcal{O}}^{m-1}} g$. By Proposition 8.2.5, the image of h is $V^{\#}_m$. This is clearly an ind-uniform parameterization. By Lemma 8.2.8, it is strong. By

Lemma 8.2.9 the kernel of h is definable on definable pieces, and so a strict ind-definable structure is induced. Uniqueness similarly follows by comparing to another parameterization, which will also be strong by Lemma 8.2.8, and so isomorphic to the given one by Lemma 8.2.9 and Remark 8.2.10. □

As $V^{\#}$ is the disjoint union of $V^{\#}_m$ over $m \le \dim(V)$, Lemma 2.2.13 endows $V^{\#}$ with a strict ind-definable structure; it is the unique such structure such that the dimension $\dim(p)$ is an ind-definable function and $V^{\#}$ is ind-uniformly strongly stably dominated. From now on this will be the way we shall view $V^{\#}$ as ind-definable.

8.3 Γ-INTERNAL SETS AND STRONGLY STABLY DOMINATED POINTS

Let V be a variety over a valued field and let W be an iso-definable Γ-internal subset of $\widehat{V}$. By the o-minimal dimension $\dim(W)$ of W we mean the dimension of any definable subset of Γ^m, for some $m \ge 0$, pro-definably isomorphic to W. Note that by Lemma 6.1.2, $\dim(W) \le \dim V$. If W' is an iso-∞-definable subset of W, we set $\dim(W') = \inf \dim(Z)$, where Z ranges over all iso-definable Γ-internal subsets containing W'. Note that if $\dim(W') = n$ then W' extends to a complete type of dimension n over any model over which V and W' are defined.

For a point x of $\widehat{V}$, we define $\dim_x(W)$ to be the infimum of $\dim(W \cap O)$, where O runs over all relatively definable neighborhoods of x. Assume that $\dim_x(W) = n \ge 0$ and that V and W are defined over some base structure A. Then there exists a complete type q over A, whose solution set is a subset $W' \subset W$, such that x lies in the closure $cl(W')$ of W' (i.e. every definable neighborhood of x intersects W') and W' has o-minimal dimension n. Indeed, the collection $C_A(x)$ of A-definable subsets W'' of W such that $x \notin cl(W'')$ is closed under finite unions. By assumption, for $W'' \in C_A(x)$, $\dim(W \smallsetminus W'') \ge n$. Hence $C'_A(x) = \{W'' \cup W''' : W'' \in C_A(x), W''' \subset W, \dim(W''') < n\}$ is also closed under finite unions and does not include W. So there exists a type over A, on W, avoiding each element of $C'_A(x)$.

We shall say W is of *pure dimension* n if it has o-minimal dimension n at every point.

In Theorem 11.1.1 (7) we will prove the existence of skeleta of pure dimension n for varieties of pure dimension n. By Theorem 11.1.1 (5) (or by Theorem 8.4.2 (3)) the skeleton points will be strongly stably dominated. The following proposition will permit us to find homotopies fixing such a given skeleton; the idea is roughly that when the skeleton already has dimension n, there is no room for the homotopy to move things around.

PROPOSITION 8.3.1. *Let V be a variety over a valued field and let $W \subset \widehat{V}$ be iso-definable and Γ-internal. Assume V is of dimension n.*

(1) *Away from a countable union of iso-definable subsets of dimension* $< n$, *all points of* W *are strongly stably dominated (see* Theorem 8.4.2 (3) *for a stronger statement).*

(2) *Let* $\phi : V \to \Gamma_\infty^r$ *be a definable function inducing a finite-to-one map* $W \to \Gamma_\infty^r$. *Let* $p \in \widehat{V}$ *with* $\dim_p W = n$, *and let* $h : I \times \widehat{V} \to \widehat{V}$ *be a homotopy respecting* ϕ. *Then* h *fixes* p. *In particular if* W *has pure dimension* n, *then* h *fixes pointwise* W.

Proof. (1) For $\alpha \in W$, let p_α denote the associated stably dominated definable type. Let A be a countable base model such that V and W are defined over A, and there exist A-definable functions $\phi_i : V \to \Gamma_\infty$, $1 \leq i \leq r$, such that the restriction of $(\phi_1, \dots, \phi_r) : V \to \Gamma_\infty^r$ to W is finite-to-one, cf. Proposition 6.2.7.

Claim. *Let* $W' \subset W$ *be the solution set of a type over* A *with* $\dim(W') = n$. *Then for any* $\alpha \in W'$, p_α *is strongly stably dominated.*

Proof of the claim. Pick $\alpha \in W'$. Let M be a maximally complete model containing A. There exists $\alpha' \models \mathrm{tp}(\alpha/A)$ with $\mathrm{tp}(\alpha'/M)$ of o-minimal dimension n. Without loss of generality (applying an automorphism of the universal domain, say) we may assume $\alpha = \alpha'$; so $\mathrm{tp}(\alpha/M)$ has o-minimal dimension n. Let $c \models p_\alpha|M(\alpha)$. Let β be a basis for $\Gamma(M(c))$ over $\Gamma(M)$. So $\beta \in M(\alpha)$. Also, as M is maximally complete, $\mathrm{tp}(c/M(\beta))$ extends to a stably dominated type r; so $r|M(\beta)$ generates a complete type over $M(\beta) \cup \Gamma$, and in particular over $M(\alpha)$. It follows that $r|M(\alpha) = p_\alpha|M(\alpha)$, so $r = p_\alpha$. Thus $\alpha \in M(\beta) \subset M(c)$. After multiplying each α_i by some positive integer, we can write $\phi_i(\alpha) = \mathrm{val}(f_i(c))$, where f_i is a rational function over M. Reordering if necessary, we may assume $\phi_1(\alpha), \dots, \phi_n(\alpha)$ are $\mathbb{Q}$-linearly independent. Let $\phi = (\phi_1, \dots, \phi_n)$, $\gamma_i := \phi_i(\alpha), \gamma = \phi(\alpha)$. Let $\psi_i(x) = \mathrm{rv}(f_i(x))$, $\psi = (\psi_1, \dots, \psi_n)$; note $\mathrm{rv}(u)$ lies in a $\mathrm{val}(u)$-definable stable sort. As $(\gamma_1, \dots, \gamma_n)$ are linearly independent modulo $\Gamma(M)$, the type of $f_1(c), \dots, f_n(c)$ over M is determined; in particular, $\mathrm{rv}(f_1(\alpha)), \dots, \mathrm{rv}(f_n(\alpha))$ are algebraically independent over $M(\gamma)$. By Proposition 8.1.2 (4), p_α is strongly stably dominated; in fact p_α is dominated by $\psi_* p_\alpha$, over $A(\gamma)$. □

Thus all points of W are strongly stably dominated, apart from ones lying in an A-definable $n-1$-dimensional set. As there are only countably many such A-definable sets, this proves (1).

(2) Let $h : I \times \widehat{V} \to \widehat{V}$ be a homotopy respecting the ϕ_i. Let $W' \subset W$ be the solution set of a complete type over A, with $\dim(W') = n$, such that p lies in the closure of W'. It suffices to prove that the elements of W' are fixed by h. Pick $\alpha \in W'$, let M be a maximally complete model containing A and set $\gamma_i = \phi_i(\alpha)$. As above, we may assume $(\gamma_1, \dots, \gamma_n)$ are linearly independent modulo $\Gamma(M)$. Let $t \in I$ be non-algebraic over $M(\gamma_1, \dots, \gamma_n)$ and set $\alpha' = h_t(\alpha)$. Since h respects the levels

of the ϕ_i, we have $\phi_i(\alpha') = \gamma_i$ for each i. Again by the linear independence of $(\gamma_1, \ldots, \gamma_n)$ over $\Gamma(M)$, $\mathrm{rv}(f_1(\alpha')), \ldots, \mathrm{rv}(f_n(\alpha'))$ are algebraically independent over $M(\gamma)$. So $\psi_* p_{\alpha'} = \psi_* p_\alpha$ is the generic type of $\mathrm{RV}(\gamma) = \Pi_i \mathrm{RV}(\gamma_i)$ (which is the unique type over $M(\gamma)$ in $\mathrm{RV}(\gamma)$). As above it follows that $p_{\alpha'}$ is defined over $M(\gamma)$, and so does not depend on t. Thus for non-algebraic t, $h_t(\alpha)$ takes a constant value; since non-algebraic values of t are dense, and h_t is continuous, this constant value must be α, and we must have $h_t(\alpha) = \alpha$ for all $t \in I$. □

REMARK 8.3.2. The proof of Proposition 8.3.1 (1) shows also the following. Let V be a quasi-projective variety of dimension n over a valued field and let $\rho : W \to \widehat{V}$ be pro-definable, continuous, and injective (or finite-to-one) where $W \subset \Gamma_\infty^m$ is a definable set of pure dimension n. Then almost all points of $\rho(W)$ are strongly stably dominated.

On the other hand, a non-strongly stably dominated point p can always be deformed in at least one direction, at least in the weaker sense of the existence of a path from p to a strongly stably dominated one. If p cannot be moved by a homotopy, it must belong to (every) skeleton in the sense of Definition 11.1.2 and is thus strongly stably dominated after all by Theorem 11.1.1.

8.4 TOPOLOGICAL PROPERTIES OF $V^\#$

Let us call a pro-definable subset X of $\widehat{V}$ *pro-Γ-internal* if the image of X in each definable quotient of $\widehat{V}$ is Γ-internal. For $X \subset \widehat{\mathbb{A}^n}$, this is equivalent to the statement that for each m, letting H_m be the space of polynomials on $\mathbb{A}^n$ of degree $\leq m$ modulo the polynomials that vanish on X, there exists a finite set F_m of bases of H_m such that for any $p \in X$, the lattice $\Lambda_m(p)$ corresponding to p in H_m is diagonal with respect to one of the bases in F_m. A Γ-parameterized pro-definable set, in particular an iso-definable Γ-internal one, is pro-Γ-internal.

PROPOSITION 8.4.1. *Let P be a pro-Γ-internal subset of $\widehat{V}$. Then the closure of $P \cap V^\#$ is contained in $V^\#$. In particular the closure of a Γ-parameterized subset of $V^\#$ is contained in $V^\#$.*

Proof. Let $q \in \widehat{V}$ and assume every neighborhood of q contains a point of $P \cap V^\#$. Say P and q are N-definable with $N \models \mathrm{ACVF}$ somewhat saturated. Find a net $p_i \in (P \cap V^\#)(N)$ with $p_i \to q$. We have to show that $q \in V^\#$.

We may assume the dimension of the Zariski closure of p_i is a fixed integer $n_0 \leq \dim(V)$, and that the stable dimension of all p_i, i.e. the maximal dimension of an image in a stable sort, is a fixed number $d_0 \leq n_0$. By Proposition 8.1.2 (4) we have in fact $d_0 = n_0$, and it suffices to show the analogous fact for q. It is enough to prove that $d(q) \geq d_0$ and $n(q) \leq n_0$.

We may assume V is affine, and even $V = \mathbb{A}^n$. Let H_d be the vector space of polynomials in n-variables of degree $\leq d$; so $\cup_d H_d$ is a K-algebra. For $p \in \widehat{V}$, let $J_d(p) = \{h \in H_d : p_*(\mathrm{val}(h)) \geq 0\}$. Then $J_d(p)$ is a lattice in $H_d/K_d(p)$, where $K_d(p)$ is the maximal K-space contained in $J_d(p)$. Since all definable maps from Γ to varieties are piecewise constant, the set of possibilities for $K_d(p)$ is an ∞-definable set of bounded cardinality, so it is finite; we may assume that $K_d(p_i)$ is constant for large i, say equal to K_d. Since $p_i \to q$, we have $K_d \subset K_d(q)$. It follows that $n(q) \leq n_0$.

For some d, we may find $x_1, \ldots, x_{d(q)}$ in H_d whose q-residues are algebraically independent elements of the residue field over $\mathbf{k}(N)$, and thus form a transcendence basis. For $p \in \widehat{V}$, let $\gamma_\nu(p) = p_*(\mathrm{val}(x_\nu))$. Thus, $\gamma_\nu(p_i) \to \gamma_\nu(q) =: \gamma_\nu$. Let $r_\nu(p)$ the image of p under $\mathrm{rv}(x_\nu)$ in $\mathrm{RV}_{\gamma_\nu(p)} = \{u : \mathrm{val}(u) = \gamma_\nu(p)\}/(1+\mathcal{M})$. Then for any $y \in H_{d'}$, $d' \geq d$, $y + \mathcal{M}J_d(q)$ depends algebraically on $r_1(q), \ldots, r_{d(q)}(q)$. For a fixed such y, the dependence is witnessed by a strict valuation inequality of the form

$$\mathrm{val}(f(y, x_1 \ldots, x_{d(q)}))_*(q) > \min(\mathrm{val}(c_\mu) + \mu \cdot \gamma_\nu)$$

for some polynomial $f = \sum c_\mu x^\mu \in N[x_0, x_1, \ldots, x_{d(q)}]$ with coefficients c_μ in $\mathcal{O}$. Thus, for large enough i, one has

$$\mathrm{val}(f(y, x_1 \ldots, x_{d(q)}))_*(p_i) > \min(\mathrm{val}(c_\mu) + \mu \cdot \gamma_\nu(p_i)).$$

This shows that the $r_i(p)$ algebraically span the image of p in the stable sorts too. Thus $d_0 = d(p_i) \leq d(q)$. □

We borrow from Definition 11.1.2 the notion of a skeleton of $\widehat{V}$. It is an iso-definable Γ-internal subset Υ of $\widehat{V}$, definably homeomorphic to a definable subset of Γ_∞^w, for some finite definable set w, such that there exists a definable deformation retraction $h : I \times \widehat{V} \to \widehat{V}$ with image Υ, and such that for each irreducible component W of V, $\Upsilon \cap W$ is of pure dimension $\dim(W)$. In particular skeleta are topologically Γ-internal. Let us conclude our study of $V^\#$ with the following theorem. Note that the proof of (4) relies on Theorem 11.1.1, that we permit ourselves to quote here.

THEOREM 8.4.2. *Let V be a quasi-projective variety over a valued field.*

(1) *Let $X \subset V^\#$ be iso-definable and Γ-internal. Then the closure of X in $\widehat{V}$ is contained in $V^\#$.*

(2) *In any iso-definable Γ-internal subset of $\widehat{V}$, the strongly stably dominated points form a* closed *iso-definable subset.*

(3) *Let X be an iso-definable Γ-internal subset of $\widehat{V}$ of pure dimension $n = \dim(V)$. Then $X \subset V^\#$.*

(4) *The set $V^\#$ is exactly the union of all skeleta of $\widehat{V}$.*

Proof. Clause (1) follows directly from Proposition 8.4.1. Iso-definability in (2) follows from 8.2.11 and closedness follows from (1).

For (3), observe that by (2), if X is an iso-definable Γ-internal subset of $\widehat{V}$, $X \cap V^{\#}$ is closed. But it follows from Proposition 8.3.1 (1) that if X is of pure dimension $n = \dim(V)$, then $X \cap V^{\#}$ is dense in X.

Let us prove (4). The fact that any skeleton of $\widehat{V}$ is contained in $V^{\#}$ follows from (3) (note that it is enough to consider the case when V is irreducible). For the converse, we shall use that $V^{\#}$ has a canonical strict ind-definable structure. Let a be a point of $V^{\#}$, and fix $M \models \mathrm{ACVF}$, over which a is realized. Since we may require that our retractions are Zariski generalizing in the sense of Theorem 11.1.1 (3), there is no harm in assuming that $a \in \widehat{W}$ for any irreducible component W of V. Set $n = \dim(V)$. We shall prove by descending induction on $k \leq n$ that if $b \in V^{\#}$ is such that $P = \mathrm{tp}(b/M)$ has o-minimal dimension k, then b belongs to some skeleton. For $k = 0$ this includes the case of $b = a$. Let $\alpha : V \to \Gamma^{\ell}$ be an M-definable function which is injective on P as provided by Theorem 6.2.8. Let $h : I \times \widehat{V} \to \widehat{V}$ be an M-definable deformation retraction as in Theorem 11.1.1, preserving the levels of α. When $k = n$, it follows from Proposition 8.3.1 (2) that all points of P are fixed by h hence belong to the image of h which is a skeleton, and thus b too. Suppose now $k < n$. For $c \in P$, let $\tau(c)$ be the maximal point $\tau \in I$ such that $h(c, t) = c$ for $t < \tau$. If $\tau(c)$ is the final point of I for one (hence for all) $c \in P$, then all $c \in P$ are contained in the final image of h and so (4) holds. Otherwise, let q_c be the type over $M(c)$ of elements of I just greater than $\tau(c)$. Consider the set $S = \{h(c, t) : c \in P, t \in q_c\}$. This is a type-definable subset of some Γ-internal definable subset of $V^{\#}$. If $\dim(S) \leq k = \dim(P)$, find an M-definable set S' of dimension k containing $S \cup P$; for $c \in P$ and $t - \tau(c)$ sufficiently small, $h(c, t)$ lies in S'. Note that P is open in S' since P is a complete type, hence for $t - \tau(c) > 0$ small enough, $h(c, t)$ must still lie in P. However, as the levels of α are preserved, $\alpha(h(c, t)) = \alpha(c)$ so $h(c, t) = c$, contradicting the definition of τ. Thus $\dim(S) = k + 1$. Clearly each point of P lies in the closure of Q (consider a path reversing the homotopy). By induction, any point of Q lies in the closure of an iso-definable Γ-internal set of dimension n; hence so does each point of P. Thus, it follows from Theorem 11.1.1 together with Proposition 8.3.1 (2), similarly as when $k = n$, that b lies on some skeleton. □

REMARK 8.4.3. Let W be an o-minimal subset of $\widehat{V}$ of pure dimension n. The fact that every point of W is strongly stably dominated also follows from Theorem 11.1.1. Indeed by Theorem 11.1.1 (1) and (5) and by Proposition 8.3.1 (2) we can find a homotopy fixing W and with strongly stably dominated final image.

REMARK 8.4.4. Modulo Theorem 11.1.1, Theorem 8.4.2 (4) is equivalent to a converse to (3) that does not mention retractions, namely that the local o-minimal dimension of $V^\#$ is everywhere equal to the local dimension of V: e.g. if V has pure dimension n, then every point p of $V^\#$ is contained in a Γ-internal set of local dimension n at p. However we do not know how to prove this local statement without using Theorem 11.1.1.

REMARK 8.4.5. It would be natural to consider $V^\#$ with the *direct limit* topology, rather than the topology induced from $\widehat{V}$. We saw that $V^\#$ has a canonical ind-definable structure; we topologize each definable subset according to the embedding in $\widehat{V}$, but then topologize $V^\#$ as a direct limit. This is another, and probably better, *canonical* topology on $V^\#$. Theorem 8.4.2 (1) implies that any ind-o-minimal subset of $V^\#$ becomes an ind-o-minimal space, i.e. a direct limit of o-minimal spaces under a system of closed embeddings.

Chapter Nine

Specializations and $\mathrm{ACV^2F}$

Summary. We introduce the theory $\mathrm{ACV^2F}$ of iterated places in 9.3. It provides us with algebraic criteria for v- and g-continuity. Some applications of the continuity criteria are given in 9.7 and 9.8. The result on definability of v- and g-criteria in 9.9 will be used in 11.7 to handle uniformity with respect to parameters. Compare to [28].

9.1 G-TOPOLOGY AND SPECIALIZATION

Let F be a valued field, and consider pairs (K, Δ), with (K, v_K) a valued field extension of F, and Δ a proper convex subgroup of $\Gamma(K)$, with $\Delta \cap \Gamma(F) = (0)$. Let $\pi : \Gamma(K) \to \Gamma(K)/\Delta$ be the quotient homomorphism. We extend π to $\Gamma_\infty(K)$ by $\pi(\infty) = \infty$. Let $\mathbf{K}$ be the field K with valuation $\pi \circ v_K$. We will refer to pairs $(K, \mathbf{K})$ as *g-pairs* over F.

The convention of 2.1 shall be in use: any ACVF_F-definable set or function will be assumed to be defined by a quantifier-free formula. This will allow us to evaluate them on g-pairs. Note that if F has characteristic $(0, p)$, i.e. $0 < v(p) < \infty$, then as $p \in F$, $v(p) \notin \Delta$, so $\mathbf{K}$ has characteristic $(0, p)$ as well. The residue field of $\mathbf{K}$ is thus a valued field of characteristic (p, p), with the same residue field as the one of K.

LEMMA 9.1.1. *Let F be a valued field, V an F-variety, and let U and X be ACVF_F-definable subsets with $U \subset X \subset V$. Then the following conditions are equivalent:*

1. *U is g-open in X;*
2. *U is the intersection of X with a positive Boolean combination of Zariski closed and open sets defined over F and sets of the form $\{w \in W : \mathrm{val}(f(w)) > \mathrm{val}(g(w))\}$, with f ang g regular functions on a Zariski open W in U, all defined over F;*
3. *for any g-pair $(K, \mathbf{K})$ over F, we have $U(\mathbf{K}) \cap X(K) \subset U(K)$;*
4. *same as (3), with K (as a field) of the form $F(a)$, with $a \in U$.*

Proof. One verifies immediately that each of the conditions is satisfied if and only if it holds on every F-definable Zariski open subset of V. So we may assume V is affine.

Let us prove that (1) implies (3). Assume U is g-open in X, and let $(K, \mathbf{K})$ be a g-pair over F. If $a \in X(K)$ and $a \in U(\mathbf{K})$, we have to show that $a \in U(K)$. We may pass for this to algebraic closures of K and $\mathbf{K}$; thus we may assume $K = K^{\text{alg}}$. Let U' be g-open, with $a \in U'$ and such that $\text{ACVF}_F \models U' \cap X \subset U$. As U' is g-open, it is defined by a positive Boolean combination of strict inequalities $\text{val}(f) < \text{val}(g)$, with f and g regular functions on V and algebraic equalities and inequalities. Since K is a model, all these data can be chosen to be defined over K. Since π is order-preserving on Γ_∞, if $\pi \circ v_K(f) < \pi \circ v_K(g)$ then $v_K(f) < v_K(g)$. The algebraic equalities and inequalities are preserved since the fields are the same. Hence $U'(\mathbf{K}) \subset U'(K)$, so $a \in U(K)$.

Since trivially (2) implies (1) and (3) implies (4), it remains to prove that (4) implies (2). Let $W = X \smallsetminus U$. Then $W \subset \text{VF}^n$ is ACVF_F-definable, and for any g-pair $(K, \mathbf{K})$ over F, $X(\mathbf{K}) \cap W(K) \subset W(\mathbf{K})$. We have to show that W is cut out of X by a finite disjunction of finite conjunctions of weak valuation inequalities $\text{val}(f) \leq \text{val}(g)$, equalities $f = g$ and inequalities $f \neq g$ involving regular functions defined over F. It suffices to show that any complete quantifier-free type q over F extending W implies a finite conjunction of this form, which in turn implies W. Let q' be the set of all equalities, inequalities and weak valuation inequalities in q, along with the formula defining X. By compactness, it suffices to show that q' implies W. Let $a \models q'$, and let $\mathbf{K}$ be the valued field $F(a)$. We have $a \in X$, and we are done if $a \in W$; so suppose $a \in U$. Let $b \models q$, and let $K = F(b)$. Since q' is complete inasfar as ACF formulas go, $F(a), F(b)$ are F-isomorphic, and we may assume $a = b$ and K and $\mathbf{K}$ coincide as fields. Any element c of K can be written as $f(a)/g(a)$ for some polynomials f, g. Let $c, c' \in K$; say $c = f(a)/g(a), c' = f'(a)/g'(a)$. If $v_K(c) \geq v_K(c')$ then $v_K(f(a)g'(a)) \geq v_K(f'(a)g(a))$; the weak valuation inequality $v_K(f(x)g'(x)) \geq v_K(f'(x)g(x))$ is thus in q, hence in q', so $v_{\mathbf{K}}(f(a)g'(a)) \geq v_{\mathbf{K}}(f'(a)g(a))$, and hence $v_{\mathbf{K}}(c) \geq v_{\mathbf{K}}(c')$. It follows that the map $v_K(c) \mapsto v_{\mathbf{K}}(c)$ is well-defined, and weak order-preserving; it is clearly a group homomorphism $\Gamma(K) \to \Gamma(\mathbf{K})$, and is the identity on $\Gamma(F)$. By the hypothesis, $W(K) \cap X(\mathbf{K}) \subset W(\mathbf{K})$. Since $b \in W(K)$, we have $a \in W(\mathbf{K})$. But a was an arbitrary realization of q', so q' implies W. □

REMARK 9.1.2. We can now see that the family of g-open sets is *definable in definable families.* In other words, if $\{U_a : a \in P\}$ is an F-definable family of definable subsets of V, and C is the set of elements $a \in P$ with U_a g-open, then C is a definable subset of P. (We may take P to be affine k-space.) Indeed it is clear from the definition that C is a union of definable sets; so it suffices to show that if $a \notin C$, then for some formula $\phi \in \text{tp}(a/F)$, any realization of ϕ is not in C. Recall the theory ACV^2F (cf. 9.3). Here we take the sorts to be the valued field sort, and the value group; the latter is enriched with a predicate for a convex subgroup $\Delta \leq \Gamma$. If (K, Δ) is the data for a g-pair, with $\text{val} : K \to \Gamma$ surjective and K algebraically closed, then $(K, \Delta) \models \mathrm{ACV^2F}$. Let $T = \mathrm{Th}(K, \Delta, c)_{c \in F}$. The complete T-type of a/F is then generated by the $ACVF_F$-diagram D of a, along with the set S of sentences: $\mathrm{val}(f(a)) > 0 \to f(a) \notin \Delta$ (for every rational function f over F, defined at a). Equivalently, we can take S to be the set of sentences: $\mathrm{val}(f(a)) > \mathrm{val}(g(a)) \to \mathrm{val}(f(a)) - \mathrm{val}(g(a)) \notin \Delta$, with f, g polynomials in k variables over F. (This makes it clear that S is independent of the type of a.) By Lemma 9.1.1, as $a \notin C$, $T + D + S \vdash U_a(\mathbf{K}) \not\subset U_a(K)$. So for some ACVF-formula $\psi \in D$, already $T + \psi(a) + S \vdash U_a(\mathbf{K}) \not\subset U_a(K)$. Hence again by the criterion, as soon as $\psi(a', b)$ holds, $a' \notin C$.

LEMMA 9.1.3. *Let F_0 be a valued field, V an F_0-variety, and let $W \subset V$ be ACVF_{F_0}-definable. Then W is g-closed if and only if for any $F \geq F_0$ with F maximally complete and algebraically closed, and any g-pair $(K, \mathbf{K})$ over F such that $\Gamma(K) = \Gamma(F) + \Delta$ with Δ convex and $\Delta \cap \Gamma(F) = (0)$, we have $W(K) \subset W(\mathbf{K})$.*

When V is an affine variety, W is g-closed iff $W \cap E$ is g-closed for every bounded, g-closed, definable subset E of V.

Proof. The "only if" direction follows from Lemma 9.1.1. For the "if" direction, suppose W is not g-closed. By Lemma 9.1.1 there exists a g-pair $(K, \mathbf{K})$ over F_0 with $W(K) \not\subset W(\mathbf{K})$; furthermore, one may assume K is finitely generated over F_0, so that $\Gamma(K) \otimes \mathbb{Q}$ is finitely generated over $\Gamma(F_0) \otimes \mathbb{Q}$ as a $\mathbb{Q}$-space. Let $c_1, \ldots, c_k \in K$ be such that $\mathrm{val}(c_1), \ldots, \mathrm{val}(c_k)$ form a $\mathbb{Q}$-basis for $\Gamma(K) \otimes \mathbb{Q}/(\Delta + \Gamma(F_0)) \otimes \mathbb{Q}$. Let $F = F_0(c_1, \ldots, c_k)$. Then $(K, \mathbf{K})$ is a g-pair over F, $\Gamma(K) = \Gamma(F) + \Delta$, and $W(K) \not\subset W(\mathbf{K})$. We continue to modify F, K, and $\mathbf{K}$. As above we may replace F by F^{alg}. Next, let K' be a maximally complete immediate extension of K, F' a maximally complete immediate extension of F, and embed F' in K' over F. Let $\mathbf{K}'$ be the same field as K', with valuation obtained by composing $\mathrm{val} : K' \to \mathrm{val}(K') = \mathrm{val}(K)$ with the quotient map $\mathrm{val}(K) \to \mathrm{val}(K)/\Delta$. Then $\mathbf{K}$ embeds in $\mathbf{K}'$ as a valued field. We have now the same situation but with F maximally complete. This proves the criterion.

For the statement regarding bounded sets, suppose again that W is not g-closed; let $(K, \mathbf{K})$ be a g-pair as above, $a \in W(K)$, $a \notin W(\mathbf{K})$. Then $a \in V \subset \mathbb{A}^n$; say $a = (a_1, \ldots, a_n)$ and let $\gamma = \max_{i \leq n} - \mathrm{val}(a_i)$. Then $\gamma \in \Delta + \Gamma(F)$ so $\gamma \leq \gamma'$ for some $\gamma' \in \Gamma(F)$. Let $E = \{(x_1, \ldots, x_n) \in V : \mathrm{val}(x_i) \geq -\gamma'\}$. Then E is F-definable, bounded, g-closed, and $W \cap E$ is not g-closed, by the criterion. $\square$

As pointed out by an anonymous referee, if W is not g-closed, there may still be no bounded subset E defined over F_0 with $W \cap E$ non-g-closed; for instance this happens when F_0 is trivially valued and $W = \{x : \mathrm{val}(x) < 0\}$. On the other hand since the family of g-closed sets is definable in definable families, if F_0 is nontrivially valued, then such a set W will be definable over

F_0^a (a model of ACVF); and it follows that one will also be definable over F_0.

COROLLARY 9.1.4. *Let W be a definable subset of a variety V. Assume whenever a definable type p on W, viewed as a set of (simple) points on $\widehat{W}$, has a limit point $p' \in \widehat{V}$, then $p' \in \widehat{W}$. Then W is g-closed.*

Proof. We will verify the criterion of Lemma 9.1.3. Let (K, Δ) give rise to a g-pair $(K, \mathbf{K})$ over F with K finitely generated over F, and $\Gamma(K) = \Delta + \Gamma(F)$, F maximally complete. Let $a \in W(K)$. Let a' be the same point a, but viewed as a point of $V(\mathbf{K})$. We have to show that $a' \in W(\mathbf{K})$. Let $d = (d_1, \dots, d_n)$ be a basis for Δ. Note $\mathrm{tp}(d/F)$ has $0 = (0, \dots, 0)$ as a limit point, in the sense of Lemma 4.2.12. Hence $\mathrm{tp}(d/F)$ extends to an F-definable type q. Now $\mathrm{tp}(a/F(d))$ is stably dominated by Theorem 2.9.2 (2), so in particular definable; hence $p = \mathrm{tp}(a/F)$ is definable. Since F is maximally complete and $\Gamma(\mathbf{K}) = \Gamma(F)$, $p' = \mathrm{tp}(a'/F)$ is stably dominated by Theorem 2.9.2. Furthermore, p' is a limit of p. To check this, since F is an elementary submodel and p, p' are F-definable, it suffices to consider F-definable open subsets of $\widehat{V}$, of the form $\mathrm{val}(g) < \infty$, $\mathrm{val}(g) < 0$ or $\mathrm{val}(g) > 0$ with g a regular function on a Zariski open subset of V. If p' belongs to such an open set, the strict inequality holds of $g(a')$, and hence clearly of $g(a)$; so p belongs to it too. By assumption, $p' \in \widehat{W}$, so $a' \in W$. □

LEMMA 9.1.5. *Let F be a valued field, V an F-variety, and let $Z \subset V \times \Gamma^\ell$ be ACVF_F-definable. Then Z is g-closed if and only if for any g-pair $(K, \mathbf{K})$ over F, $\pi(Z(K)) \subset Z(\mathbf{K})$.*

Proof. If Z is g-closed then the condition on g-pairs is also clear, since π is order-preserving. In the other direction, let $\widetilde{Z}$ be the pullback of Z to $V \times \mathrm{VF}^\ell$. Then Z is g-closed if and only if $\widetilde{Z}$ is g-closed. The condition $\pi(Z(K)) \subset Z(\mathbf{K})$ implies $\widetilde{Z}(K) \subset \widetilde{Z}(\mathbf{K})$. By Lemma 9.1.1, since this holds for any g-pair $(K, \mathbf{K})$, $\widetilde{Z}$ is indeed g-closed. □

9.2 V-TOPOLOGY AND SPECIALIZATION

Let F be a valued field, and consider pairs (K, Δ), with (K, v_K) a valued field extension of F, and Δ a proper convex subgroup of $\Gamma(K)$, with $\Gamma(F) \subset \Delta$. Let $R = \{a \in K : v_K(a) > 0 \text{ or } v_K(a) \in \Delta\}$. Then $M = \{a \in R : v_K(a) \notin \Delta\}$ is a maximal ideal of R and we may consider the field $\widetilde{K} = R/M$, with valuation $v_{\widetilde{K}}(r) = v_K(a)$ for nonzero $r = a + M \in \widetilde{K}$. We will refer to $(K, \widetilde{K})$ and the related data as a *v-pair* over F. For an affine F-variety $V \subset \mathbb{A}^n$, let $V(R) = V(K) \cap R^n$. If $h : V \to V'$ is an isomorphism between F-varieties, defined over F, then since $F \subset R$ we have $h(V(R)) = V'(R)$. Hence $V(R)$ can be defined independently of the embedding in $\mathbb{A}^n$, and the notion can be extended to an arbitrary F-variety. We have a residue map

$\pi : V(R) \to V(\widetilde{K})$. We will write $\pi(x') = x$ to mean: $x' \in V(R)$ and $\pi(x') = x$, and say: x' specializes to x. Note that $\Gamma(\widetilde{K}) = \Delta$. If $\gamma = v_K(x)$ with $x \in R$, we also write $\pi(\gamma) = \gamma$ if $v_K(x) \in \Delta$, and $\pi(\gamma) = \infty$ if $\gamma > \Delta$. Note also that if F has characteristic $(0, p)$, i.e. $p \neq 0$ but $v(p) > 0$ in F, then $v(p) \in \Delta$, so $p \notin M$, and hence $\widetilde{K}$ also has characteristic $(0, p)$.

LEMMA 9.2.1. *Let V be an F-variety, W an ACVF_F-definable subset of V. Let $(K, \widetilde{K})$ be any v-pair over F, with $\widetilde{K} \models \mathrm{ACVF}$. Then W is v-closed if and only if $\pi(W(R)) \subset W(\widetilde{K})$.*

Proof. Since ACVF_F is complete and eliminates quantifiers, we may assume W is defined without quantifiers. By the discussion above, we may take V to be affine; hence we may assume $V = \mathbb{A}^n$.

Assume the criterion holds. Let $b \in V(\widetilde{K}) \smallsetminus W(\widetilde{K})$. If $a \in V(R), b = \pi(a)$, then $a \notin W$. Thus there exists a K^{alg}-definable open ball containing a and disjoint from W. Since $F \subset \widetilde{K}$, we may view $\widetilde{K}$ as embedded in R, hence take $a = b$. It follows that the complement of W is v-open, so W is v-closed.

Conversely, assume W is v-closed, and let $a \in W(R)$, $b = \pi(a)$. Then $b \in V(\widetilde{K})$. If $b \notin W$, there exists $\gamma \in \Gamma(F)$ such that, in ACVF_F, the γ-polydisc $D_\gamma(b)$ is disjoint from W. However we have $a \in D_\gamma(b)$, and $a \in W$, a contradiction. □

LEMMA 9.2.2. *Let U be a variety over a valued field F, let $f : U \to \Gamma_\infty$ be an F-definable function, and let $e \in U(F)$. Then f is v-continuous at e if and only if for any v-pair $(K, \widetilde{K})$ over F and any $e' \in U(R)$, with $\pi(e') = e$, we have $f(e) = \pi(f(e'))$. Furthermore, if F is nontrivially valued, one can take $\widetilde{K} = F$, and if $f(e) \in \Gamma$ then in fact f is v-continuous at e if and only if it is constant on some v-neighborhood of e.*

Proof. Embed U in affine space; then we have a basis of v-neighborhoods $N(e, \delta)$ of e in U parameterized by elements of Γ, with $\delta \to \infty$.

First suppose $\gamma = f(e) \in \Gamma$. Assume for some nontrivial v-pair (K, F) and for every $e' \in U(R)$ with $\pi(e') = e$, we have $f(e) = \pi(f(e'))$. To show that $f^{-1}(\gamma)$ contains an open neighborhood of e, it suffices, since $f^{-1}(\gamma)$ is a definable set, to show that it contains an open neighborhood defined over some set of parameters. Now if we take $\delta > \Gamma(F)$, $\delta \in \Gamma(K)$, then any element e' of $N(e, \delta)$ specializes to e, i.e. $\pi(e') = e$, hence $f(e) = f(e')$ and $f^{-1}(\gamma)$ contains an open neighborhood.

Conversely if $f^{-1}(\gamma)$ contains an open neighborhood of e, this neighborhood can be taken to be $N(e, \delta)$ for some $\delta \in \mathbb{Q} \otimes \Gamma(F)$. It follows that the criterion holds, i.e. $\pi(e') = e$ implies $e' \in N(e, \delta)$ so $f(e') = f(e)$, for any v-pair $(K, \widetilde{K})$.

Now suppose $\gamma = \infty$. Assume for some nontrivial v-pair (K, F) and for every $e' \in U(R)$ with $\pi(e') = e$, we have $f(e) = \pi(f(e'))$. We have to show that for any γ', $f^{-1}((\gamma', \infty])$ contains an open neighborhood of e. In case

F is nontrivially valued, it suffices to take $\gamma' \in \Gamma(F)$. Indeed as above, any element e' of $N(e,\delta)$ must satisfy $f(e') > \gamma'$, since $\pi(f(e')) = \infty$. Conversely, if continuity holds, then for some definable function $h : \Gamma_{>0} \to \Gamma_{>0}$, if $e' \in N(e, h(\gamma'))$ then $f(e') > \gamma'$; so if $\pi(e') = e$, i.e. $e' \in N(e,\delta)$ for all $\delta > \Gamma(F)$, then $f(e') > \Gamma(F)$ so $\pi(f(e')) = \infty$. □

REMARK 9.2.3. Let $f : U \to \Gamma$ be as in Lemma 9.2.2, but suppose it is merely (v-to-g-)-continuous at e, i.e. the inverse image of any interval around $\gamma = f(e) \in \Gamma$ contains a v-open neighborhood of e. Then f is v-continuous at e.

Proof. It is easy to verify that, under the conditions of the remark, the criterion holds: $\pi(f(e'))$ will be arbitrarily close to $f(e)$, hence they must be equal. (Let us also sketch a more geometric proof. We have to show that $f^{-1}(\gamma)$ contains an open neighborhood of e. If not then there are points u_i approaching e with $f(u_i) \neq \gamma$. By curve selection we may take the u_i along a curve; so we may replace U by a curve. By pulling back to the resolution, it is easy to see that we may take U to be smooth. By taking an étale map to $\mathbb{A}^1$ we find an isomorphism of a v-neighborhood of e with a neighborhood of 0 in $\mathbb{A}^1$; so we may assume $e = 0 \in U \subset \mathbb{A}^1$. For some neighborhood U_0 of 0 in U, and some rational function F, we have $f(0) = \mathrm{val}(F)$ for $u \in U_0 \setminus 0$. By (v-to-g-)-continuity we have $f(0) = \infty$ or $f(0) = \mathrm{val}(F) \neq \infty$ also. But by assumption $\gamma \neq \infty$. Now $f = \mathrm{val}(F)$ is v-continuous, a contradiction.) □

LEMMA 9.2.4. *Let V be an F-variety with F algebraically closed, $W' \subset W$ two ACVF_F-definable subsets of V. Then W' is v-dense in W if and only if for any $a \in W(F)$, for some v-pair (K,F) and $a' \in W'(K)$, $\pi(a') = a$.*

Proof. Straightforward, but this and Lemma 9.2.5 will not be used and are left as remarks. □

LEMMA 9.2.5. *Let U be an algebraic variety over a valued field F, and let Z be an F-definable family of definable functions $U \to \Gamma$. Then the following are equivalent:*

(1) *There exists an ACVF_F-definable, v-dense subset U' of U such that each $f \in Z$ is v-continuous at each point;*

(2) *for any $K, \widetilde{K}$ such that $(\widetilde{K}, F)$ and $(K, \widetilde{K})$ are both v-pairs over F, for any $e \in U(F)$, for some $e' \in U(\widetilde{K})$ specializing to e, for any $f \in Z(\widetilde{K})$ and any $e'' \in U(K)$ specializing to e', we have $f(e'') = f(e')$.*

Proof. Let U' be the set of points where each $f \in Z$ is v-continuous. Then U' is ACVF_F-definable, and by Lemma 9.2.2, for $\widetilde{K} \models \mathrm{ACVF}_F$ we have that $e' \in U'(\widetilde{K})$ if and only if for any $f \in Z(\widetilde{K})$, any v-pair $(K, \widetilde{K})$ and any $e'' \in U(K)$ specializing to e', $f(e'') = f(e')$. Thus (2) says that for any v-pair $(\widetilde{K}, F)$, and any $e \in U(F)$, some $e' \in U'(\widetilde{K})$ specializes to e. By Lemma 9.2.4 this is equivalent to U' being dense. □

Let U be an F-definable v-open subset of a smooth quasi-projective variety V over a valued field F, let W be an F-definable open subset of Γ^m, let Z be an algebraic variety over F, and let $f : U \times W \to \widehat{Z}$ or $f : U \times W \to \Gamma_\infty^k$ be an F-definable function. We consider Γ^m and Γ_∞^k with the order topology. We say f is *(v,o)-continuous* at $(a, b) \in U \times W$ if the preimage of every open set containing $f(a, b)$ contains the product of a v-open containing a and an open containing b.

LEMMA 9.2.6. *Let U be an F-definable v-open subset of a smooth quasi-projective variety V over a valued field F, let W be an F-definable open subset of Γ^m, let Z be an algebraic variety over F, and let $f : U \times W \to \widehat{Z}$ or $f : U \times W \to \Gamma_\infty^k$ be an F-definable function. Then f is (v,o)-continuous if and only if it is continuous separately in each variable. More precisely f is (v,o)-continuous at $(a, b) \in U \times W$ provided that $f(x, b)$ is v-continuous at a, and $f(a', y)$ is continuous at b for any $a' \in U$, or dually that $f(a, y)$ is continuous at b, and $f(x, b')$ is v-continuous at a for any $b' \in W$.*

Proof. Since a base change will not affect continuity, we may assume $F \models$ ACVF. The case of maps into $\widehat{Z}$ reduces to the case of maps into Γ_∞, by composing with continuous definable maps into Γ_∞, which determine the topology on $\widehat{Z}$. For maps into Γ_∞^k, since the topology on Γ_∞^k is the product topology, it suffices also to check for maps into Γ_∞. So assume $f : U \times W \to \Gamma_\infty$ and $f(a, b) = \gamma_0$. Suppose f is not continuous at (a, b). So for some neighborhood N_0 of γ_0 (defined over F) there exist (a', b') arbitrarily close to (a, b) with $f(a', b') \notin N_0$. Fix a metric on V near a, and write $\nu(u)$ for the valuative distance of u from a. Also write $\nu'(v)$ for $\min |v_i - b_i|$, where $v = (v_1, \dots, v_m), b = (b_1, \dots, b_m)$. For any $F' \supset F$, let $r_0^+|F'$ be the type of elements u with $\mathrm{val}(a) < \mathrm{val}(u)$ for every nonzero a in F', and let $r_1^-|F'$ be the type of elements v with $0 < \mathrm{val}(v) < \mathrm{val}(b)$ for every b in F' with $\mathrm{val}(b) > 0$. Then r_0^+, r_1^- are definable types, and they are orthogonal to each other, that is, $r_0^+(x) \cup r_1^-(y)$ is a complete definable type. Consider $u, v \in \mathbb{A}^1$ with $u \models r_0^+|F, v \models r_1^-|F$. Since $F(u, v)^{\mathrm{alg}} \models$ ACVF, there exist $a' \in U(F(u, v)^{\mathrm{alg}})$ and $b' \in W(F(u, v)^{\mathrm{alg}})$ such that $\nu(a') \geq \mathrm{val}(u)$, $\nu'(b') \leq \mathrm{val}(v)$, and $f(a', b') \notin N_0$. Note that any nonzero coordinate of $a' - a$ realizes r_0^+; since r_0^+ is orthogonal to r_1^- and $v \models r_1^-|F(u)$, we have $a' - a \in F(u)^{\mathrm{alg}}$, so $a' \in F(u)^{\mathrm{alg}}$. Similarly $b' \in \Gamma(F(v)^{\mathrm{alg}})$. Say two points of Γ_∞ are very close over F if the interval between them contains no point of $\Gamma(F)$. By the continuity assumption (say the first version), $f(a', b')$ is very close to $f(a', b)$ (even over $F(u)$) and $f(a', b)$ is very close to $f(a, b)$ over F. So $f(a', b')$ is very close to $f(a, b)$ over F. But then $f(a', b') \in N_0$, a contradiction. □

Corollary 9.2.7. *More generally, let* $f : U \times \Gamma_\infty^\ell \times W \to \widehat{Z}$ *be* F*-definable, and let* $a \in U \times \Gamma_\infty^\ell$, $b \in W$. *Then* f *is (v,o)-continuous at* (a, b) *if* $f(a, y)$ *is continuous at* b, *and* $f(x, b')$ *is (v,o)-continuous at* a *for any* $b' \in W$.

Proof. Pre-compose with $\mathrm{Id}_U \times \mathrm{val} \times \mathrm{Id}_W$. □

Remark 9.2.8. It can be shown that a definable function $f : \Gamma^n \to \Gamma$, continuous in each variable, is continuous. But this is not the case for Γ_∞. For instance, $|x - y|$ is continuous in each variable, if it is given the value ∞ whenever $x = \infty$ or $y = \infty$. But it is not continuous at (∞, ∞), since on the line $y = x + \beta$ it takes the value β. By pre-composing with $\mathrm{val} \times \mathrm{Id}$ we see that Lemma 9.2.6 cannot be extended to $W \subset \Gamma_\infty^m$.

9.3 ACV^{2}F

We consider the theory ACV^{2}F of triples (K_2, K_1, K_0) of fields with surjective, non-injective places $r_{ij} : K_i \to K_j$ for $i > j$, $r_{20} = r_{10} \circ r_{21}$, such that K_2 is algebraically closed. We shall denote by ACV^{2}F$_{p_2,p_1,p_0}$ the theory of such triples with K_i of characteristic p_i. We will work in ACV^{2}F$_{F_2}$, i.e. over constants for some subfield of K_2, but will suppress F_2 from the notation. The lemmas below should be valid over imaginary constants too, at least from Γ.

We let Γ_{ij} denote the value group corresponding to r_{ij} and we write $\Gamma_{ij,\infty}$ for Γ_{ij} with an element ∞ added with usual conventions. Then we have a natural exact sequence

$$0 \to \Gamma_{10} \to \Gamma_{20} \to \Gamma_{21} \to 0.$$

The inclusion $\Gamma_{10} \to \Gamma_{20}$ is given as follows: for $a \in \mathcal{O}_{21}$, $\mathrm{val}_{10}(r_{21}(a)) \mapsto \mathrm{val}_{20}(a)$. Note that if $\mathrm{val}_{10}(r_{21}(a)) = 0$ then $a \in \mathcal{O}_{20}^*$ so $\mathrm{val}_{20}(a) = 0$. The surjection on the right is $\mathrm{val}_{20}(a) \mapsto \mathrm{val}_{21}(a)$.

Note that (K_2, K_1, K_0) is obtained from (K_2, K_0) by expanding the value group Γ_{20} by a predicate for Γ_{10}. On the other hand it is obtained from (K_2, K_1) by expanding the residue field K_1.

We will use characteristics $(0, 0, 0)$, resp. (p, p, p), when starting with a value field of characteristic $(0, 0)$, resp. (p, p); when the field we start with has characteristic $(0, p)$, we will use ACV^{2}F$_{0,p,p}$ for the g-criterion, and ACV^{2}F$_{0,0,p}$ for the v-criterion.

Lemma 9.3.1.

(1) *The theory* ACV^{2}F$_{p_2,p_1,p_0}$ *is complete.*

(2) *The induced structure on* (K_1, K_0) *is just the valued field structure; moreover* (K_1, K_0) *is stably embedded.*

(3) *The set of stably dominated types* $\widehat{V}$ *is unambiguous for* V *over* K_1, *whether interpreted in* (K_1, K_0) *or in* (K_2, K_1, K_0).

(4) *The sorts* (K_2, Γ_{20}) *admit quantifier elimination in the language with the ring operations on* K_2, *the valuation map into* Γ_{20}, *the group operations on* Γ_{20} *and the predicate* Γ_{10} *on* Γ_{20}.

Proof. (1) and (2) are special cases of Lemma 2.4.1. Indeed, take T to be ACVF_{p_2,p_1}, with sort D denoting the residue field, and take T_D^* to be ACVF_{p_1,p_0}. (3) is a consequence of (2). For (4), we use the quantifier elimination statement of Lemma 2.4.1 applied to $T = \mathrm{ACVF}_{p_2,p_0}$, $D = \Gamma$ (which we will refer to as Γ_{20}), T_D^* the expansion of $\mathrm{Th}(\Gamma_{2,0})$ by a predicate for a nontrivial convex subgroup Γ_{10}. Quantifier-elimination for ACVF as well as the stable embeddedness of Γ via term functions is well-known, cf. 2.7; quantifier elimination for $(\Gamma_{20}, \Gamma_{21}, +, -, 0, <)$ is easy and left to the reader. □

LEMMA 9.3.2. *Let* W *be a definable set in* (K_2, K_1) *(possibly in an imaginary sort).*

(1) *Let* $f : W \to \Gamma_{2,1}$ *be a definable function in* (K_2, K_1, K_0). *Then there exist* (K_2, K_1)*-definable functions* $f_1, \ldots, f_k$ *such that on any* $a \in \mathrm{dom}(f)$ *we have* $f_i(a) = f(a)$ *for some* i.

(2) *Let* $f : \Gamma_{21} \to W$ *be a* (K_2, K_1, K_0)*-definable function. Then* f *is* (K_2, K_1)*-definable (with parameters; see remark below on parameters).*

In fact this is true for any expansion of (K_2, K_1) *by relations* $R \subset K_1^m$.

Proof. We may assume (K_2, K_1, K_0) is saturated. We shall use some basic properties of stably embedded sets for which we refer to the appendix of [8].

(1) It suffices to show that for any $a \in W$ we have $f(a) \in \mathrm{dcl}_{21}(a)$, where dcl_{21} refers to the structure $M_{21} = (K_2, K_1)$. We have at all events that $f(a)$ is fixed by $\mathrm{Aut}(M_{21}/K_1, a)$. By stable embeddedness of K_1 in M_{21}, we have $f(a) \in \mathrm{dcl}_{21}(e, a)$ for some $e \in K_1$. But by orthogonality of Γ_{21} and K_1 in M_{21} we have $f(a) \in \mathrm{dcl}_{21}(a)$.

(2) Let A be a base structure, and consider a type p over A of elements of Γ_{21}. Note that the induced structure on Γ_{21} is the same in (K_2, K_1, K_0) as in (K_2, K_1), and that Γ_{21} is orthogonal to K_1 in both senses. For $a \models p$, $b = f(a)$, let $g(b)$ be an enumeration of the (K_2, K_1)-definable closure of b within K_1 (over A). By orthogonality, $g \circ f$ must be constant on p; say it takes value e on p. Now $\mathrm{tp}_{21}(ab/A, e) \models \mathrm{tp}_{21}(ab/A, K_1)$ by stable embeddedness of K_1 within (K_2, K_1). By considering automorphisms it follows that $\mathrm{tp}_{21}(ab/A, e) \models \mathrm{tp}_{210}(ab/A, e)$, so $\mathrm{tp}_{21}(ab/A, e)$ is the graph of a function on p; this function must be $f|p$. By compactness, f is (K_2, K_1)-definable. □

REMARK 9.3.3. Let D be definable in (K_2, K_1, K_0) over an algebraically closed substructure (F_2, F_1, F_0) of constants. In particular $\mathrm{res}_{21}(F_2) \subset F_1$ and $\mathrm{res}_{10}(F_1) \subset F_0$, but we do not assume equality. If D is (K_2, K_1)-definable with additional parameters, then D is (K_2, K_1)-definable over (F_2, F_1).

Proof. We may take (K_2, K_1, K_0) saturated. Let e be a canonical parameter for D as a (K_2, K_1)-definable set. Note that e is fixed by the group $\mathrm{Aut}(K_2, K_1/F_2, K_1)$. Hence by stable embeddedness of (K_1, K_0), we have $e \in \mathrm{dcl}_{K_2,K_1}(F_2, F_1')$ for some (small) $F_1' \subset K_1$. As K_1 is a pure algebraically closed field stably embedded in (K_2, K_1) and has elimination of imaginaries, $\mathrm{dcl}_{K_2,K_1}(F_2, e) = \mathrm{dcl}_{K_2,K_1}(F_2, c)$ for some tuple $c = (c_1, \dots, c_m)$ of elements of K_1. Now each c_i is fixed by $\mathrm{Aut}(K_2, K_1, K_0/F_2, F_1, F_0)$, hence by $\mathrm{Aut}(K_1, K_0/F_1, F_0)$; it follows easily that $c_i \in F_1$ (since non-algebraic elements of a valued field cannot be definable over residue field elements). □

LEMMA 9.3.4. *Let W be a (K_2, K_1)-definable set and let I be a definable subset of Γ_{21} and let $f : I \times W \to \Gamma_{21,\infty}$ be a (K_2, K_1, K_0)-definable function such that for fixed $t \in I$, $f_t(w) = f(t, w)$ is (K_2, K_1)-definable. Then f is (K_2, K_1)-definable.*

Proof. Applying compactness to the hypothesis, we see that there exist finitely many functions g_k, h_k such that g_k is (K_2, K_1)-definable, h_k is definable, and that for any $t \in I$ for some k we have $f(t, w) = g_k(h_k(t), w)$. Now by Lemma 9.3.2 (2), h_k is actually (K_2, K_1)-definable too. So we may simplify to $f(t, w) = G_k(t, w)$ with G_k a (K_2, K_1)-definable function. But every definable subset of I is (K_2, K_1)-definable, in particular $\{t : (\forall w)(f(t, w) = G_k(t, w))\}$. From this it follows that $f(t, w)$ is (K_2, K_1)-definable. □

LEMMA 9.3.5. *Let T be any theory, T_0 the restriction to a sublanguage L_0, and let $\mathbb{U} \models T$ be a saturated model, $\mathbb{U}_0 = \mathbb{U}|L_0$. Let V be a definable set of T_0. Let $\widehat{V}$, $\widehat{V_0}$ denote the spaces of generically stable types in V of T, T_0 respectively. Then there exists a map $r_0 : \widehat{V} \to \widehat{V_0}$ such that $r_0(p)|\mathbb{U}_0 = (p|\mathbb{U})|L_0$. If $A = \mathrm{dcl}(A)$ (in the sense of T) and p is A-definable, then $r_0(p)$ is A-definable.*

Proof. In general, a definable type p of T over $\mathbb{U}$ need not restrict to a definable type of T_0. However, when p is generically stable, for any formula $\phi(x, y)$ of L_0 the p-definition $(d_p x)\phi(x, y)$ is equivalent to a Boolean combination of formulas $\phi(x, b)$. Hence $(d_p x)\phi(x, y)$ is $\mathbb{U}_0$-definable. The statement on the base of definition is clear by Galois theory. □

REMARK 9.3.6. The same holds of course when T_0 is interpreted in T (not necessarily as a reduct).

Returning to $\mathrm{ACV}^2\mathrm{F}$, we have:

LEMMA 9.3.7. *Let V be an algebraic variety over K_1. Then the restriction map of* Lemma 9.3.5 *from the stably dominated types of V in the sense of (K_2, K_1, K_0) to those in the sense of (K_1, K_0) is a bijection.*

Proof. This is clear since (K_1, K_0) is embedded and stably embedded in (K_2, K_1, K_0). ("Embedded" means that the induced structure on (K_1, K_0) is just the ACV^{2}F-structure.) □

We can thus write unambiguously $\widehat{V}_{10}$ for V an algebraic variety over K_1.

Now let V be an algebraic variety over K_2. Note that K_1 may be interpreted in $(K_2, K_0, \Gamma_{20}, \Gamma_{10})$ (the enrichment of (K_2, K_0, Γ_{20}) by a predicate for Γ_{10}).

LEMMA 9.3.8. *Any stably dominated type of (K_2, K_0) in V over $\mathbb{U}$ generates a complete type of (K_2, K_1, K_0). More generally, assume T is obtained from T_0 by expanding a linearly ordered sort Γ of L_0, and that p_0 is a stably dominated type of T_0. Then p_0 generates a complete definable type of T; over any base set $A = \mathrm{dcl}(A) \leq M \models T$, $p_0|A$ generates a complete T-type over A.*

Proof. We may assume T has quantifier elimination. Then $\mathrm{tp}(c/A)$ is determined by the isomorphism type of $A(c)$ over A. Now since $\Gamma(A(c)) = \Gamma(A)$, any L_0-isomorphism $A(c) \to A(c')$ is automatically an L-isomorphism. □

LEMMA 9.3.9. *Assume T is obtained from T_0 by expanding a linearly ordered sort Γ of L_0, and that in T_0, a type is stably dominated if and only if it is orthogonal to Γ. Let V be an L_0-definable set. Then the following properties of a type on V over $\mathbb{U}$ are equivalent:*

(1) *p is stably dominated;*

(2) *p is generically stable;*

(3) *p is orthogonal to Γ;*

(4) *the restriction p_0 of p to L_0 is stably dominated.*

Proof. The implication (1) to (2) is true in any theory, and so is (2) to (3) given that Γ is linearly ordered. Also in any theory (3) implies that p_0 is orthogonal to Γ, so by the assumption on T_0, p is stably dominated, hence (4). Finally, let p_0 be stably dominated and generating a type p of T (Lemma 9.3.8), let us prove this type is also stably dominated. Using the terminology from [20] p. 37, say p is dominated via some $*$-definable functions $f : V \to D$, with D a stable ind-definable set of T_0.

Since T is obtained by expanding Γ, which is orthogonal to D, the set D remains stable in T. Now for any base A of T we have that $p|A$ is implied by $p_0|A$, hence by $(f_*(p_0)|A)(f(x))$, hence by $(f_*(p)|A)(f(x))$. So (4) implies (1). □

It follows from Lemmas 9.3.8 and 9.3.9 that for any definable set V in M_{20}^{eq}, the restriction map $\widehat{V}_{210} \to \widehat{V}_{20}$ is a bijection.

LEMMA 9.3.10. *For* $\mathrm{ACV^2F}$, *the following properties of a type on* V *over* $\mathbb{U}$ *are equivalent:*

(1) *p is stably dominated;*

(2) *p is generically stable;*

(3) *p is orthogonal to Γ_{20};*

(4) *the restriction p_{20} of p to the language of (K_2, K_0) is stably dominated.*

Proof. Follows directly from Lemma 9.3.9 upon letting T_0 be the theory of (K_2, K_0). □

9.4 THE MAP $R^{20}_{21} : \widehat{V}_{20} \to \widehat{V}_{21}$

Let V be an algebraic variety over K_2. We write $V_{210}, V_{20}, V_{21}, V_2$, etc., when we wish to view V as a definable set for (K_2, K_1, K_0), (K_2, K_0), (K_2, K_1), or just the field K_2, respectively.

We have on the face of it three spaces: $\widehat{V}_{2j}$ the space of stably dominated types for (K_2, K_j) for $j = 0$ and 1, and $\widehat{V}_{210}$ the space of stably dominated types with respect to the theory (K_2, K_1, K_0). But in fact $\widehat{V}_{20}$ can be identified with $\widehat{V}_{210}$, as Lemma 9.3.8 and Lemma 9.3.9 show. We thus identify $\widehat{V}_{210}$ with $\widehat{V}_{20}$. In particular we use this identification to define a topology on $\widehat{V}_{210}$.

By Lemma 9.3.5, we have a restriction map $R^{20}_{21} : \widehat{V}_{20} = \widehat{V}_{210} \to \widehat{V}_{21}$. If a stably dominated type over a model M is viewed as a sequence of functions into Γ (sending an M-definable function into Γ to its generic value), then R^{20}_{21} is just composition with the natural homomorphism $\Gamma_{20} \to \Gamma_{21}$. Note that R^{20}_{21} is the identity on simple points and that R^{20}_{21} is continuous.

The following Lemma 9.4.1 will not be used in the rest of the paper. Note that in (2) and (3) of Lemma 9.4.1, it is important that V be allowed to be made of imaginaries of (K_2, K_1). (In (1) this is irrelevant, since ACF eliminates imaginaries.) This allows applying them to stable completions in (4).

LEMMA 9.4.1.

(1) *Let U be a variety (or constructible set) over K_1. Let $\widehat{U}_1$ be the space of stably dominated types of U within* ACF. *Then the restriction map $\widehat{U}_{10} \to \widehat{U}_1$ is surjective.*

(2) *Let V be a pro-definable set over (K_2, K_1). Then the restriction $\widehat{V}_{210} \to \widehat{V}_{21}$ is surjective. (The same is true rationally over any algebraically closed substructure of (K_2, K_1, K_0).)*

(3) *Let V be a pro-definable set over (K_2, K_1). Then any definable type q on V_{21} extends to a definable type q' of V_{210} (moreover, with q' orthogonal to Γ_{10}).*

(4) *Let V be a quasi-projective variety over K_2. Then R^{20}_{21} is surjective and definably closed.*

(5) *The topology on $\widehat{V}_{21}$ is the quotient topology from $\widehat{V}_{210}$.*

Proof. (1) $\widehat{U}_1$ is also the space of definable types of U_1, or again the space of generics of irreducible subvarieties of U. Let W be an absolutely irreducible variety over K_1. We have to show that the generic type of W expands to a stably dominated type of (K_1, K_0). Let $\mathcal{W}$ be a scheme over $\mathcal{O}_1$ with generic fiber W, and with special fiber of dimension equal to $\dim(W)$. Then there are finitely many types q over K_1 of elements of $\mathcal{W}(\mathcal{O})$ whose residues have transcendence degree equal to $\dim(W)$, and all of them are stably dominated and have Zariski closure equal to W.

(2) Let p be a stably dominated type of (K_2, K_1); it is dominated via some definable map f to a finite-dimensional vector space over K_1. So f_*p is a definable type of K_1. By the previous paragraph, f_*p expands to a stably dominated type q of (K_1, K_0). It is now easy to see (as in Remark 9.3.3) that q dominates a unique definable type r of (K_2, K_1, K_0) via f; and clearly $R^{20}_{21}(r) = p$.

(3) For types on Γ_{21} this is easy and left to the reader; in this case, note that the type of n $\mathbb{Q}$-linearly independent elements over Γ_{21} actually generates a complete type over Γ_{210}. Now any definable type r on V_{21} is the integral over some definable q on Γ_{21} of a definable map into $\widehat{V}_{21}$; i.e. for any M (over which r is defined), $r = \mathrm{tp}(c/M)$ where $a \models q|M$ and $s = \mathrm{tp}(c/M(a))_{21}$ is stably dominated. Let q' be an expansion of q to Γ_{210}; we may assume $a \models q'|M$. By (2), there exists a (K_2, K_1, K_0) expansion s' of s to a stably dominated type s' over $M(a) = \mathrm{acl}(M(a))$. Integrating s' over q' we obtain a definable type of V_{210} restricting to r.

(4) Since V itself is open in some projective variety, we may assume V is projective. Let X be a closed pro-definable subset of $\widehat{V}_{210}$ and let q be a definable type on $\mathbf{X} = R^{20}_{21}(X) \subset \widehat{V}_{21}$. By (3), q extends to a definable type q' on $\mathbf{X}_{210}$ (the same pro-definable set $\mathbf{X}$, now viewed within the structure (K_2, K_1, K_0)). Using Remark 4.2.8, q' lifts to a definable type $\tilde{q}$ on X. Let $\tilde{c} \in X$ be a limit point of $\tilde{q}$; it exists by definable compactness of $\widehat{V}_{210} = \widehat{V}_{20}$. Let $c = R^{20}_{21}(\tilde{c})$; by continuity it is a limit point of $\mathbf{X}$.

(5) Follows from (2) and (4). □

We move towards the (K_2, K_1)-definability of the image of (K_2, K_1, K_0)-definable paths in $\widehat{V}$.

LEMMA 9.4.2. *Let $f : \Gamma_{20,\infty} \to \widehat{V}_{20}$ be (K_2, K_1, K_0)-(pro)-definable. Assume $R^{20}_{21} \circ f = \bar{f} \circ \pi$ for some $\bar{f} : \Gamma_{21,\infty} \to \widehat{V}_{21}$ with $\pi : \Gamma_{20,\infty} \to \Gamma_{21,\infty}$ be the natural projection. Then $\bar{f}$ is (K_2, K_1)-(pro)-definable.*

Proof. Let U be a (K_2, K_1)-definable set, and let $g : V \times U \to \Gamma_{21,\infty}$ be definable. We have to prove the (K_2, K_1)-definability of the map: $(\gamma, u) \mapsto g(\bar{f}(\alpha), u)$, where $g(q, u)$ denotes here the q-generic value of $g(v, u)$. For fixed γ, this is just $u \mapsto g(q, u)$ for a specific $q = R^{20}_{21}(p)$, which is certainly (K_2, K_1)-definable. By Lemma 9.3.4, the map : $(\gamma, u) \mapsto g(\bar{f}(\alpha), u)$ is (K_2, K_1)-definable. □

LEMMA 9.4.3. *Let $f : \Gamma_{20,\infty} \to \widehat{V}_{20}$ be a (K_2, K_1, K_0)-(pro)-definable path. Then there exists a path $\bar{f} : \Gamma_{21,\infty} \to \widehat{V}_{21}$ such that $R^{20}_{21} \circ f = \bar{f} \circ \pi$.*

Proof. Let us first prove the existence of $\bar{f}$ as in Lemma 9.4.2. Fixing a point of $\Gamma_{21,\infty}$, with a preimage a in $\Gamma_{20,\infty}$, it suffices to show that $R^{20}_{21} \circ f$ is constant on $\{\gamma + a : \gamma \in \Gamma_{10,\infty}\}$. Hence, for any definable family of test function $\phi(x, y) : V \to \Gamma_{20,\infty}$ we need to show that $\gamma \mapsto \pi(f(\gamma + a)_* \phi)$ is constant in γ; or again that for any b, the map $\gamma \mapsto \pi(f(\gamma + a)_* \phi(b))$ is constant in γ. This is clear since any definable map $\Gamma_{10} \to \Gamma_{21}$ has finite image (due to orthogonality of Γ_{21} and K_1 inside M_{210}, and since $\Gamma_{10} \subset K_1^{eq}$), and by continuity. By Lemma 9.4.2 $\bar{f}$ is definable, it remains to show it is continuous. This amounts, as the topology on $\widehat{V}$ is determined by continuous functions into $\Gamma_{20,\infty}$, to checking that if $g : \Gamma_{20,\infty} \to \Gamma_{20,\infty}$ is continuous and (K_2, K_1, K_0)-definable, then the induced map $\Gamma_{21,\infty} \to \Gamma_{21,\infty}$ is continuous, which is easy. □

EXAMPLE 9.4.4. Let $a \in \mathbb{A}^1$ and let $f_a : [0, \infty] \to \widehat{\mathbb{A}^1}$ be the map with $f_a(t) =$ the generic of the closed ball around a of valuative radius t. Then $R^{20}_{21} \circ f_a(t) = f_a(\pi(t))$, where on the right f_a is interpreted in (K_2, K_1) and on the left in (K_2, K_0). Also, if $f_a^\gamma(t)$ is defined by $f_a^\gamma(t) = f_a(\max(t, \gamma))$ for then $R^{20}_{21} \circ f_a^\gamma(t) = f_a^{\pi(\gamma)}(\bar{t})$.

Let $\mathbb{P}^1$ be endowed with the standard metric of Lemma 3.10.1. Given a Zariski closed set $D \subset \mathbb{P}^1$ of points, recall the standard homotopy $\psi_D : [0, \infty] \times \mathbb{P}^1 \to \widehat{\mathbb{P}^1}$ defined in 7.5.

LEMMA 9.4.5. *For every (t, a) we have $R^{20}_{21} \circ \psi_D(t, a) = \psi_D(\pi(t), a)$, where on the right ψ is interpreted in (K_2, K_1) and on the left in (K_2, K_0).*

Proof. Clear, since $\pi(\rho(a, D)) = \rho_{21}(a, D)$. □

LEMMA 9.4.6. *Let $f : V \to V'$ be an* ACF*-definable map of varieties over K_2. Then f induces $f_{20} : \widehat{V}_{20} \to \widehat{V'}_{20}$ and also $f_{21} : \widehat{V}_{21} \to \widehat{V'}_{21}$. We have $R^{20}_{21} \circ f_{20} = f_{21} \circ R^{20}_{21}$.*

Proof. Clear from the definition of R^{20}_{21}. □

9.5 RELATIVE VERSIONS

Let V be an algebraic variety over U, with U an algebraic variety over K_2, that is, a morphism of algebraic varieties $f : V \to U$ over K_2. We have already defined the relative space $\widehat{V/U}$. It is the subset of $\widehat{V}$ consisting of types $p \in \widehat{V}$ such that $\widehat{f}(p)$ is a simple point of $\widehat{U}$. A map $h : W \to \widehat{V/U}$ will be called pro-definable (or definable) if the composite $W \to \widehat{V}$ is. We endow $\widehat{V/U}$ with the topology induced by the topology of $\widehat{V}$. In particular one can speak of continuous, v-, g-, or v+g-continuous maps with values in $\widehat{V/U}$. Exactly as above we obtain $R^{20}_{21} : \widehat{V/U}_{20} \to \widehat{V/U}_{21}$. Thus, for any $u_0 \in U$, the map R^{20}_{21} restricts to the previous map $R^{20}_{21} : \widehat{V_{u_0}}_{20} \to \widehat{V_{u_0}}_{21}$ between the respective fibers over u_0.

The relative version of all the above lemmas holds without difficulty:

Lemma 9.5.1. *Let $f : U\times\Gamma_{20,\infty} \to \widehat{V/U}_{20}$ be a (K_2, K_1, K_0)-(pro)-definable map commuting with the structural maps to U. Assume $R^{20}_{21} \circ f = \bar{f} \circ \pi$ for some $\bar{f} : U\times\Gamma_{21,\infty} \to \widehat{V/U}_{21}$. Then $\bar{f}$ is (K_2, K_1)-(pro)-definable.*

Proof. Same proof as Lemma 9.4.2, or by restriction. □

Lemma 9.5.2. *Let $f : U\times\Gamma_{20,\infty} \to \widehat{V/U}_{20}$ be a (K_2, K_1, K_0)-(pro)-definable map commuting with the structural maps to U. Then the assumption that $R^{20}_{21} \circ f$ factors through $U\times\Gamma_{21,\infty}$ is automatically verified.*

Proof. This follows from Lemma 9.4.3 since a function on $U \times \Gamma_{20,\infty}$ factors through $U \times \Gamma_{21,\infty}$ if and only if this is true for the section at a fixed u, for each u. □

Example 9.4.4 goes through for the relative version $\widehat{\mathbb{A}^1 \times U/U}$, where now a may be taken to be a section $a : U \to \mathbb{A}^1$.

The standard homotopy on $\mathbb{P}^1$ defined in 7.5 may be extended fiberwise to a homotopy $\psi : [0, \infty] \times \mathbb{P}^1 \times U \to \widehat{\mathbb{P}^1 \times U/U}$, which we still call standard. Consider now an ACF-definable (constructible) set $D \subset \mathbb{P}^1 \times U$ whose projection to U has finite fibers. One may consider as above the standard homotopy with stopping time defined by D at each fiber $\psi_D : [0, \infty] \times \mathbb{P}^1 \times U \to \widehat{\mathbb{P}^1 \times U/U}$.

In this framework Lemma 9.4.5 still holds, namely:

Lemma 9.5.3. *For every (t, a) we have $R^{20}_{21} \circ \psi_D(t, a) = \psi_D(\pi(t), a)$, where on the right ψ is interpreted in (K_2, K_1) and on the left in (K_2, K_0).*

Finally Lemma 9.4.6 also goes through in the relative setting:

Lemma 9.5.4. *Let $f : V \to V'$ be an ACF-definable map of varieties over U (and over K_2). Then f induces $f_{20} : \widehat{V/U}_{20} \to \widehat{V'/U}_{20}$ and also $f_{21} : \widehat{V/U}_{21} \to \widehat{V'/U}_{21}$. We have $R^{20}_{21} \circ f_{20} = f_{21} \circ R^{20}_{21}$.* □

9.6 G-CONTINUITY CRITERION

Let $F \leq K_2$. Assume $v_{20}(F) \cap \Gamma_{10} = (0)$; so $(F, v_{20}|F) \cong (F, v_{21}|F)$ and $((K_2, v_{20}), (K_2, v_{21}))$ is a g-pair over F. In this case any ACVF_F-definable object ϕ can be interpreted with respect to $(K_2, K_1)_F$ or to $(K_2, K_0)_F$. We refer to ϕ_{20}, ϕ_{21}. In particular if V is an algebraic variety over F, then $V_{20} = V_{21} = V$; $\widehat{V}$ is ACVF_F-pro-definable, and $\widehat{V}_{20}, \widehat{V}_{21}$ have the meaning considered above. If $f : W \to \widehat{V}$ is a definable function with W a g-open ACVF_F-definable subset of V, we obtain $f_{2j} : W \to \widehat{V}_{2j}$, $j = 0, 1$. Let W_{21}, W_{20} be the interpretations of W in (K_2, K_1), (K_2, K_0). By Lemma 9.1.1 we have $W_{21} \subset W_{20}$.

PROPOSITION 9.6.1. *Let V be an algebraic variety over F and W be a g-open ACVF_F-definable subset of V. Assume $v_{20}(F) \cap \Gamma_{10} = (0)$.*

(1) *An ACVF_F-definable map $g : W \to \Gamma_\infty$ is g-continuous if and only if $g_{21} = \pi \circ g_{20}$ on W_{21}.*

(2) *An ACVF_F-definable map $g : W \times \Gamma_\infty^n \to \Gamma_\infty$ is g-continuous if and only if $g_{21} \circ \pi_2 = \pi \circ g_{20}$ on $W_{21} \times \Gamma_{20,\infty}$, where $\pi_2(u,t) = (u, \pi(t))$, π being the projection $\Gamma_{20} \to \Gamma_{21}$.*

(3) *An ACVF_F-definable map $f : W \to \widehat{V}$ is g-continuous if and only if $f_{21} = R^{20}_{21} \circ f_{20}$ on W_{21}.*

(4) *An ACVF_F-definable map $f : W \times \Gamma_\infty^n \to \widehat{V}$ is g-continuous if and only if $f_{21} \circ \pi_2 = R^{20}_{21} \circ f_{20}$ on $W_{21} \times \Gamma_{20,\infty}$.*

Proof. (1) Recall that g-continuity of maps to Γ_∞ was defined with respect to the g-topology on Γ_∞ (as well as on W). The function g is g-continuous with respect to ACVF_F if and only if $g^{-1}(\infty)$ is g-open and for any open interval I of Γ_{21}, $g^{-1}(I)$ is g-open.

Let us start with an interval of the form $I_a = \{x : x > \mathrm{val}_{21}(a)\}$, with $a \in K_2$.

By increasing F we may assume $a \in F$. (We may assume $F = F^{\mathrm{alg}}$. There is no problem replacing F by $F(a)$ unless $v_{20}(F(a)) \cap \Gamma_{10} \neq (0)$. In this case it is easy to see that $v_{21}(a) = v_{21}(a')$ for some $a' \in F$, so we may replace a by a'.)

We view $U_a = g^{-1}(I_a)$ as defined by $\infty > g(u) > \mathrm{val}(a)$ in ACVF_F. By Lemma 9.1.1, U_a is g-open if and only if $(U_a)_{21} \subset (U_a)_{20}$, that is, $\infty > g_{21}(u) > \mathrm{val}_{21}(a)$ implies $\infty > g_{20}(u) > \mathrm{val}_{20}(a)$. Thus, $g^{-1}(I_a)$ is g-open for every a if and only if $g_{21}(u) \leq \pi(g_{20}(u))$ and $g_{20}(u) < \infty$ whenever $g_{21}(u) < \infty$. Let $I'_a = \{x : x < \mathrm{val}_{21}(a)\}$. One gets similarly that $g^{-1}(I'_a)$ is g-open for every a if and only if $g_{21}(u) \geq \pi(g_{20}(u))$ whenever $g_{21}(u) < \infty$. Again by Lemma 9.1.1, $g^{-1}(\infty)$ is g-open if and only if $g_{20}(u) = \infty$ whenever $g_{21}(u) = \infty$. The statement follows.

(2) Let $G(u, a) = g(u, \mathrm{val}(a))$. Then g is g-continuous if and only if G is g-continuous. The statement follows from (1) applied to G.

For (3) and (4), we pass to affine V, and consider a regular function H on V. Let $g(u) = f(u)_*(\mathrm{val}(H))$. Then $f_{21} = R^{20}_{21} \circ f_{20}$ if and only if for each such H we have $g_{21} = \pi \circ g_{20}$; and f is g-continuous if and only if, for each such H, g is g-continuous. Thus (3) follows from (1), and similarly (4) from (2). □

REMARK 9.6.2. A similar criterion is available when W is g-closed rather than g-open; in this case we have $W_{20} \subset W_{21}$, and the equalities must be valid on W_{20}. In practice we will apply the criterion only with g-clopen W.

9.7 SOME APPLICATIONS OF THE CONTINUITY CRITERIA

As an example of using the continuity criteria, assume $h : V \to W$ is a finite surjective morphism of separable degree n between algebraic varieties of pure dimension d, with W normal. For $w \in W$, one may endow $h^{-1}(w)$ with the structure of a multi-set (i.e. a finite set with multiplicities assigned to points) of constant cardinality n as follows. One considers a pseudo-Galois covering $h' : V' \to W$ of separable degree n' with Galois group G factoring as $h' = h \circ p$ with $p : V' \to V$ finite of separable degree m. If $y' \in V'$, one sets $m(y') = |G|/|\mathrm{Stab}(y')|$ and for $y \in V$, one sets $m(y) = 1/m \sum_{p(y')=y} m(y')$. The function m on V is independent from the choice of the pseudo-Galois covering h' (if h'' is another pseudo-Galois covering, consider a pseudo-Galois covering dominating both h' and h''). Also, the function m on V is ACF-definable. Let R be a regular function on V and set $r = \mathrm{val} \circ R$. More generally, R may be a tuple of regular functions $(R^1, \dots, R^m)$, and $r = (\mathrm{val} \circ R^1, \dots, \mathrm{val} \circ R^m)$. The pushforward $r(h^{-1}(w))$ is also a multi-set of size n, and a subset of Γ^m_∞. Given a multi-set Y of size n in a linear ordering, we can uniquely write $Y = \{y_1, \dots, y_n\}$ with $y_1 \leq \dots \leq y_n$ and with repetitions equal to the multiplicities in Y. Thus, using the lexicographic ordering on Γ^m_∞, we can write $r(h^{-1}(w)) = \{r_1(w), \dots, r_n(w)\}$; in this way we obtain definable functions $r_i : W \to \Gamma_\infty$, $i = 1, \dots, n$. In this setting we have:

LEMMA 9.7.1. *The functions r_i are v+g-continuous.*

Proof. Note that if $g : A \to B$ is a weakly order-preserving map of linearly ordered set, X is a multi-subset of A of size n and $Y = g(X)$, then $g(x_i) = y_i$ for $i \leq n$. It follows that both the v-criterion Lemma 9.2.2 and the g-criterion Proposition 9.6.1 (1) hold in this situation. □

COROLLARY 9.7.2. *Let $h : V \to W$ be a finite surjective morphism between algebraic varieties of pure dimension d over a valued field, with W normal. Then $\widehat{h} : \widehat{V} \to \widehat{W}$ is an open map.*

Proof. We may assume that W and hence V are affine. A basic open subset of $\widehat{V}$ may be written as $G = \{p : (r(p)) \in U\}$ for some $r = (\mathrm{val} \circ R^1, \dots, \mathrm{val} \circ R^m)$, R^i regular functions on V, and some v+g-open definable subset U of Γ_∞^n. Consider the functions r_i as in Lemma 9.7.1. By Lemma 9.7.1 they are v+g-continuous. By Lemma 3.8.4, they extend to continuous functions $\widehat{r_i} : \widehat{W} \to \Gamma_\infty$. Since $w \in \widehat{h}(G)$ if and only if for some i we have $\widehat{r_i}(w) \in U$, it follows that $\widehat{h}(G)$ is open. □

Note the necessity of the assumption of normality. If h is a a pinching of $\mathbb{P}^1$, identifying two points $a \neq b$, the image of a small valuative neighborhood of a is not open.

We also have:

LEMMA 9.7.3. *Let U and V be algebraic varieties over a valued field and let $p : U \times V \to U$ be the projection. Then $\widehat{p}$ is open.*

Proof. By taking open covers, we may assume U, and then V, are affine. Embedding V in $\mathbb{A}^n$, so that an open subset of $\widehat{U \times V}$ is the restriction of an open subset of $\widehat{U \times \mathbb{A}^n}$, we may assume $V = \mathbb{A}^n$. By induction on n, we reduce to the case $V = \mathbb{A}^1$. It suffices to consider open subsets $\widehat{H}$ of $\widehat{U \times V}$ cut out by inequalities $\mathrm{val}(F_i) > 0, \mathrm{val}(G_j) < 0$ where F_i, G_j are regular functions on $U \times V$. By Lemma 4.2.6, $\widehat{p}(\widehat{H}) = \widehat{p(H)}$. Since F_i, G_j are continuous in the valuation topology, it is clear that $p(H)$ is v-open. To see that it is g-open, it suffices by Lemma 9.1.1 to show that for any g-pair $(K, \mathbf{K})$ over the base field, $p(H)(\mathbf{K}) \subset p(H)(K)$. This is clear since $H(\mathbf{K}) \subset H(K)$ (strict inequalities being stronger for $\mathbf{K}$), and since $K, \mathbf{K}$ have the same underlying set. □

COROLLARY 9.7.4. *Let $h : V \to W$ be a morphism between algebraic varieties over a valued field, with W normal. Assume W and V are of pure dimension m and $m + n$ and that $h = f \circ g$ where $f : V \to W \times \mathbb{P}^n$ is a finite surjective morphism, g is the projection map $W \times \mathbb{P}^n \to W$. Then $\widehat{h} : \widehat{V} \to \widehat{W}$ is an open map.*

Proof. Clear from Corollary 9.7.2 and Lemma 9.7.3. □

COROLLARY 9.7.5. *Let $h : V \to W$ be a finite morphism of algebraic varieties of pure dimension d over a valued field, with W normal and V quasi-projective. Let $\xi : V \to \Gamma_\infty^n$ be a definable function. Then there exists a definable function $\xi' : W \to \Gamma_\infty^m$ such that for any path $p : I \to \widehat{V}$, still denoting by ξ and ξ' their canonical extensions to $\widehat{V}$ and $\widehat{W}$, if $\xi' \circ h \circ p$ is constant on I, then so is $\xi \circ p$.*

Proof. By Lemma 9.7.6 we may assume ξ is continuous. Also, we can treat the coordinate functions separately, so we may as well take $\xi : V \to \Gamma_\infty$.

Let $d = \deg(h)$, and define $\xi_1, \dots, \xi_d$ on W as above, so that the canonical extension of ξ_i (still denoted by ξ_i) is continuous on $\widehat{W}$ and $\xi(v) \in \{\xi_1(h(v)), \dots, \xi_d(h(v))\}$. Let $\xi' = (\xi_1, \dots, \xi_d)$. Now if $\xi' \circ h \circ p$ is constant on I, then $\xi \circ p$ takes only finitely many values, so by definable connectedness of I, cf. 10.4, it must be constant too. □

LEMMA 9.7.6. *Let V be a quasi-projective variety over a valued field and let $\xi : V \to \Gamma_\infty^n$ be a definable function. Then there exists a v+g-continuous definable function $\xi^* : V \to \Gamma_\infty^N$ and a definable function $d : \Gamma_\infty^N \to \Gamma_\infty^n$ such that $\xi = d \circ \xi^*$.*

Proof. We may assume $V = \mathbb{P}^m$. The statement follows from the following remark: if f/g is a rational function on $\mathbb{P}^m$ with f and g homogeneous of the same degree, the map $x \mapsto \mathrm{val}((f/g)(x))$ factors through the maps $x \mapsto \max(0, \mathrm{val}(f(x)) - \mathrm{val}(g(x)))$ and $x \mapsto \max(0, \mathrm{val}(g(x)) - \mathrm{val}(f(x)))$. □

9.8 THE V-CRITERION ON $\widehat{V}$

Let V be an algebraic variety defined over a field $F_2 \subset \mathcal{O}_{21}$. This means that $v_{21}(a) \geq 0$ for $a \in F_2$, so $v_{21}(a) = 0$ for $a \in F_2$, equivalently $v_{20}(F_2^\times) \subset \Gamma_{10}$. This is the condition considered in relation with the v-criterion in 9.2. The place r_{21} induces a field isomorphism $\mathrm{res}_{21} : F_2 \to F_1$. Let V_1 be the conjugate of V under this field isomorphism, so $(F_2, V) \cong (F_1, V_1)$. We can also view V_1 as the special fiber of the $\mathcal{O}_{21}$-scheme $V_2 \otimes_{F_2} \mathcal{O}_{21}$. As noted earlier, $\widehat{V_1}$ is unambiguous for varieties over F_1.

Recall $\widehat{V}_{20} = \widehat{V}_{210}$. Now $\widehat{V}_{210}$ has a subset $\widehat{V}_{\mathcal{O}} = \widehat{V(\mathcal{O}_{21})}$ consisting of types concentrating on $V(\mathcal{O}_{21})$. We have a definable map $\mathrm{res} : V(\mathcal{O}_{21}) \to V(K_1)$. This induces a map

$$\mathrm{res}_{21*} : \widehat{V}_{\mathcal{O}} \to \widehat{V}_1.$$

Let $\Gamma_{20}^+ = \{x \in \Gamma_{20,\infty} : x \geq 0 \vee x \in \Gamma_{10}\}$. Define a retraction $\pi : \Gamma_{20}^+ \to \Gamma_{10,\infty}$ by letting $\pi(x) = \infty$ for $x \in \Gamma_{20}^+ \smallsetminus \Gamma_{10}$. Note that this is the same as the map π in 9.2.

LEMMA 9.8.1. *Let V be an algebraic variety over F_2, let W be an ACVF_{F_2}-definable subset of $\mathbb{P}^n \times \Gamma_\infty^m$ and consider an ACVF_{F_2}-definable map $f : V \to \widehat{W}$.*

(1) *Let x be a point in $V(\mathcal{O}_{21})$. Then f is v-continuous at x if and only if $(\mathrm{res}_{21*} \circ f_{20})(x) = (f_{10} \circ \mathrm{res}_{21})(x)$.*

(2) *Let X be an ACVF_{F_2}-definable subset of V and assume $\mathrm{res}_{21*} \circ f_{20} = f_{10} \circ \mathrm{res}_{21}$ at x whenever $x \in V(\mathcal{O}_{21})$ and $\mathrm{res}_{21}(x) \in X$. Then f is v-continuous at each point of X. In particular, if f is also g-continuous, then the canonical extension $F : \widehat{V} \to \widehat{W}$ is continuous at each point of $\widehat{X}$.*

Proof. Let x be a point in $V(\mathcal{O}_{21})$. As in the proof of Lemma 3.8.2, f is v-continuous at x if and only if for every continuous definable function $c : \widehat{W} \to \Gamma_\infty^n$, $c \circ f$ is v-continuous at x. On the other hand, by the "only if" direction in Lemma 9.2.2, the other condition holds for f at x if and only if it holds for $c \circ f$, for any continuous definable function $c : \widehat{W} \to \Gamma_\infty^n$. Thus, in the proof of (1), we may assume $f : V \to \Gamma_\infty$, in which case the statement follows from Lemma 9.2.2. (2) follows directly from (1) and Lemma 3.8.2. □

REMARK 9.8.2. Let $F(X) \in \mathcal{O}_{21}[X]$ be a polynomial in one variable, and let $f(X)$ be the specialization to $K_1[X]$. Assume $f \neq 0$. Then the map r_{21} takes the roots of F onto the roots of f. Indeed, consider a root of f; we may take it to be 0. Then the Newton polygon of f has a vertical edge. So the Newton polygon of F has a very steep edge compared to Γ_{10}. Hence it has a root of that slope, specializing to 0.

The following lemma states that a continuous map on X remains continuous relative to a set U that it does not depend on; i.e. viewed as a map on $X \times U$ with dummy variable U, it is still continuous. This sounds trivial, and the proof is indeed straightforward if one uses the continuity criteria; it seems curiously nontrivial to prove directly.

For U a variety and $b \in U$, let s_b denote the corresponding simple point of $\widehat{U}$, i.e. the definable type $x = b$. For V a variety and $q \in \widehat{V}$, let $q \otimes s_b$ denote the unique definable type $q(v, u)$ extending $q(v)$ and $s_b(u)$.

LEMMA 9.8.3. *Let U, V and V' be varieties over a valued field. Assume U and V are quasi-projective and X be a v+g-open definable subset of V', or of $V' \times \Gamma_\infty^N$. Let $f : X \to \widehat{V}$ be v+g-continuous, and let $\bar{f}(x, u) = f(x) \otimes s_u$. Then $\bar{f} : X \times U \to \widehat{V \times U}$ is v+g-continuous.*

Proof. For g-continuity, we use Proposition 9.6.1 (3) and (4). We have $f_{21} = R_{21}^{20} \circ f_{20}$ on X_{21}. Also for $x \in X_{21}, u \in U_{21}$, we have $\bar{f}_{21}(x, u) = f_{21}(x) \otimes s_u$, and $\bar{f}_{20}(x, u) = f_{20}(x) \otimes s_u$. Moreover we noted that R_{21}^{20} is the identity on simple points, so $R_{21}^{20}(p \otimes s_b) = R_{21}^{20}(p) \otimes s_b$ in the natural sense. The criterion follows.

For v-continuity, Lemma 9.8.1 applies. Assume $\mathrm{res}_{21}(x) \in X$, so $x \in X$. Let $u \in U(\mathcal{O}_{21})$. We have $\mathrm{res}_{21*} \circ f_{20}(x) = f_{10} \circ \mathrm{res}_{21}(x)$. Now $\mathrm{res}_{21*}(q \otimes s_u) = \mathrm{res}_{21*}(q) \otimes s_{\bar{u}}$, where $\bar{u} = \mathrm{res}_{21}(u)$, and $\mathrm{res}_{21}(x, u) = (\mathrm{res}_{21}(x), \bar{u})$, so the criterion follows. □

REMARK 9.8.4. In the context of the previous lemma, recall that the map $\otimes : \widehat{U} \times \widehat{V} \to \widehat{U \times V}$ is well-defined but not continuous in general. If $f : I \times \widehat{V} \to \widehat{V}$ is a homotopy, let $\phi : I \times V \to \widehat{V}$ be the restriction to simple points, and let $(\phi \otimes \mathrm{Id})(t, v, u) = \phi(t, v) \otimes u$. By Lemma 9.8.3, $(\phi \otimes \mathrm{Id})$ is v+g-continuous. By Lemma 3.8.5, it extends to a homotopy $I \times \widehat{V \times U} \to \widehat{V \times U}$, which we denote $\widehat{f \times \mathrm{Id}}$. We easily compute: $\widehat{f \times \mathrm{Id}}(t, p \otimes q) = f(t, p) \otimes q$.

Corollary 9.8.5. *Let U and V be quasi-projective varieties over a valued field and let X and Y be definable subsets of U and V. Let $f : I \times \widehat{X} \to \widehat{X}$ and $g : I' \times \widehat{Y} \to \widehat{Y}$ two definable deformation retractions onto iso-definable Γ-internal subsets S and T, respectively. Assume f and g are restrictions of homotopies $F : I \times \widehat{U} \to \widehat{U}$ and $G : I' \times \widehat{V} \to \widehat{V}$, respectively. Then there exists a definable deformation retraction $h : (I+I') \times \widehat{X \times Y} \to \widehat{X \times Y}$ whose image is equal to $S \otimes T$.*

Proof. Recall $I + I'$ is obtained from the disjoint union of I and I' by identifying the endpoint e_I of I with the initial point of I'. The homotopy $\widehat{F \times \mathrm{Id}}$ restricts to a homotopy $\widehat{f \times \mathrm{Id}} : I \times \widehat{X \times Y} \to \widehat{X \times Y}$ and similarly $\widehat{\mathrm{Id} \times G}$ restricts to a homotopy $\widehat{\mathrm{Id} \times g} : I' \times \widehat{X \times Y} \to \widehat{X \times Y}$. Let h be the concatenation of $\widehat{f \times \mathrm{Id}}$ with $\widehat{\mathrm{Id} \times g}$, that is, defined by

$$h(t,z) = \widehat{f \times \mathrm{Id}} \text{ for } t \in I, \quad h(t,z) = \widehat{\mathrm{Id} \times g}(t, \widehat{f \times \mathrm{Id}}(e_I, z)) \text{ for } t \in I'.$$

So $h(t, p \otimes q) = f(t,p) \otimes q$ for $t \in I$, and $= f(e_I, p) \otimes g(t,q)$ for $t \in I'$. In particular, $h(e_{I'}, p \otimes q) = f(e_I, p) \otimes g(e_{I'}, q)$.

Since any simple point of $\widehat{X \times Y}$ has the form $a \otimes b$, we see that $h(e_{I'}, X \times Y) \subset S \otimes T$. Hence for any $r \in \widehat{X \times Y}$, $h(e_{I'}, r)$ is an integral over r of a function into $S \otimes T$. But as $S \otimes T$ is iso-definable and Γ-internal, and r is stably dominated, this function is generically constant on r, and the integral is an element of $S \otimes T$. Thus the final image of h is contained in $S \otimes T$.

Using again the expression for $h(t, p \otimes q)$ we see that if $f(t,s) = s$ for $s \in S$ and $g(t,y) = y$ for $y \in T$, then $h(t,z) = z$ for all t and all $z \in S \otimes T$. So the final image is exactly equal to $S \otimes T$. □

The following statement is a consequence of Corollary 9.8.5 and Theorem 11.1.1.

Corollary 9.8.6. *Let U and V be quasi-projective varieties over a valued field and let X and Y be definable subsets of U and V. The canonical map $\pi : \widehat{X \times Y} \to \widehat{X} \times \widehat{Y}$ is a homotopy equivalence.*

Proof. We may assume U and V are projective. By Theorem 11.1.1, there exists definable deformation retractions $F : I \times \widehat{U} \to \widehat{U}$ and $G : I' \times \widehat{V} \to \widehat{V}$, leaving X and Y invariant, whose images Σ and Θ are iso-definable and Γ-internal. Since Σ and Θ are continuous definable images of $\widehat{U}$ and $\widehat{V}$, they are definably compact. The map $\pi_\Sigma \times \pi_\Theta : \Sigma \otimes \Theta \to \Sigma \times \Theta$ is continuous and injective, hence a homeomorphism. Thus the inverse map $\otimes : \Sigma \times \Theta \to \Sigma \otimes \Theta$ is continuous.

Let $f : I \times \widehat{X} \to \widehat{X}$, $g : I' \times \widehat{Y} \to \widehat{Y}$ be the restrictions of F and G, respectively, with images iso-definable and Γ-internal subsets S and T. Being the restriction of a continuous map, $\otimes : S \times T \to S \otimes T$ is continuous, thus

$\pi_S \times \pi_T : S \otimes T \to S \times T$ is a homeomorphism. Denote by e and e' the endpoints of I and I'.

By Corollary 9.8.5, we have a homotopy equivalence $h_{e'} : \widehat{X \times Y} \to S \otimes T$ such that the following diagram is commutative:

$$\begin{array}{ccc} \widehat{X \times Y} & \xrightarrow{h_{e'}} & S \otimes T \\ {\scriptstyle \pi_X \times \pi_Y} \downarrow & & \downarrow {\scriptstyle \pi_S \times \pi_T} \\ \widehat{X} \times \widehat{Y} & \xrightarrow[f_e \times g_{e'}]{} & S \times T. \end{array}$$

Since $f_e \times g_{e'}$ is a homotopy equivalence and $\pi_S \times \pi_T$ is a homeomorphism, $\pi_X \times \pi_Y$ is a homotopy equivalence. □

9.9 DEFINABILITY OF V- AND G-CRITERIA

We shall consider $V^{\#}$ with its canonical strict ind-definable structure defined in 8.2.

PROPOSITION 9.9.1. *Let V and W be varieties over a valued field and let C be the set of definable functions $V \to W^{\#}$ that extend to continuous functions $\widehat{V} \to \widehat{W}$. Assume W is quasi-projective. Then C is (strict) ind-definable. If V and W depend on a parameter t, then this is uniform in the parameter.*

Proof. We will use the v- and g-criteria to show that for each definable set of definable functions $V \to W^{\#}$ the subset of those that are v-continuous, resp. g-continuous, is definable.

We begin with v-continuity. Let V and W be defined over a field $F_2 \subset \mathcal{O}_{21}$. Let $f = f_b : V \to W^{\#} \subset \widehat{W}$ be a definable map, with parameter $b \in F_2$. By Lemma 9.8.1 (1), f is v-continuous iff the equation:

$$\mathrm{res}_{21*} \circ f_{20} = f_{10} \circ \mathrm{res}_{21} \qquad (*)$$

holds on $V(\mathcal{O}_{21})$.

There is no harm in assuming that W is projective, so as to simplify notation: $W(\mathcal{O}_{21}) = W(K_2)$. Now the map

$$\mathrm{res}_{21} : W(K_2) = W(\mathcal{O}_{21}) \to W(K_1)$$

is $\mathrm{ACV^2F}$-definable. It induces a map

$$\mathrm{res}_{21*} : \widehat{W_{20}} \to \widehat{W_{10}}.$$

It is easy to see that $\mathrm{res}_{21*}(W^{\#}) \subset {W_1}^{\#}$. For instance, the argument for $\widehat{W_{20}} = \widehat{W_{210}}$ shows also that the strongly stably dominated types of these structures coincide; and the image of a strongly stably dominated type under a definable map is strongly stably dominated in K_{210}, and hence in K_{10} which is stably embedded; see Proposition 2.6.12 (2).

The restriction r of res_{21*} to $W^{\#}$ is $\mathrm{ACV^2F}$-piecewise definable (i.e. definable on definable pieces), since res_{21*} itself is pro-definable. Now the set $f_{20}(V(\mathcal{O}_{21}))$ is contained in a definable subset of $W^{\#}$ which does not depend on b. Hence the displayed equation $(*)$ is an $\mathrm{ACV^2F}$-definable property of b. Now any $\mathrm{ACV^2F}$-definable subset X of K_2^n is defined by quantifier-free formulas in the field language along with valuation maps v_{20}, v_{21}, and the group operations on Γ_{20}; this is Lemma 9.3.1 (4). But on F_2, the valuation v_{21} is trivial; hence $X \cap F_2^n$ is cut out by quantifier-free formulas in the field language along with v_{20} alone, so it is already cut out by (K_2, K_0)-formula. Thus the set of b from F_2 for which f_b is v-continuous is ACVF-definable.

Similarly, we use the g-criterion Proposition 9.6.1 (3) for proving that for each definable set of definable functions $V \to W^{\#}$ the subset of those that are g-continuous is definable. Here the defining equation is

$$f_{21} = R_{21}^{20} \circ f_{20} \text{ on } V,$$

R_{21}^{20} is the composition of the equality $W_{210}^{\#} = W_{20}{}^{\#}$ with the restriction map $W_{210}^{\#} \to W_{21}^{\#}$, and is clearly piecewise definable in $\mathrm{ACV^2F}$. Once again the quantifier-free induced structure on F_2 is the same as the v_{20}-ACVF-structure, which implies the statement. □

Chapter Ten

Continuity of homotopies

Summary. This chapter consists mostly of preliminary material useful for the proof of the main theorem in Chapter 11. In 10.1 and 10.2 we use the continuity criteria of Chapter 9 to prove the continuity of functions and homotopies used in Chapter 11. In 10.3, we construct inflation homotopies, which are a key tool in our approach. Finally, in 10.4 we prove GAGA type results for connectedness and prove additional results regarding the Zariski topology.

10.1 PRELIMINARIES

The following lemma will be used both for the relative curve homotopy, and for the inflation homotopy.

LEMMA 10.1.1. *Let $f : W \to U$ be a morphism of quasi-projective varieties over some valued field F. Let $h : [0, \infty] \times U \to \widehat{U}$ be F-definable. Let $H : [0, \infty] \times W \to \widehat{W}$ be an F-definable lifting of h. Let $H_w(t) = H(t, w)$ and $h_u(t) = h(t, u)$. Assume for all $w \in W$, H_w and $h_{f(w)}$ are paths and that H_w is the unique path lifting $h_{f(w)}$ with $H_w(\infty) = w$. Let X be a g-open definable subset of U. Assume h is g-continuous, and v-continuous on (respectively, at each point of) $[0, \infty] \times X$. Then H is g-continuous, and is v-continuous on (respectively, at each point of) $[0, \infty] \times f^{-1}(X)$ (we say a function is v-continuous on a subset, if its restriction to that subset is v-continuous).*

Proof. We first use the criterion of Proposition 9.6.1 (4) to prove g-continuity. We may assume the data are defined over a subfield F of K_2, such that $v_{20}(F) \cap \Gamma_{10} = (0)$; so $(F, v_{20}) \cong (F, v_{21})$.

To show that $H_{21} \circ \pi_2 = R^{20}_{21} \circ H_{20}$, we fix $w \in W$. By Lemma 9.4.3, $R^{20}_{21} \circ H_{20}(w, t) = H'_w \circ \pi$ for some path H'_w. To show that $H_{21}(w, t) = H'_w(t)$, it is enough to show that $f_{21} \circ H'_{w,21} = h_{f(w),21}$. It is clear that $H'_w(\infty) = H_{20}(\infty) = w$ since R^{20}_{21} preserves simple points. To see that $f_{21} \circ H'_{w,21} = h_{f(w),21}$ it suffices to check that $f \circ H'_w \circ \pi = h_{f(w)} \circ \pi$, i.e. $f \circ R^{20}_{21} \circ H_{20}(w, t) = h_{f(w)}(\pi(t))$. Now $f \circ R^{20}_{21} \circ H_{20} = R^{20}_{21} \circ h_{20} = h_{21} \circ \pi_2$. It follows that the g-continuity criterion for H is satisfied.

Let us now use the v-continuity criterion in Lemma 9.8.1 above X, $(\mathrm{res}_{21*} \circ H_{20})(t, v) = (H_{10} \circ \mathrm{res}_{21})(t, v)$ whenever $(f \circ \mathrm{res}_{21})(v) \in X$. Fixing $w = \mathrm{res}_{21}(v)$, $H_{10}(t, w)$, for $t \in \Gamma_{10}$, is the unique path lifting $h_{f(w)}$ and starting at w, hence to conclude it is enough to prove that $\mathrm{res}_{21*} \circ H_{20}(t, v)$ also has these properties. But continuity follows from Lemma 10.1.2 and the lifting property from Lemma 10.1.3. □

In the next two lemmas we shall use the notations and assumptions in 9.8. In particular we will assume that $v_{20}(F_2^{\times}) \subset \Gamma_{10,\infty}$.

LEMMA 10.1.2. *Let V be a quasi-projective variety over F_2. Let $f : [0, \infty] \subset \Gamma_{20,\infty} \to \widehat{V}_{20}$ be a (K_2, K_{O})-definable path defined over F_2, with $f(\infty)$ a simple point p_{O} of $\widehat{V}_{\mathcal{O}}$. Then:*

(1) *For all t, $f(t) \in \widehat{V}_{\mathcal{O}}$.*

(2) *We have $\mathrm{res}_{21*}(f(t)) = \mathrm{res}_{21*}(p_{\mathrm{O}})$ for positive $t \in \Gamma_{20} \smallsetminus \Gamma_{10}$.*

(3) *The restriction of $\mathrm{res}_{21*} \circ f$ to $[0, \infty] \subset \Gamma_{10,\infty}$ is a continuous (K_1, K_{O})-definable path $[0, \infty] \subset \Gamma_{10,\infty} \to \widehat{V}_1$.*

Proof. Using base change if necessary and Lemma 6.2.1 we may assume $V \subset \mathbb{A}^n$ is affine. So $f : [0, \infty] \subset \Gamma_{20,\infty} \to \widehat{\mathbb{A}^n}_{20}$ and we may assume $V = \mathbb{A}^n$.

To prove (1) and (2), by using the projections to the coordinates, one reduces to the case $V = \mathbb{A}^1$. Let $\rho(t) = v(f(t) - p_{\mathrm{O}})$. Then ρ is a continuous function $[0, \infty] \to \Gamma_\infty$, which is F_2-definable (in (K_2, K_{O})), and sends ∞ to ∞. If ρ is constant, there is nothing to prove, since f is constant, so suppose not. As Γ is stably embedded, it follows that there is $\alpha \in \Gamma_{20}(F_2) \subset \Gamma_{10}$ such that for all $t \in [0, +\infty], \alpha \leq \rho(t)$. Hence, if $t \in [0, +\infty]_{20}$, then $v_{20}(f(t) - p_{\mathrm{O}}) \geq \alpha$, which implies that $f(t) \in \widehat{O_{21}}$ as desired, and gives (1). Again, by F_2-definability and since f is not constant, for some $\mu > 0$ and $\beta \in \Gamma_{20}(F_2)$, if $t > \beta$, then $\rho(t) > \mu t$. Thus, when $t > \Gamma_{10}$, then $\pi(\rho(t)) = 0$, i.e., $\mathrm{res}_{21*}(f(t)) = \mathrm{res}_{21*}(p_{\mathrm{O}})$.

(3) Definability of the restriction of $\mathrm{res}_{21*} \circ f$ to $[0, \infty] \subset \Gamma_{10,\infty}$ follows directly from Lemma 9.3.1. For continuity, note that if h is a polynomial on $V = \mathbb{A}^n$, over K_1 and if H is a polynomial over $\mathcal{O}_{21}$ lifting h, then $v_{20}(H(a)) = v_{10}(h(\mathrm{res}(a)))$. It follows that for $t \neq \infty$ in $[0, \infty] \subset \Gamma_{10,\infty}$ continuity of f at t implies continuity of $\mathrm{res}_{21*} \circ f$.

In fact since $(\mathrm{res}_{21*} \circ f(t))_* h$ factors through $\pi_{10}(t)$ as we have shown in (2), the argument in (3) shows continuity at ∞ too. To see this directly, one may again consider a polynomial h on $V = \mathbb{A}^n$ over K_1 and a lift H over $\mathcal{O}_{21}$, and also lift an open set containing $\mathrm{res}_{21*}(p_{\mathrm{O}})$ to one defined over a subfield F_2' contained in $\mathcal{O}_{21}$. The inverse image contains an interval (γ, ∞), and since γ is definable over F_2' we necessarily have $\gamma \in \Gamma_{10}$. The pushforward by π_{10} of (γ, ∞) contains an open neighborhood of ∞. □

LEMMA 10.1.3. *Let $f : V \to V'$ be a morphism of varieties defined over F_2. Then f induces $f_{20} : \widehat{V}_{20} \to \widehat{V'}_{20}$ and also $f_{10} : \widehat{V}_{10} \to \widehat{V'}_{10}$. We have $\mathrm{res}_{21*} \circ f_{20} = f_{10} \circ \mathrm{res}_{21*}$ on $\widehat{V}_{\mathcal{O}}$.*

Proof. In fact f_{20}, f_{10} are just induced from restriction of the morphism $f \otimes_{F_2} \mathcal{O}_{21} : V \times_{F_2} \text{Spec } \mathcal{O}_{21} \to V' \times_{F_2} \text{Spec } \mathcal{O}_{21}$, to the general and special fiber respectively, and the statement is clear. □

LEMMA 10.1.4. *Let U be a projective variety over a valued field, D a divisor. Let m be a metric on U, cf.* Lemma 3.10.1. *Then the function $u \mapsto \rho(u, D) = \sup\{m(u,d) : d \in D\}$ is v+g-continuous on U.*

Proof. By Lemma 4.2.29, the supremum is attained. Let $\rho(u) = \rho(u, D)$. It is clearly v-continuous. Indeed, if $\rho(u, D) = \alpha \in \Gamma$, then $\rho(u', D) = \alpha$ for any u' with $m(u, u') > \alpha$. If $\rho(u, D) = \infty$ then $\rho(u', D) > \alpha$ for any u' with $m(u, u') > \alpha$. Let us show g-continuity by using the criterion in Proposition 9.6.1. Let (K_2, K_1, K_0) and F be as in that criterion. Let $u \in U(K_2)$. We have to show that $\rho_{21}(u) = (\pi \circ \rho_{20})(u)$. Say $\rho_{20}(u) = m(u, d)$ with $d \in D(K_2)$. Then $m_{21}(u, d) = \pi(m(u, d))$ by g-continuity of m. Let $\alpha = \pi(m(u, d))$ and suppose for contradiction that $\rho_{21}(u) \neq \alpha$. Then $m_{21}(u, d') > \alpha$ for some d'. We have again $m_{21}(u, d') = \pi(m_{20}(u, d'))$ so $m_{20}(u, d') > m_{20}(u, d)$, a contradiction. □

REMARK 10.1.5. In the proof of Lemma 10.1.4, semi-continuity can be seen directly as follows. Indeed, $\rho^{-1}(\infty) = D$ which is g-clopen. It remains to show $\{u : \rho(u, D) \geq \alpha\}$ and $\{u : \rho(u, D) \leq \alpha\}$ are g-closed. Now $\rho(u, D) \geq \alpha$ if and only if $(\exists y \in D)(\rho(u, y) \geq \alpha)$; this is the projection of a v+g-closed subset of U, hence v+g-closed. The remaining inequality seems less obvious without the criterion, which serves in effect as a topological refinement of quantifier elimination.

LEMMA 10.1.6. *Let U be an algebraic variety over a valued field or a definable subset of such an algebraic variey. Let $h : I \times \widehat{U} \to \widehat{U}$ (resp. $h : I \times U \to \widehat{U}$) be a homotopy. Let $\gamma : \widehat{U} \to I$ be a definable continuous function (resp. $\gamma : U \to I$ be a definable v+g-continuous function). Let $h[\gamma]$ be the cut-off, defined by $h[\gamma](t, u) = h(\max(t, \gamma(u)), u)$. Then $h[\gamma]$ is a homotopy. Also, if h satisfies $(*)$ of 3.9, then so does $h[\gamma]$.*

Proof. Clear. □

Let U be a quasi-projective variety, Z a definable subset of U, $f : Z \to \Gamma$ a definable function. We say f is *locally bounded* if every point $p \in Z$ has a neighborhood, in the valuation topology, on which f is bounded. Say f is *U-locally bounded* if every point $p \in U$ has a neighborhood O in the valuation topology, with $f|O$ bounded. Note that when Z is v-closed, these two notions coincide.

LEMMA 10.1.7. *Let U be a quasi-projective variety over a valued field, Z a U-definable subset of U, $f : Z \to \Gamma$ a definable function. Then f is U-locally bounded on Z if and only if for any bounded definable subset W of U, $f|(W \cap Z)$ is bounded.*

Proof. Assume U, Z and f are defined over K with $K \models$ ACVF. We may also assume U is affine. It is enough to prove that if f is locally bounded on U, then for every v-closed bounded K-definable subset W of Z, $f(W)$ is bounded. Suppose this does not hold. Then there would exist such a W such that, for some elementary extension $K^* \geq K$, there exists $a \in W(K^*)$ with $f(a) > \Gamma(K)$. Consider the valuation ring

$$R = \{x \in K^* : \exists b \in K \operatorname{val}(x) \geq \operatorname{val}(b)\}.$$

Since W is bounded, $a \in W(R)$. The residue field K' is an elementary extension of K. Denote by $\pi : R \to K'$ the canonical projection and set $b = \pi(a)$. Since W is v-closed, $b \in W(K')$ by Lemma 9.2.1. We claim that f is not locally bounded at b. Otherwise, there would exist $\gamma, \delta \in \Gamma(K')$ such that, denoting by $B_\gamma(b)$ the open polydisc of polyradius $(\gamma, \ldots, \gamma)$ around b, for every y in $Z \cap B_\gamma(b)$, $f(y) \leq \delta$. After increasing γ and δ we may assume they belong to $\Gamma(K)$. Now consider a Hahn field extension $L = K'((t^{\mathbb{Q}}))$ with $\operatorname{val}(t) > \Gamma(K)$. Let k' denote the residue field of K'. Since (K^*, K', k') and (L, K', k') are models of ACV^{2}F with the same characteristics, they are elementary equivalent with parameters in K' by Lemma 9.3.1. It follows there exists $a' \in W(L)$ such that $a' \in B_\gamma(b)$ and $f(a') > \delta$, leading to a contradiction with the definition of δ and γ, since L is an elementary extension of K'. □

LEMMA 10.1.8. *Let V be a projective variety over some valued field F, V' a Zariski locally closed subset, U a v-closed definable subset of V', $f : U \to \Gamma$ be an F-definable function. Assume f is locally bounded on U. Then there exists a v+g-continuous F-definable function $G : V \to \Gamma_\infty$ such that $f(x) \leq G(x) \in \Gamma$ for $x \in U$.*

Proof. By embedding V as a closed subset of projective space, we can find a v+g-continuous function $g : V \to [0, \infty]$ (distance to the boundary), such that g is finite on V' and for $\alpha \in \Gamma$,

$$V_\alpha = \{x \in V : g(x) \leq \alpha\}$$

is a v+g-closed and bounded subset of V'. Let $U_\alpha = V_\alpha \cap U$. Then f is bounded on U_α by Lemma 10.1.7; let $f_1(\alpha)$ be the least upper bound. Since f_1 is a piecewise affine function, one can find $m \in \mathbb{N}$ and $c_0 \in \Gamma$ such that $f_1(\alpha) \leq m\alpha + c_0$ for all $\alpha \geq 0$ and the function $G(x) = mg(x) + c_0$ does the job. □

10.2 CONTINUITY ON RELATIVE $\mathbb{P}^1$

We fix three points 0, 1, ∞ in $\mathbb{P}^1$. In particular, the notion of a ball and the standard homotopy are well-defined, cf. Lemma 3.10.1, Theorem 7.5.1. Let

U_0 be a normal variety and set $E_0 = U_0 \times \mathbb{P}^1$. In practice, U_0 will be a dense open subset of $U = \mathbb{P}^{n-1}$. Let D be a divisor on E_0 containing the divisor at ∞ at each fiber.

Write $D = D' \cup D''$, with D' finite over U_0 and D'' the preimage of a closed divisor Z in U_0. Set $U_0' = U_0 \smallsetminus Z$ and $E_0' = E_0 \smallsetminus D''$. Let $\psi_{D'} : [0, \infty] \times E_0' \to \widehat{E_0'/U_0'}$ be the standard homotopy with stopping time defined by D' at each fiber, as defined above Lemma 9.5.3. We extend $\psi_{D'}$ to a map $\psi_D : [0, \infty] \times E_0 \to \widehat{E_0/U_0} \subset \widehat{E_0}$ by $\psi_D(t, x) = x$ for $x \in D''$.

LEMMA 10.2.1. *Assume D is finite over U_0 (thus D'' is empty). Then the pro-definable map $\psi_D : [0, \infty] \times E_0 \to \widehat{E_0/U_0}$ is v+g-continuous.*

Proof. Thanks to the g-criterion in Proposition 9.6.1, one deduces from Lemma 9.5.3 and Lemma 9.5.4, that ψ_D is g-continuous.

We clearly have v-continuity for the basic homotopy on $\mathbb{P}^1$, applied fiberwise on $\mathbb{P}^1 \times U_0$. Let $\varrho_D^F : \mathbb{P}^1 \times U_0 \to \Gamma_\infty$ be the fiberwise distance to D: $\varrho_D^F(y, u)$ is the maximum of all $d(y, z)$ with $(z, u) \in D$, with d the metric on $\mathbb{P}^1$. Let us check ϱ_D^F is v-continuous. There is no harm in assuming U_0 is projective. Fix $(y, u) \in \mathbb{P}^1 \times U_0$ and let $\alpha = \varrho_D^F(y, u)$. Fix $\varepsilon > 0$ in Γ and set $W_\varepsilon = \{x \in \mathbb{P}^1 : d(y, x) \geq \alpha + \varepsilon\}$. Fix a metric on $\mathbb{P}^1 \times U_0$ and consider as in Lemma 10.1.4 the distance function to D. By Lemma 10.1.4 it is v+g-continuous on $\mathbb{P}^1 \times U_0$. Thus, by Lemma 4.2.29, on the bounded v+g-closed definable set $W_\varepsilon \times \{u\}$ its maximum is attained. Since $D \cap (W_\varepsilon \times \{u\}) = \varnothing$, this maximum is finite. It follows that on some definable v-open set containing (y, u), $\varrho_D^F \leq \alpha + \varepsilon$, proving upper semi-continuity. Lower semi-continuity follows from the fact that the morphism $\widehat{D} \to \widehat{U_0}$ is an open map by Corollary 9.7.2, since D has pure codimension 1 in E_0. Thus, by Lemma 10.1.6, ψ_D is v-continuous on $[0, \infty] \times E_0$. □

LEMMA 10.2.2. *Let $\xi : \mathbb{P}^1 \times U \to \Gamma_\infty$ be a definable map, with U an algebraic variety over a valued field. Then there exists a divisor D_ξ on $\mathbb{P}^1 \times U$ such that, for any divisor D containing D_ξ, the standard homotopy with stopping time defined by D preserves ξ.*

Proof. If U is not affine, there exists a divisor D_0 in U whose complement is affine. By making $\mathbb{P}^1 \times D_0$ a component of D_ξ, we reduce to the case that U is affine. Write $\mathbb{P}^1 = \mathbb{A}^1 \cup \{\infty\}$; by adding $\{\infty\} \times U$ to D_ξ we can ensure that ξ is preserved there, and so it suffices to preserve $\xi | \mathbb{A}^1 \times U$. Since $\xi | \mathbb{A}^1 \times U$ factorizes through a finite number of functions of the form $\mathrm{val}(g)$, with g regular on $\mathbb{A}^1 \times U$, we may assume $\xi | \mathbb{A}^1 \times U$ is actually of the form $\xi(u) = \mathrm{val}(g)$ with g regular on $\mathbb{A}^1 \times U$. Write $g = g(x, u)$, so for fixed $u \in U$ we have a polynomial $g(x, u)$; let D_ξ include the divisor of zeroes of g. Now it suffices to see for each fiber $\mathbb{P}^1 \times \{u\}$ separately that the standard homotopy with stopping time defined by a divisor containing the roots of g must preserve $\mathrm{val}(g)$. This is clear since this standard homotopy fixes any ball containing a root of g; while on a ball containing no root of g, $\mathrm{val}(g)$ is constant. □

LEMMA 10.2.3. *Let $f : W \to U$ be a generically finite morphism of varieties over a valued field F, with U a normal variety, and $\xi : W \to \Gamma_\infty$ an F-definable map. Then there exists a divisor D on U and F-definable maps $\xi_1, \dots, \xi_n : U \to \Gamma_\infty$ such that any homotopy $I \times W \to \widehat{W}$ lifting a homotopy of $I \times U \to \widehat{U}$ fixing pointwise D and the levels of the functions ξ_i also preserves ξ.*

Proof. There exists a divisor D_0 of U such that f is finite above the complement of D_0, and such that $U \setminus D_0$ is affine. By making D_0 a component of D, we reduce to the case that U is affine, and f is finite. So W is also affine, and ξ factorizes through functions of the form $\mathrm{val}(g)$, with g regular. We may thus assume ξ is of this form and in particular that it is v+g-continuous, so that it induces a continuous function on $\widehat{W}$. Let $\xi_i(u)$, $i = 1, \dots, n$, list the values of ξ on $f^{-1}(u)$ and let also ξ_{n+1} be the function given by the valuation of the characteristic function of D. Let h be a homotopy of W lifting a homotopy of U fixing D and the levels of the ξ_i. Then for fixed $w \in W$, $\xi(h(t, w))$ can only take finitely many values as t varies. On the other hand $t \mapsto \xi(h(t, w))$ is continuous, so it must be constant. □

10.3 THE INFLATION HOMOTOPY

LEMMA 10.3.1. *Let V be a quasi-projective variety over a valued field F and let W be a closed and bounded F-pro-definable subset of $\widehat{V}$. Let D and D' be closed F-subvarieties of V, and suppose $W \cap \widehat{D'} \subset \widehat{D}$. Then there exists a v+g-closed, bounded F-definable subset Z of V with $Z \cap D' \subset D$, and $W \subset \widehat{Z}$.*

Proof. We may assume V is affine. Indeed, we may assume $V = \mathbb{P}^n$; then find finitely many affine open $V_i \subset V$ and closed bounded definable subsets $B_i \subset V_i$ such that $W = \cup_i B_i$; given Z_i solving the problem for V_i, set $Z = \cup_i (B_i \cap Z_i)$.

Choose a finite generating family (f_i) of the ideal of regular functions vanishing on D and set $d(x, D) = \inf \mathrm{val}(f_i(x))$ for x in V. Similarly, fixing a finite generating family of the ideal of regular functions vanishing on D', one defines a distance function $d(x, D')$ to D'. Note that the functions $d(x, D)$ and $d(x, D')$ may be extended to $x \in \widehat{V}$.

For $\alpha \in \Gamma$, let V_α be the set of points x of V with $d(x, D) \leq \alpha$. Let $W_\alpha = W \cap \widehat{V_\alpha}$. Then $W_\alpha \cap \widehat{D'} = \varnothing$. So $d(x, D') \in \Gamma$ for $x \in W_\alpha$. By Lemma 4.2.29 there exists $\delta(\alpha) \in \Gamma$ such that $d(x, D') \leq \delta(\alpha)$ for $x \in W_\alpha$. We may take $\delta : \Gamma \to \Gamma$ to be a continuous non-decreasing definable function. Since any such function $\Gamma \to \Gamma$ extends to a continuous function $\Gamma_\infty \to \Gamma_\infty$, we may extend δ to a continuous function $\delta : \Gamma_\infty \to \Gamma_\infty$. Also, since any such function is bounded by a continuous function with value ∞ at ∞ we may assume $\delta(\infty) = \infty$. Let

$$Z_1 = \{x \in V : d(x, D') \leq \delta(d(x, D))\}.$$

This is a v+g-closed set. Let c be a realization of $p \in W$. We have $c \in Z_1$ and $Z_1 \cap D' \subset D$. Since, by Lemma 4.2.10, W is contained in $\widehat{Z_2}$ with Z_2 a bounded v+g-closed definable subset of V, we may set $Z = Z_1 \cap Z_2$. □

LEMMA 10.3.2. *Let D be a closed subvariety of a projective variety V over a valued field F, and assume there exists an étale map $e : V \setminus D \to U$, with U a Zariski open subset of $\mathbb{A}^n$. Then there exists an F-definable homotopy $H : [0, \infty] \times \widehat{V} \to \widehat{V}$ fixing $\widehat{D}$ (that is, such that $H(t, d) = d$ for $t \in [0, \infty]$ and $d \in \widehat{D}$), with image $\mathbf{Z} = H(0, \widehat{V})$, such that for any subvariety D' of V of dimension $< \dim(V)$ we have $\mathbf{Z} \cap \widehat{D'} \subset \widehat{D}$. Moreover given a finite family of F-definable v-continuous functions $\xi_i : V \setminus D \to \Gamma$, $i \in I$, one can choose the homotopy such that the levels of the ξ_i are preserved. The same statement remains true if instead of being F-definable, the functions ξ_i are only assumed to be F'-definable, with F' a finite Galois extension of F, and the functions ξ_i are permuted by the action of $\mathrm{Gal}(F'/F)$. If a finite group G acts on V over U, inducing a continuous action on $\widehat{V}$ and leaving D and the fibers of e invariant, then H may be chosen to be G-equivariant.*

Proof. Let $I = [0, \infty]$ and let $h_0 : I \times \mathbb{A}^n \to \widehat{\mathbb{A}^n}$ be the standard homotopy sending (t, x) to the generic type of the closed polydisc of polyradius $(t, \ldots, t)$ around x. Denote by $H_0 : I \times \widehat{\mathbb{A}^n} \to \widehat{\mathbb{A}^n}$ its canonical extension (cf. Lemma 3.8.5). Note the following fundamental inflation property of H_0: if W is closed subvariety of $\mathbb{A}^n$ of dimension $< n$, then, for any (t, x) in $I \times \widehat{\mathbb{A}^n}$, if $t \neq \infty$, then $H_0(t, x) \notin \widehat{W}$.

By Lemma 7.3.1 and Lemma 7.3.4, for each $u \in U$ there exists $\gamma_0(u) \in \Gamma$ such that $h_0(t, u)$ lifts uniquely to $V \setminus D$ beginning with any $v \in e^{-1}(u)$, up to $\gamma_0(u)$. By Lemma 10.1.8 we can take γ_0 to be v+g-continuous. For $t \geq \gamma_0(u)$, let $h_1(t, v)$ be the unique continuous lift.

Since ξ_i is v-continuous outside D, $\xi_i^{-1}(\xi_i(v))$ contains a v-neighborhood of v. So for some $\gamma_1(u) \geq \gamma_0(u)$, for all $t \geq \gamma_1(u)$ we have $\xi_i(h_1(t, v)) = \xi_i(v)$. We may take $\gamma_1(u) = \min\{\alpha \in \Gamma_{\geq 0} : \xi_i(h_1(t, v)) = \xi_i(v), \forall t \in [\alpha, \infty), \forall v \in e^{-1}(u), \forall i\}$, which is locally bounded and F-definable, not only when the functions ξ_i are assumed to be F-definable, but also when they are only assumed to be F'-definable, with F' a finite Galois extension of F, and permuted by the action of $\mathrm{Gal}(F'/F)$. Thus, we may use Lemma 10.1.8 again to replace γ_1 by a v+g-continuous F-definable function. By Lemma 10.1.6, the cut-off $h_0[\gamma_1]$ is v+g-continuous, and by Lemma 10.1.1, $h_1[\gamma_1 \circ e]$ is v+g-continuous on $V \setminus D$. However we would like to fix D pointwise and have continuity on D.

Let m be a metric on V, as provided by Lemma 3.10.1. Given $v \in V$ let $\rho(v) = \sup\{m(d,v) : d \in D\}$. By Lemma 4.2.29 we have $\rho : V \setminus D \to \Gamma$. Let $\gamma_2 : U \to \Gamma$, $\gamma_2 \geq \gamma_1$, such that for $t \geq \gamma_2(u)$ we have $m(h_1(t,v),v) \geq \rho(v)$ for each v with $e(v) = u$. By Lemma 10.1.8 we can take γ_2 to be v+g-continuous.

Let H the canonical extension of $h_1[\gamma_2 \circ e]$ to $\widehat{V \setminus D} \times I$ provided by Lemma 3.8.5. We extend H to $\widehat{V} \times I$ by setting $H(t,x) = x$ for $x \in \widehat{D}$. We want to show that H is continuous on $\widehat{V}$. Since we already know it is continuous at each point of the open set $\widehat{(V \setminus D)} \times I$, it is enough to prove H is continuous at each point of $\widehat{D} \times I$.

Let $d \in \widehat{D}$, $t \in I$. Then $H(t,d) = d$. Let $\mathbf{G}$ be an open neighborhood of d. One may assume $\mathbf{G}$ to have the form $\{x \in G_0 : \mathrm{val}(r(x)) \in J\}$, with J open in Γ_∞, and r a regular function on a Zariski open neighborhood G_0 of d (which is just a Zariski open subset of V supporting d). So $\mathbf{G} = \widehat{G}$ where G is a v+g-open subset of V.

We have to find an open neighborhood W of (t,d) such that $H(W) \subset \mathbf{G}$. We may take $W \subset \mathbf{G} \times \Gamma_\infty$, so we have $H(W \cap \widehat{D}) \subset \mathbf{G}$. Since the simple points of $W \setminus \widehat{D}$ are dense in $W \setminus \widehat{D}$ and by construction of the canonical extensions in 3.8, it suffices to show that for some neighborhood W, the simple points are mapped to $\widehat{G}$.

View d as a type (defined over some model M_0); if $z \models d|M_0$, then for some $\varepsilon \in \Gamma$, $H(B(z;m,\varepsilon)) \subset \mathbf{G}$. Fix ε, independently of z. The set

$$W_0 = \{v \in V : B(v;m,\varepsilon) \subset G\}$$

is v+g-open since its complement is

$$\{v \in V : (\exists y) m(x,y) \leq \varepsilon \wedge y \in (V \setminus G)\}.$$

Now the projection of a (bounded) v+g-closed set is also v+g-closed.

If there is no neighborhood W as desired, there exist a net (t_i, v_i) with $t_i \to t$ and $v_i \in V \setminus D$ simple points with $v_i \to d$, such that $H(t_i, v_i) \notin G$.

Since

$$H(t_i, v_i) = h_1(\max(\gamma_2(e(v_i)), t_i), v_i),$$

we have $m(H(t_i,v_i),v_i) \geq \rho(v_i)$. As $\rho(v_i) \to \rho(d) = m(d,D) = \infty$, it follows that $m(H(t_i,v_i),v_i) \to \infty$. So, for large i, we have $H(t_i,v_i) \in \widehat{B(v_i;m,\varepsilon)}$, and also $v_i \subset W_0$. So $B(v_i,m,\varepsilon) \subset G$, hence $H(t_i,v_i) \in \widehat{G} = \mathbf{G}$, a contradiction. This shows that H is continuous.

It remains to prove that if $\mathbf{Z} = H(0,\widehat{V})$, then, for any subvariety D' of V of dimension $< \dim(V)$, we have $\mathbf{Z} \cap \widehat{D'} \subset \widehat{D}$. This follows from the inflation property of H_0 stated at the beginning, applied to $e(D' \cap (V \setminus D))$.

The statement on the group action follows from the uniqueness of the continuous lift. □

REMARK 10.3.3. Lemma 10.3.2 remains true if one supposes only that D contains the singular points of V. Indeed, one can find divisors D_i with

$D = \cap_i D_i$, and étale morphisms $h_i : V \setminus D_i \to \mathbb{A}^n$, and iterate the lemma to obtain successively $\mathbf{Z} \cap \widehat{D'} \subset \widehat{D_1} \cap \ldots \cap \widehat{D_i}$. In particular, when V is smooth, Lemma 10.3.2 is valid for $D = \varnothing$.

10.4 CONNECTEDNESS AND THE ZARISKI TOPOLOGY

Let V be an algebraic variety over some valued field. We say a strict pro-definable subset Z of $\widehat{V}$ is *definably connected* if it contains no clopen strict pro-definable subsets other than $\varnothing$ and Z. We say that Z is *definably path connected* if for any two points a and b of Z there exists a definable path in Z connecting a and b. Clearly definable path connectedness implies definable connectedness. When V is quasi-projective and $Z = \widehat{X}$ with X a definable subset of V, the reverse implication will eventually follow from Theorem 11.1.1.

If X is a definable subset of V, $\widehat{X}$ is definably connected if and only if X contains no v+g-clopen definable subsets, other than X and $\varnothing$. Indeed, if U is a clopen strict pro-definable subset of $\widehat{X}$, the set $U \cap X$ of simple points of U is a v+g-clopen definable subset of X, and U is the closure of $U \cap X$. When X is a definable subset of V, we shall say $\widehat{X}$ has a finite number of connected components if X may be written as a finite disjoint union of v+g-clopen definable subsets U_i with each $\widehat{U_i}$ definably connected. The $\widehat{U_i}$ are called *connected components* of $\widehat{X}$.

LEMMA 10.4.1. *Let V be a smooth quasi-projective variety over a valued field and let Z be a nowhere dense Zariski closed subset of V. Then $\widehat{V}$ has a finite number of connected components if and only if $\widehat{V \setminus Z}$ has a finite number of connected components. Furthermore, if $\widehat{V}$ is a finite disjoint union of connected components $\widehat{U_i}$ then the $\widehat{U_i} \setminus \widehat{Z}$ are the connected components of $\widehat{V \setminus Z}$.*

Proof. By Remark 10.3.3, there exists a homotopy $H : I \times \widehat{V} \to \widehat{V}$ such that its final image Σ is contained in $\widehat{V \setminus Z}$. Also, by construction of H, the simple points of $V \setminus Z$ move within $\widehat{V \setminus Z}$, and so H leaves $\widehat{V \setminus Z}$ invariant. Thus, we have a continuous morphism of strict pro-definable spaces $\varrho : \widehat{V} \to \Sigma$. If V is a finite disjoint union of v+g-clopen definable subsets U_i with each $\widehat{U_i}$ definably connected, note that each $\widehat{U_i}$ is invariant by the homotopy H. Thus, $\varrho(\widehat{U_i}) = \Sigma \cap \widehat{U_i}$ is definably connected. Since $\Sigma \cap \widehat{U_i} = \Sigma \cap (\widehat{U_i} \setminus \widehat{Z})$ and any simple point of $U_i \setminus Z$ is connected via H within $\widehat{U_i \setminus Z}$ to $\Sigma \cap \widehat{U_i}$ it follows that $\widehat{U_i} \setminus \widehat{Z}$ is definably connected. For the reverse implication, assume $V \setminus Z$ is a finite disjoint union of v+g-clopen definable subsets V_i with each $\widehat{V_i}$ definably connected. Then $\varrho(\widehat{V_i}) = \Sigma \cap \widehat{V_i}$ is definably connected. Let U_i denote the set of simple points in $\varrho^{-1}(\Sigma \cap \widehat{V_i})$. Then $\widehat{U_i}$ is definably connected. □

THEOREM 10.4.2. *Let V be an algebraic variety over a valued field F. Assume V is geometrically connected for the Zariski topology. Then $\widehat{V}$ is definably connected.*

Proof. We may assume F is algebraically closed and V is irreducible. It follows from the version of Bertini's theorem given in [31] p. 56, that any two points of V are contained in an irreducible curve C on V. So, since simple points are dense, the lemma reduces to the case of irreducible curves, and by normalization, to the case of normal irreducible curves C. As in the beginning of Theorem 7.1.1, one may thus assume C is smooth and irreducible. By Lemma 10.4.1 one may assume that C is also projective. The case of genus 0 is clear using the standard homotopies of $\mathbb{P}^1$. So assume C has genus $g > 0$. By Theorem 7.5.1 there is a retraction $\varrho : \widehat{C} \to \Upsilon$ with Υ an iso-definable Γ-internal subset. It follows from Theorem 6.2.8 that Υ is a finite disjoint union of connected iso-definable Γ-internal subsets Υ_i. Denote by C_i the set of simple points in C mapping to Υ_i. Each C_i is a v+g-clopen definable subset of C and $\widehat{C_i}$ is definably connected, thus $\widehat{C}$ has a finite number of connected components. Assume this number is > 1. Then $\widehat{C^g}/\mathrm{Sym}(g)$ has also a finite number > 1 of connected components, since $\widehat{C^g}$ may be written as a disjoint union of the definably connected sets $\widehat{C_{i_1} \times \cdots \times C_{i_g}}$.

Let J be the Jacobian variety of C. There exist proper subvarieties W of C^g and V of J, with W invariant under $\mathrm{Sym}(g)$, and a biregular isomorphism of varieties $(C^g \smallsetminus W)/\mathrm{Sym}(g) \to J \smallsetminus V$. By Lemma 10.4.1, $\widehat{(C^g \smallsetminus W)}/\mathrm{Sym}(g)$ has a finite number > 1 of connected components, hence also $\widehat{J \smallsetminus V}$. By Lemma 10.4.1 again, $\widehat{J}$ would have a finite number > 1 of connected components. The group of simple points of J acts by translation on $\widehat{J}$, homeomorphically, and so acts also on the set of connected components. Since it is a divisible group, the action must be trivial. On the other hand, it is transitive on simple points, which are dense, hence on connected components. This leads to a contradiction, hence $\widehat{C}$ is connected, which finishes the proof. □

LEMMA 10.4.3. *Let V be an algebraic variety over a valued field F and let $f : V \to \Gamma_\infty$ be a v+g-continuous F-definable function. Then $f^{-1}(\infty)$ is a subvariety of V.*

Proof. Note that, for constructible sets, the Zariski closure and the v-closure coincide. Hence, since $f^{-1}(\infty)$ is v-closed, it suffices to show that it is constructible. We may assume F is algebraically closed. By noetherian induction we may assume $f^{-1}(\infty) \cap W$ is a subvariety of W, for any proper subvariety W of V. So it suffices to show that $f^{-1}(\infty) \cap V'$ is an algebraic variety, for some Zariski open $V' \subset V$. In particular we may assume V is affine, smooth and irreducible. Since any definable set is v-open away from some proper subvariety, we may also assume that $f^{-1}(\infty)$ is v-open. On the other hand $f^{-1}(\infty)$ is v-closed. The point ∞ is an isolated point in the g-topology, so $f^{-1}(\infty)$ is g-closed and g-open. By Lemma 3.7.7 it follows that $\widehat{f^{-1}(\infty)}$ is a clopen subset of $\widehat{V}$. Since $\widehat{V}$ is definably connected by Theorem 10.4.2, one deduces that $f^{-1}(\infty) = V$ or $f^{-1}(\infty) = \varnothing$, proving the lemma. □

Let w be a finite definable set. It will be convenient to use the following terminology. By a *z-closed* subset of Γ_∞^w we mean one of the form $[x_i = \infty]$, an intersection of such sets, or a finite union of such intersections. Note that such sets are not automatically defined over the given base (but some of them are). Let $Y \subset \Gamma_\infty^w$ be a definable set. A z-closed subset of Y is the intersection with Y of a z-closed subset. (If Y is A-definable, an A-definable z-closed subset of Y can be written as $Y \cap Z$, where Z is z-closed and A-definable; this can be done by taking unions of Galois conjugates.) By a *z-irreducible* subset we mean a z-closed subset which cannot be written as the union of two proper z-closed subsets. Any z-closed set can be written as a finite union of z-closed z-irreducible sets; these will be called *z-components*. A z-open set is the complement of a z-closed set Z. A z-open set is *dense* if its complement does not contain any z-component of Y.

Let Y be a definable subset of Γ_∞^w. Define a *Zariski closed* subset of Y (resp. *Zariski open*) to be a clopen definable subset of a z-closed subset of Y (resp. z-open). By o-minimality, there are finitely many such clopen subsets, the unions of the definably connected components. A definable set X thus has only finitely many Zariski closed subsets; if X is connected and z-irreducible, there is a maximal proper one.

This has nothing to do with the topology on Γ^n generated by translates of subspaces defined by $\mathbb{Q}$-linear equations, for which the name Zariski would also be natural. We will use this latter topology little, and will refer to it as the linear Zariski topology on Γ^n, when required.

Lemma 10.4.3 can be strengthened as follows:

LEMMA 10.4.4. *Let V be an algebraic variety over a valued field F, let w be a finite F-definable set and let $f : V \to Y \subset \Gamma_\infty^w$ be a v+g-continuous F-definable function. Then $f^{-1}(U)$ is Zariski open (resp. closed) in V, whenever U is Zariski open (resp. closed) in Y.*

Proof. It suffices to prove the statement about closed sets. We may assume F is algebraically closed. So U is a clopen subset of U', with U' z-closed. By Lemma 10.4.3, $f^{-1}(U')$ is Zariski closed; write $f^{-1}(U') = V_1 \cup \ldots \cup V_m$ with V_i Zariski irreducible. It suffices to prove the lemma for $f|V_i$, for each i; so we may assume $V_i = V$ is Zariski irreducible. By Theorem 10.4.2, $f^{-1}(U) = V$. □

Here is a converse:

LEMMA 10.4.5. *Let* $X \subset \Gamma_\infty^w$ *and let* $\beta : X \to \widehat{V}$ *be a continuous, pro-definable map. Let* W *be a Zariski closed subset of* $\widehat{V}$. *Then* $\beta^{-1}(W)$ *is Zariski closed in* X.

Proof. Let $F_1, \ldots, F_\ell$ be the nonempty, proper Zariski closed subsets of X. Removing from X any F_i with $F_i \subset \beta^{-1}(W)$, we may assume no such F_i exist. By working separately in each component, we may assume X is connected, and in fact z-irreducible. Moreover by induction on z-dimension, we can assume the lemma holds for proper z-closed subsets of X.

CLAIM. $\beta^{-1}(W) \cap F_i = \varnothing$ *for each* i.

Proof of the claim. Otherwise, let P be a minimal F_i with nonempty intersection with $\beta^{-1}(W)$. Let Q be the z-closure of P; then $Q \neq X$. As Zariski closed in Q implies Zariski closed in X, $Q \cap \beta^{-1}(W) = \varnothing$. (Thanks to Z. Chatzidakis for this argument.) □

Say $\beta^{-1}(W) \subset \Gamma_\infty^{w_1} \times \{\infty\}^{w_2}$ with (w_1, w_2) a partition of w and $|w_1|$ minimal. Then $\beta^{-1}(W) \cap (x_i = \infty) = \varnothing$ for $i \in w_1$, i.e. $\beta^{-1}(W) \subset \Gamma^{w_1} \times \{\infty\}^{w_2}$. Projecting homeomorphically to Γ^{w_1}, we may assume $w_1 = w$ and $X \subset \Gamma^w$. However, W is of the form $\widehat{F}$ with F g-clopen, so $\beta^{-1}(W)$ is g-clopen. Since any g-clopen subset of Γ_∞^w which is also closed and contained in Γ^w is clopen, it follows that $\beta^{-1}(W)$ is clopen, which implies that it is after all Zariski closed in X. □

COROLLARY 10.4.6. *Let* Υ *be an iso-definable subset of* $\widehat{V}$, X *a definable subset of* Γ_∞^w, *and let* $\alpha : \Upsilon \to X$ *be a pro-definable homeomorphism. Then* α *takes the Zariski topology on* Υ *to the Zariski topology on* X.

Proof. Follows from Lemma 10.4.4 and Lemma 10.4.5. □

Chapter Eleven

The main theorem

Summary. The main theorem is stated in 11.1 and several preliminary reductions are performed in 11.2 that allow us to essentially reduce to a curve fibration. We construct a relative curve homotopy in 11.3 and a liftable base homotopy in 11.4. In 11.5 a purely combinatorial homotopy is constructed in the Γ-world. Finally in 11.6 we end the proof of the main theorem; the homotopy retraction is constructed by concatenating the previous three homotopies together with an inflation homotopy. The chapter ends with 11.7 which is devoted to the relative version of the main theorem.

11.1 STATEMENT

THEOREM 11.1.1. *Let V be a quasi-projective variety over a valued field F and let X be a definable subset of $V \times \Gamma_\infty^\ell$ over some base set $A \subset \mathrm{VF} \cup \Gamma$, with $F = \mathrm{VF}(A)$. Then there exists an A-definable deformation retraction $h : I \times \widehat{X} \to \widehat{X}$ with image an iso-definable subset Υ definably homeomorphic to a definable subset of Γ_∞^w, for some finite A-definable set w.*

One can furthermore require the following additional properties for h to hold simultaneously:

(1) *Given finitely many A-definable functions $\xi_i : X \to \Gamma_\infty$, with canonical extension $\xi_i : \widehat{X} \to \Gamma_\infty$ as in 3.8, one can choose h to respect the ξ_i, i.e. to satisfy $\xi_i(h(t,x)) = \xi_i(x)$ for all $(t,x) \in I \times \widehat{X}$. In particular, finitely many definable subsets U of X can be preserved, in the sense that the homotopy restricts to one of $\widehat{U}$.*

(2) *Assume given, in addition, a finite algebraic group G acting on V and leaving X globally invariant. Then the retraction h can be chosen to be equivariant with respect to the G-action.*

(3) *Assume $\ell = 0$. The homotopy h is Zariski generalizing, i.e. for any Zariski open subset U of V, $\widehat{U} \cap \widehat{X}$ is invariant under h.*

(4) *The homotopy h satisfies condition $(*)$ of 3.9, i.e.: $h(e_I, h(t,x)) = h(e_I, x)$ for every t and x.*

(5) *The homotopy h restricts to $h^\# : I \times X^\# \to X^\#$, cf.* Definition 2.6.9 *and* 8.1.

(6) *One has* $h(e_I, X) = \Upsilon$, *i.e.* Υ *is the image of the simple points.*

(7) *Assume* $\ell = 0$ *and* $X = V$. *Given a finite number of closed irreducible subvarieties* W_i *of* V, *one can demand that* $\Upsilon \cap \widehat{W_i}$ *has pure dimension* $\dim(W_i)$.

Definition 11.1.2. Let V be a quasi-projective variety, X be a definable subset of V over some base set $A \subset \mathrm{VF} \cup \Gamma$. Let Υ be an A-iso-definable subset of $\widehat{X}$. We call Υ a *skeleton* of $\widehat{X}$ if it is definably homeomorphic to a definable subset of Γ_∞^w, for some finite A-definable set w, there exists an A-definable deformation retraction $h : I \times \widehat{X} \to \widehat{X}$ with image Υ, and in addition (7) holds for each irreducible component W of the Zariski closure of X.

The last condition may look inelegant, but will allow us to prove that any two skeleta are contained in a third, and more generally that the homotopy in Theorem 11.1.1 can be required to fix pointwise any given skeleton. A possible alternate definition could be to replace the last condition by the condition that Υ is contained in $V^\#$. By Theorem 8.4.2 any such skeleton is contained in a skeleton in the sense of Definition 11.1.2, and any skeleton in the sense of Definition 11.1.2 lies in $V^\#$.

Remark 11.1.3.

(1) Without parameters, one cannot expect in general Υ to be definably homeomorphic to a subset of Γ_∞^n, because of the existence of Berkovich analytifications for which the Galois group acts nontrivially on the cohomology, cf. the earlier observation in 6.1.

(2) Let $\pi : V' \to V$ be a finite surjective morphism of F-varieties with V normal, and $\xi' : V' \to \Gamma_\infty^m$ be an A-definable morphism. Then, when $X = V$ one can find h as in the theorem lifting to $h' : I \times \widehat{V'} \to \widehat{V'}$ respecting ξ'. To see this, let $V'' \to V'$ be such that $V'' \to V$ admits a finite group action G, and V' is the quotient variety of some subgroup. An equivariant homotopy of $\widehat{V''}$ will induce homotopies on $\widehat{V'}$ and on $\widehat{V}$. The continuity of the induced homotopies follows from Lemma 3.9.6 and Remark 3.9.7 and the iso-definability of their image from Lemma 2.2.5.

(3) In Theorem 11.1.1 (1), one may demand that the homotopy preserve a given A-definable map $\xi : X \to \Gamma_\infty^w$ with w a finite A-definable set. Indeed, let $\xi' : X \to \Gamma_\infty^m$ (where $m = |w|$) be a map such that for $v \in X$, $\xi'(v)$ is an m-tuple in non-decreasing order, enumerating the underlying set of the w-tuple $\xi(v)$. There exist definable sets U_i such that $\xi | U_i$ is continuous. We can ask that the homotopy h preserves the U_i and ξ'. Then along the homotopy h, ξ is preserved up to a permutation of w, hence by continuity it is preserved.

(4) Property (3) in Theorem 11.1.1 implies that, for any irreducible component W of V, $\Upsilon \cap \widehat{W}$ is Zariski dense in $X \cap \widehat{W}$ in the sense of 3.11 and that $X \cap \widehat{W}$ is invariant under h. For the first assertion note that one cannot have $\Upsilon \cap \widehat{W} \subset \widehat{Z}$, for some proper Zariski closed subset Z of W, since then a point in $W \smallsetminus Z$ would have its final image in $\widehat{Z}$. For the second one, let W_0 be the complement in W of the other components. By (3) $\widehat{W_0 \cap X}$ is invariant under h and the invariance of $\widehat{W} \cap X$ follows by continuity.

(5) Assume $\ell = 0$. The retraction $\widehat{X} \to \Upsilon$ can be taken to be *definably proper*, i.e. so that the pullback of a definably compact set is definably compact. Indeed V embeds in some projective variety V', in an G-equivariant way as in the beginning of 11.2. We can use the theorem to find a homotopy $\widehat{V'} \to \Upsilon'$, preserving the data, and also preserving $V' \smallsetminus X$ and X. The retraction $\widehat{X} \to \Upsilon$ is just the restriction of $\widehat{V'} \to \Upsilon'$, and hence also definably proper.

It is worth pointing out that the fibers of $\widehat{X} \to \widehat{Y}$, over an element $\mathbf{y} \in \widehat{Y}$, for a definable map $X \to Y$, are not in general spaces of the form $\widehat{U}$. The fiber $\widehat{X}_{\mathbf{y}}$ over an element $\mathbf{y} \in \widehat{Y}$ does contain a subset $IX_{\mathbf{y}}$ accessible in our language, namely $\{\int_{\mathbf{y}} g\}$ for $g : Y \to \widehat{X/Y}$ a definable section. But this does not exhaust the fiber. Nonetheless, the proof of Theorem 11.1.1 is inductive, using appropriate fibrations. What permits this is that our homotopy is determined by its restriction to the simple points, cf. Lemma 3.8.5. Given relative homotopies of the fibers, on the simple points of X one obtains a map into $\widehat{X}$ whose image, over a fiber $\mathbf{y}$, does fall into the "inductive" subset $IX_{\mathbf{y}}$ mentioned above. In addition, under appropriate circumstances, a homotopy of $\widehat{Y}$ can be extended to a homotopy of $\widehat{X}$. Though the methods can be applied more generally, it is worth pointing out that the homotopy restricts to a homotopy of $X^{\#}$; and that the fibers of $X^{\#} \to Y^{\#}$ *can* all be obtained as integrals, as above.

11.2 PROOF OF THEOREM 11.1.1: PREPARATION

The theorem reduces easily to the case $\ell = 0$ (for instance, take the projection of X to V, and add ξ'_i describing the fibers, as in the first paragraph of the proof of Theorem 6.4.4). We assume $\ell = 0$ from now on.

It is enough to prove the theorem when $X = V$. Indeed consider the functions ξ'_i on V obtained by extending the functions ξ_i by 0 on $V \smallsetminus X$ together with the function given by the valuation of the characteristic function of X. The theorem for $X = V$ equipped with these functions implies the statement for X and the original functions ξ_i. We now assume $X = V$.

Let G be a finite algebraic group acting on V. We may embed equivariantly V in a projective, equidimensional variety W with G-action of the same dimension. Indeed, let $\bar{V}$ be a projective completion of V. Embed V diagonally in V^G, via $v \mapsto (gv)_{g\in G}$; this is equivariant with respect to the action of G on V^G via the regular action of G on G. Taking the Zariski closure of the image in $\bar{V}^G$ and an equidimensional $\bar{V}'$ containing $\bar{V}$ with the same irreducible components of dimension $\dim V$, and then considering $\cup_{h\in G}\bar{V}'$, we get W as required. On W we can consider the extensions by 0 of the functions ξ_i together with the functions given by the valuation of the characteristic functions of the lower dimensional components of V. It is enough to prove the theorem for W equipped with these functions to have it for V with the functions ξ_i. Thus, we may assume from now on that $X = V$ is projective and equidimensional.

At this point we note that we can take the base A to be a field. Let $F = \mathrm{VF}(A)$ be the field part. Then V and G are defined over F. Write $\xi = \xi_\gamma$ with γ from Γ. Let $\xi'(x)$ be the function: $\gamma \mapsto \xi_\gamma(x)$. Clearly if the fibers of ξ' are preserved then so is each ξ_γ (cf. Remark 11.1.3 (4)). By stable embeddedness of Γ, ξ' can be coded by a function into Γ^k for some k. And this function is F-definable. Thus all the data can be taken to be defined over F, and the theorem over F will imply the general case.

We may assume F is perfect, since this does not change the notion of definability over F.

We use induction on $n = \dim(V)$. For $n = 0$, take the identity deformation $h(t, x) = x$, $w = V$, and map $a \in w$ to $(0, \dots, 0, \infty, 0, \dots, 0)$ with ∞ in the a-th place.

We start with a hypersurface (that is, a closed subset everywhere of dimension $n-1$) D_0 of V containing the singular locus V_{sing}. We assume there exists an étale morphism $V \smallsetminus D_0 \to \mathbb{A}^n$, factoring through V/G. Such a D_0 exists using generic smoothness, after choosing a separating transcendence basis at the generic point of V/G. We also assume D_0 is nonempty of dimension $n-1$ in each irreducible component of V. Note that the functions ξ_i factor through v+g-continuous functions into Γ_∞^m. Indeed, if f and g are homogeneous polynomials of the same degree, then away from the common zero set of f and g, $\mathrm{val}(f/g)$ is a function of $\max(0, \mathrm{val}(f) - \mathrm{val}(g))$ and $\max(0, \mathrm{val}(g) - \mathrm{val}(f))$. The characteristic function of a set defined by $\mathrm{val}(f_i) \geq \mathrm{val}(f_j)$ is the composition of the characteristic function of $x_i \geq x_j$ on Γ_∞^m, with the function $(\mathrm{val}(f_1), \dots, \mathrm{val}(f_m))$. Hence taking a large enough degree, and collecting together all the polynomials mentioned, and adding more so that $f_1, \dots, f_m$ never vanish simultaneously, all ξ_i factor through the function $[\mathrm{val}(f_1) : \dots : \mathrm{val}(f_m)]$ of Remark 5.2.2. Thus we may take the ξ_i to be v+g-continuous. We denote by x_h a schematic distance function to D_0, cf. 3.12 and we shall assume x_h is one of the ξ_i.

By enlarging D_0, we may assume D_0 contains $\xi_i^{-1}(\infty) \cap U$ for any irreducible component U such that ξ_i is not identically equal to ∞ on U, cf. Lemma 10.4.3. Moreover, we can demand that D_0 is G-invariant, and that the *set* $\{\xi_i : i \in I\}$ is G-invariant, by increasing both if necessary.

Note that there exists a continuous function $m = (m_1, \ldots, m_n) : \Gamma_\infty^I \to \Gamma_\infty^n$ whose fibers are the orbits of the symmetric group acting on I, namely $m((x_i)_{i\in I}) = (y_1, \ldots, y_n)$ if $(y_1, \ldots, y_n)$ is a non-decreasing enumeration of $\{x_i\}_{i\in I}$, with appropriate multiplicities. Then $\{m \circ \xi_i\}_{i\in I}$ is G-invariant. It is clear that a homotopy preserving $m \circ \xi$ also preserves each ξ_i. Thus we may assume that each ξ_i is G-invariant.

Let E be the blowing up of $\mathbb{P}^n$ at one point. Then E admits a morphism $E \to \mathbb{P}^{n-1}$, whose fibers are $\mathbb{P}^1$. We now show one may assume V admits a finite morphism to E, with composed morphism to $\mathbb{P}^{n-1}$ finite on D_0, at least when F is infinite.

LEMMA 11.2.1. *Let V be a projective variety of dimension n over a field F. Assume F is infinite. Then there exists a finite morphism $\pi : V \to \mathbb{P}^n$ and a zero dimensional subscheme Z of V such that if $v : V_1 \to V$ denotes the blowing up at Z, there exists a finite morphism $m : V_1 \to E$ making the diagram*

$$\begin{array}{ccc} V_1 & \xrightarrow{v} & V \\ \downarrow{\scriptstyle m} & & \downarrow{\scriptstyle \pi} \\ E & \longrightarrow & \mathbb{P}^n \end{array}$$

commutative. Moreover, if a divisor D_0 on V is given in advance, we may arrange that Z is disjoint from D_0, and that the composition of m with the projection $E \to \mathbb{P}^{n-1}$ is finite on $v^{-1}(D_0)$. If a finite group G acts on V, we may take all these to be G-invariant.

Proof. Let m be minimal such that V admits a finite morphism to $\mathbb{P}^m$. If $m > n$, choose an F-rational hyperplane H inside $\mathbb{P}^m$, and an F-rational point neither on H nor on the image of V; and project the image of V to H through this point. Hence $m = n$, i.e. there exists a finite morphism $V \to \mathbb{P}^n$.

Given a divisor D_0 on V, choose an F-rational point z of $\mathbb{P}^n$ not on the image of this divisor. The projection through this point to a $\mathbb{P}^{n-1}$ contained in $\mathbb{P}^n$ and not containing z determines a morphism $E \to \mathbb{P}^{n-1}$. If V_1 is the blowing up of V at the inverse image Z of z, we find a morphism $V_1 \to E$; composing with $E \to \mathbb{P}^{n-1}$ we obtain the required morphism.

To arrange for G-invariance, we shall apply the lemma to $V' := V/G$. Let $\phi : V \to V'$ be the natural projection. Let $R \subset V'$ be the ramification locus of $V \to V'$. Assuming as we may that G acts faithfully, R is the union over $h \in G$ of the set of fixed points of h; so away from R, $V \to V'$ is Galois and étale. Let D' be a divisor containing $\phi(D_0)$ and R. Applying the lemma to (V', D'), one obtains $v' : V_1' \to V'$, $m' : V_1' \to E$, $\pi' : V' \to \mathbb{P}^n$, and Z' (so v' is an isomorphism away from Z', and $Z' \cap D' = \varnothing$). Let $V_1 = V_1' \times_{V'} V$. Then $V_1 \to V$ is a blowing up of the pullback Z of Z' under the morphism ϕ which is étale over Z, and all statements are clear. □

The next lemmas provide a variant of Lemma 11.2.1 that works over finite fields too. They provide a less detailed description of V_1, but still sufficient for our purposes; the reader who wants to assume an infinite base field may skip them. Note that non-archimedean geometry over trivially valued fields, including finite ones, may have some relevant applications, cf. [39]. We are grateful to Antoine Ducros for pointing out the need for a special argument in the case of a finite base field.

LEMMA 11.2.2. *Let V be a subvariety of dimension n of $\mathbb{P}^m$ over a finite field F. Then there exist homogenous polynomials $f_1, \ldots, f_n$ in $F[x_0, \ldots, x_m]$, of equal degree, such that $Z = V \cap (f_1 = \ldots = f_n = 0)$ is finite. Given a subvariety D of V of dimension $< n$, we may choose $f_1, \ldots, f_n$ so that Z is disjoint from D and such that $[f_1 : \ldots : f_n] : D \to \mathbb{P}^{n-1}$ is a finite morphism.*

Proof. Given any finite number k of F-irreducible projective subvarieties U_i of $\mathbb{P}^m$ of positive dimension, one can always find a homogeneous polynomial f in $F[x_0, \ldots, x_m]$ which does not vanish on any of them. Indeed, by Hilbert polynomial considerations, the codimension of the space of homogeneous polynomials of degree d vanishing on U_i grows at least linearly with d. Thus, for large enough d, this codimension is $> \log_q(k)$; in particular if the field F has cardinality q, a fraction strictly less than $1/k$ of all homogeneous polynomials of degree d in $F[x_0, \ldots, x_m]$ will vanish on U_i, implying that some will vanish on no U_i.

On the other hand, let w_0 be a finite, Galois invariant, set of points of $V(F')$, with F' a finite Galois extension of F. We lift $w_0 \subset \mathbb{P}^m$ to a finite, Galois invariant, subset w of $\mathbb{A}^{m+1}$ in such a way that each element of w has some coordinate equal to 1. Let H_d denote the space of homogeneous polynomials $h(x_0, \ldots, x_m)$ of degree d (with zero added), let $H_d(w)$, resp. $H_d(w, 1)$, denote the subspace of H_d consisting of polynomials vanishing at each element of w, resp. taking value 1 on each element of w. Thus $H_d(w, 1)$ is (empty or) a coset of $H_d(w)$. We now claim that $H_d(w, 1)$ has a point over $F(w)$, for d large enough. If this is true for w and w' such that w and w' are disjoint, then it holds also for $w \cup w'$, since $H_d/H_d(w \cup w') \cong H_d/H_d(w) \oplus H_d/H_d(w')$. So it suffices to consider a Galois orbit w. Then for the elements $c = (c_0, \ldots, c_m)$ of w, we have say $c_0 = 1$. Since over the finite field $F(w)$, every function $F(w)^m \to F(w)$ is represented by a polynomial, there exists over $F(w)$ a polynomial h, say of degree d_0, with $h(c_1/c_0, \ldots, c_m/c_0) = 1$ for $(c_0, \ldots, c_m) \in w$. So $x_0^d h(x_1/x_0, \ldots, x_m/x_0)$ is a homogeneous polynomial of degree d, for $d \geq d_0$, as required. Finally, by Hilbert 90, $H_d(w)$ has an F-basis; and since also $H^1(\mathrm{Gal}(F(w)/F), G_a^N) = (0)$, as the affine space $H_d(w, 1)$ is defined over F, it has a point in F.

We now prove the lemma. The condition on equal degree is easily arranged afterwards, by taking appropriate powers of each f_i; so we ignore it. Inductively, we need to find $f = f_k$ that vanishes on no positive-dimensional component of $V_k = V \cap (f_1 = \ldots = f_{k-1} = 0)$. Then it is clear that any component of V_k has dimension at most $n-k$. In particular for $k=n$ this proves the finiteness of Z.

Further we can insist that f_k vanishes on no positive-dimensional component of $D \cap (f_1 = \ldots = f_{k-1} = 0)$. As a result, $D \cap (f_1 = \ldots = f_{n-1} = 0)$ is finite, with points $c_1, \ldots, c_m \in V(F^{\mathrm{alg}})$. When choosing $f = f_n$, we need also to insist that $f(c_i) \neq 0$. We saw above that there exists a nonempty affine subspace of H_d of codimension bounded independently of d, whose elements satisfy $f(c_i) \neq 0$. For large d, this subspace cannot be contained in the union of the linear spaces that need to be avoided in order to avoid vanishing on the components of V_n whose codimensions grow linearly with d. The finiteness of $[f_1 : \ldots : f_n] : D \to \mathbb{P}^{n-1}$ follows from Lemma 11.2.3. □

LEMMA 11.2.3. *Let Y be an irreducible quasi-projective variety of dimension > 0 over a field k. Let $f : U \to Y$ be a dominant k-morphism with U a Zariski open subvariety of $\mathbb{P}^m$. Let X be a closed subvariety of $\mathbb{P}^m$ which is contained in U. Then $f|X$ is finite.*

Proof. We may assume k is algebraically closed and it is enough to prove $f|X$ is quasi-finite. Thus we may assume $f(X)$ is a point. Let D be a divisor in Y such that $f(X) \notin D$. Let E be the Zariski closure of $f^{-1}(D)$. We have $E \subset f^{-1}(D) \cup F$, with $F = \mathbb{P}^m \setminus U$, thus $E \cap X = \varnothing$. By Bézout's theorem, if follows that X is of dimension 0. □

LEMMA 11.2.4. *Let V be a projective variety of dimension n over a finite field F, and let D be a closed subvariety, of dimension $< n$, containing any component of V of dimension $< n$. Then, there exists a projective variety V_1, a finite closed subset Z of V, disjoint from D, a morphism $v : V_1 \to V$ which is the blowing up of an ideal supported on Z (in particular it is an isomorphism above $V \setminus Z$), and a morphism $u : V_1 \to \mathbb{P}^{n-1}$ which is finite on $v^{-1}(D) \cup v^{-1}(Z)$ such that $v^{-1}(D)$ is a Cartier divisor and there exists a Zariski dense open subset U_0 of $U = \mathbb{P}^{n-1}$ such that with $V_0 = u^{-1}(U_0)$, $u|V_0$ factors as $V_0 \to E_0 = U_0 \times \mathbb{P}^1 \to U_0$, with $V_0 \to E_0$ a finite morphism, and $E_0 \to U_0$ the projection. If a finite group G acts on V, we may take all these to be G-equivariant.*

Proof. Fix an embedding of V in $\mathbb{P}^m$. By Lemma 11.2.2 there exist homogenous polynomials $f_1, \ldots, f_n$ in $F[x_0, \ldots, x_m]$, of equal degree, such that $Z = V \cap (f_1 = \ldots = f_n = 0)$ is finite and disjoint from D and such that $[f_1 : \ldots : f_n] : D \to \mathbb{P}^{n-1}$ is a finite morphism. Let $V_1 \subset V \times \mathbb{P}^{n-1}$ be the Zariski closure of the graph $\{(v, (f_1(v) : \ldots : f_n(v))) : v \in V \setminus Z\}$. Let v be the first projection and u the second projection. Thus $v : V_1 \to V$ is the blowing up of V along the ideal $(f_1, \ldots, f_n)$. By Lemma 11.2.3, the restriction of u to $v^{-1}(Z)$ is finite. The generic fiber of the morphism $V_1 \to U = \mathbb{P}^{n-1}$ is a curve (possibly reducible, and possibly containing some isolated points, in D). Thus it admits a finite morphism to $\mathbb{P}^1$ over $F(U)$. This morphism is the generic fiber of a morphism $u : V_0 \to U_0 \times \mathbb{P}^1$, over U_0, for some Zariski dense open U_0 of U. Equivariance is arranged by applying this construction to V/G in the first place. □

Let us return to the main discussion and recall our setting. We have a projective equidimensional variety V together with a hypersurface $D_0 \subset V$ containing the singular locus of V and such that there exists an étale morphism $V \smallsetminus D_0 \to \mathbb{A}^n$, factoring through V/G. Consider $v : V_1 \to V$ as provided by Lemma 11.2.1 and Lemma 11.2.4, respectively in the infinite and finite field case. It is a G-equivariant birational morphism whose exceptional locus lies above a finite subset Z of V. By Lemma 3.9.6 and Remark 3.9.7, any deformation retraction $h_1 : I \times \widehat{V_1} \to \widehat{V_1}$ leaving the exceptional locus invariant descends to a deformation retraction $h : I \times \widehat{V} \to \widehat{V}$. Furthermore, if h_1 satisfies the theorem for $X = V_1$, so does h for $X = V$. Thus, pulling back the data of Theorem 11.1.1 to V_1, and adding the above invariance requirement, we see that it suffices to prove the theorem for V_1 (which is equidimensional of dimension n). Furthermore, setting $D'_0 = v^{-1}(D_0) \cup v^{-1}(Z)$, we have $V_1 \smallsetminus D'_0 = V \smallsetminus D_0$. In particular, $V_1 \smallsetminus D'_0$ is smooth and admits an étale equivariant morphism to $\mathbb{A}^n$. Hence, we may assume $V = V_1$ and $D_0 = D'_0$.

By construction, there is a morphism $u : V \to U = \mathbb{P}^{n-1}$, whose restriction to D_0 is finite, and a Zariski dense open subset U_0 of U such that, setting $V_0 = u^{-1}(U_0)$, $u|V_0 = q \circ f$ with $f : V_0 \to E_0 = U_0 \times \mathbb{P}^1$ a finite morphism and $q : E_0 \to U_0$ the projection. If a finite group G acts on V, we may take everything to be G-equivariant. Note that the hypotheses imply that f is surjective.

Furthermore, we may assume, after possibly shrinking U_0, that the morphism $f : V_0 \to E_0$ factors through $V_0 \xrightarrow{h} V'_0 \xrightarrow{f'} E_0$ with h finite radicial and f' satisfying the following condition: for every u in U_0, the restriction $f'_u : V'_u \to \mathbb{P}^1_u$ of $V_0 \to E_0$ over u is a generically étale morphism of curves. Indeed, such a factorization exists over the generic point ξ of U_0 and can be spread out on some dense Zariski open set U_0.

11.3 CONSTRUCTION OF A RELATIVE CURVE HOMOTOPY

We fix three points $0, 1, \infty$ in $\mathbb{P}^1$. We are now in the setting of 10.2 with $U_0 \subset U = \mathbb{P}^{n-1}$. For any divisor D on E_0 we consider $\psi_D : [0, \infty] \times E_0 \to \widehat{E_0/U_0}$ as in 10.2.

LEMMA 11.3.1. *Let W be an A-iso-definable subset of $\widehat{E_0/U_0}$ such that $W \to U_0$ has finite fibers. There exists a divisor D' on E_0, generically finite over U_0, such that for every u in U_0, for every x in W over u, the intersection of D' with the ball in $\mathbb{P}^1_u$ corresponding to x is nonempty.*

Proof. Recall we are working over a field base A. By splitting W into two parts (then taking the union of the divisors D' corresponding to each part), we may assume $W \subset \widehat{\mathcal{O}} \times U_0$ where $\mathcal{O}$ is the unit ball. Let a be a point in U_0; so $W_a \subset \widehat{\mathcal{O}}$.

We claim that there exists a finite $A(a)$-definable subset D'_a of $\mathcal{O}$ such that for every x in W_a, the intersection of D'_a with the ball in $\mathcal{O}$ corresponding to x is nonempty. Let W^1 be the set of simple points in W. Thus, W splits into two disjoint iso-definable sets W^1 and $W^2 = W \smallsetminus W^1$. Let D'^1_a be the union of the simple points in W^1_a. If $A(a)$ is trivially valued, any A-definable closed sub-ball of $\mathcal{O}$ must have valuative radius 0, i.e. must equal $\mathcal{O}$. In this case we set $D'^2_a = \{0\}$. Otherwise, $A(a)$ is a nontrivially valued field, and so $\mathrm{acl}(A(a))$ is a model of ACVF. Hence, if we denote by $\widetilde{W}_a$ the finite set of closed balls corresponding to the points in W_a, for every b in $\widetilde{W}_a$, $b \cap \mathrm{acl}(A(a)) \neq \varnothing$. Thus there exists a finite $A(a)$-definable D'^2_a set such that $D'^2_a \cap b \neq \varnothing$ for every b in $\widetilde{W}_a$. Set $D'_a = D'^1_a \cup D'^2_a$.

By compactness we get a constructible set D'' finite over U_0 with the required property. Taking the Zariski closure of D'' we get a Zariski closed set D' generically finite over U_0 with the required property. □

LEMMA 11.3.2. *There exists a divisor D' on E_0 such that, for any divisor D containing D', ψ_D lifts uniquely to an A-definable map $h : [0, \infty] \times V_0 \to \widehat{V_0/U_0}$, which is fiberwise a homotopy.*

Proof. We proceed as in the proof of Theorem 7.5.1. By assumption the morphism $f : V_0 \to E_0$ factors through $V_0 \xrightarrow{h} V'_0 \xrightarrow{f'} E_0$ with h finite radicial and for every u in U_0, the restriction $f'_u : V'_u \to \mathbb{P}^1_u$ of $V_0 \to E_0$ over u is a generically étale morphism of curves. Thus, for every u in U_0, the restriction $f_u : V_u \to \mathbb{P}^1_u$ of $V_0 \to E_0$ over u factors as $V_u \xrightarrow{h_u} V'_u \xrightarrow{f'_u} \mathbb{P}^1_u$, with h_u the restriction of h. Note that $V'_0 \to U_0$ is a relative curve so that $\widehat{V'_0/U_0}$ is iso-definable over A by Theorem 7.1.2. There is a subset W_0 of $\widehat{V'_0/U_0}$, iso-definable over A, containing, for every point u in U_0, all singular points of C'_u, all ramification points of f'_u and all forward-branching points of f'_u, and such that the fibers $W_0 \to U_0$ are all finite. Such an W_0 exists by Lemma 7.4.4 (uniform finiteness of the set of forward-branching points). Let W be the image of W_0 in E_0. Then D' provided by Lemma 11.3.1 does the job. □

Let D be a divisor on E_0 as in Lemma 11.3.2, and such that D contains the image of D_0 in E_0. Assume also D contains the infinity divisor in E_0. Then ψ_D lifts to an A-definable map $h^0_{curves} : [0, \infty] \times V_0 \to \widehat{V_0/U_0}$. By Lemma 10.2.2, after enlarging D, one can arrange that h^0_{curves} preserves the functions ξ_i. Note that G-invariance follows from uniqueness of the lift. After shrinking U_0 we may assume that the restriction of $u : E_0 \to U_0$ to D is finite, that is, that D has no vertical component over U_0.

By Lemma 10.1.1 and Lemma 10.2.1, h^0_{curves} is v+g-continuous at each point of $[0,\infty] \times V_0$. We extend h^0_{curves} to $h_{curves} : [0,\infty] \times V \to \widehat{V/U}$ by setting $h_{curves}(t,x) = x$ for every t in $[0,\infty]$ and every x in $V \smallsetminus V_0$.

LEMMA 11.3.3. *The mapping h_{curves} is g-continuous on $[0,\infty] \times V$ and v-continuous at each point of $[0,\infty] \times X$ for $X = V_0 \cup D_0$.*

Proof. Since $V \smallsetminus V_0$ is g-clopen, g-continuity may be shown separately on $V \smallsetminus V_0$ and away from $V \smallsetminus V_0$. On $V \smallsetminus V_0$ it is trivial since h_{curves} is constant there. Away from $V \smallsetminus V_0$ it was already proved.

It remains to show v-continuity at points on D_0. Let F_2, res be as in 9.8 and in the v-continuity criterion Lemma 9.8.1. Let $p \in V(F_2)$ with $\mathrm{res}(p) \in D_0$. If $p \notin V_0$ then h_{curves} fixes p, so assume $p \in V_0(F_2)$. Set $q = \mathrm{res}(p)$. Fix t in $[0,\infty]$ and let $q_t = \mathrm{res}_{21*}(h_{curves}(t,p))$. Since $h_{curves}(t,q) = q$, it is enough to prove that $q_t = q$. Recall we assume one of the ξ_i is a schematic distance function x_h to D_0, cf. 3.12. Since $x_h(h_{curves}(t,p)) = x_h(p)$, it follows that $\mathrm{res}_{21*}(x_h(h_{curves}(t,p))) = \mathrm{res}_{21*}(x_h(p)) = \infty$. Thus q_t lies in $\widehat{D_0}$. Since it lies on the fiber of u at q, and the intersection of this fiber with D_0 is a finite set D_{0q}, it follows that q_t is a simple point lying on D_{0q}. Let $q' \neq q$ be another point of D_{0q} and let ϑ be a regular function on some Zariski open set containing q and q' which vanishes at q' and not at q. Thus $\mathrm{val}(\vartheta(q))$ is equal to some finite $\gamma \in \Gamma(F_1)$ and $\mathrm{val}(\vartheta(p)) = \gamma$ also. On the other hand the set of $\mathrm{val}(\vartheta(q_t))$ is finite. By continuity of h_{curves} in the t-variable one gets that $\mathrm{val}(\vartheta(h_{curves}(t,p)))$ cannot jump and is equal to γ for all t. Hence, for every t, $q_t \neq q'$, and $q_t = q$ follows. □

By Lemma 3.8.3 the restriction of h_{curves} to $[0,\infty] \times V_0 \cup D_0$ extends to a deformation retraction $H_{curves} : [0,\infty] \times \widehat{V_0 \cup D_0} \to \widehat{V_0 \cup D_0}$. Since D_0 is finite over U, the image $\Upsilon_{curves} = h_{curves}(0, V_0 \cup D_0)$ is iso-definable over A in $\widehat{V/U}$ and relatively Γ-internal. Thus, as above Theorem 6.4.2, we can identify $\widehat{\Upsilon_{curves}}$ with its image in $\widehat{V}$. It follows that the image $H_{curves}(0, \widehat{V_0 \cup D_0})$ is equal to $\widehat{\Upsilon_{curves}}$. By construction $H_{curves}(\infty, x) = x$ for every x and H_{curves} satisfies $(*)$.

Let $x_v : U \to [0,\infty]$ be a schematic distance to the image of $V \smallsetminus V_0$ in U, cf. 3.12. We still denote by x_v its pullback to V (which is a schematic distance to $V \smallsetminus V_0$) and the corresponding extension to $\widehat{V}$. Let us check that $\widehat{\Upsilon_{curves}}$ is σ-compact via (x_h, x_v). Indeed, on $\widehat{\Upsilon_{curves}}$ the infinite locus of x_v is contained in that of x_h, and $\widehat{\Upsilon_{curves}}$ is compact at $x_h^{-1}(\infty)$ since $\{x \in \widehat{V} : x_h(x) = \infty\}$ is contained in $\widehat{\Upsilon_{curves}}$. Furthermore, since for any $\gamma \in \Gamma$, the set $\{x \in \widehat{V} : x_v(x) \leq \gamma\}$ is definably compact and preserved by H_{curves}, $\{x \in \widehat{\Upsilon_{curves}} : x_v(x) \leq \gamma\}$ is definably compact, being the image by a continuous definable map of a definably compact set.

11.4 THE BASE HOMOTOPY

By Theorem 6.4.4 there exists a finite pseudo-Galois covering U' of U and a finite number of A-definable functions $\xi_i' : U' \to \Gamma_\infty$ such that, for I a generalized interval, any A-definable deformation retraction $h : I \times U \to \widehat{U}$ lifting to a deformation retraction $h' : I \times U' \to \widehat{U'}$ respecting the functions ξ_i', also lifts to an A-definable deformation retraction $I \times \widehat{\Upsilon_{curves}} \to \widehat{\Upsilon_{curves}}$ respecting the restrictions of the functions ξ_i on Υ_{curves} and the G-action.

Now by the induction hypothesis applied to U' and $\mathrm{Gal}(U'/U)$, such a pair (h, h') does exist; we can also take it to preserve x_v, the schematic distance to $V \setminus V_0$. Set $h_{base} = h$. Hence, h_{base} lifts to a deformation retraction

$$H_{\widetilde{base}} : I \times \widehat{\Upsilon_{curves}} \to \widehat{\Upsilon_{curves}},$$

which by (2) in Theorem 6.4.4 we may assume to respect the restrictions of the functions ξ_i and the G-action.

Recall the notion of Zariski density in $\widehat{U}$, 3.11. By induction h_{base} has an A-iso-definable Γ-internal final image Υ_{base} and we may assume Υ_{base} is Zariski dense in $\widehat{U}$. By Theorem 6.4.4 we may assume $H_{\widetilde{base}}$ has an A-iso-definable Γ-internal final image equal to $\widehat{\Upsilon_{curves}} \cap \widehat{u}^{-1}(\Upsilon_{base})$ and by induction we may assume $H_{\widetilde{base}}$ satisfies $(*)$.

By composing the homotopies H_{curves} and $H_{\widetilde{base}}$ one gets an A-definable deformation retraction

$$H_{bc} = H_{\widetilde{base}} \circ H_{curves} : I' \times \widehat{V_0 \cup D_0} \longrightarrow \widehat{V},$$

where I' denotes the generalized interval obtained by gluing I and $[0, \infty]$. The image is contained in the image of $H_{\widetilde{base}}$, but contains $H_{\widetilde{base}}(e_I \times \widehat{\Upsilon_{curves}/U})$, the image over the simple points of U. As these sets are equal, the image is equal to both, and is iso-definable and Γ-internal; we denote it by Υ_{bc}. Thus, $\Upsilon_{bc} = \widehat{\Upsilon_{curves}} \cap \widehat{u}^{-1}(\Upsilon_{base})$. In general Υ_{bc} is not definably compact, but it is σ-compact via (x_h, x_v), since $H_{\widetilde{base}}$ fixes x_v and $\widehat{\Upsilon_{curves}}$ is σ-compact via the same functions. (Note that $\Upsilon_{bc} \cap x_h^{-1}(\infty) = \widehat{D_0} \cap \widehat{u}^{-1}(\Upsilon_{base})$.)

LEMMA 11.4.1.

(1) *The subset* Υ_{bc} *is a Zariski dense subset of* $\widehat{V}$.

(2) *One may choose* h_{base} *so that, for every irreducible component* V_i *of* V, $\Upsilon_{bc} \cap \widehat{V_i}$ *is of pure dimension* $n = \dim(V)$.

Proof. Let V_i denote the irreducible components of V, $u : \widehat{V/U} \to U$ and $\widehat{u} : \widehat{V} \to \widehat{U}$ denote the projections. Since H_{curves} preserves $\widehat{D_0}$, its complement (check it fiberwise) and the connected components of its complement by continuity, it preserves each of the $\widehat{V_i}$. Furthermore, there exists an open dense subset $U_1 \subset U$ such that, for every $x \in U_1$, $u^{-1}(x) \cap \Upsilon_{curves} \cap \widehat{V_i}$ is Zariski dense in $\widehat{u}^{-1}(x) \cap \widehat{V_i}$ for every i. It follows that, for every $x \in \widehat{U_1}$, $\widehat{u}^{-1}(x) \cap \widehat{\Upsilon_{curves}} \cap \widehat{V_i}$ is Zariski dense in $\widehat{u}^{-1}(x) \cap \widehat{V_i}$ for every i (recall $\widehat{\Upsilon_{curves}}$ is identified with $\int_U \Upsilon_{curves}$). Pick $x \in \Upsilon_{base}$ which is Zariski dense in $\widehat{U}$, then $\widehat{u}^{-1}(x) \cap \Upsilon_{bc}$ is Zariski dense in $\widehat{V}$.

Next, we deal with local dimension. Consider a component V_i of V. Let C be an irreducible component of a fiber of V_i above U_0. Since D_0 was chosen so that $D_0 \cap C \neq \varnothing$, it follows directly from the definition that the homotopy on C has image containing more than one point. It follows by construction that the image of each irreducible component C_ℓ of C over the algebraic closure of F by the homotopy also contains more than one point. By Theorem 10.4.2, the image of each C_ℓ under that homotopy is necessarily connected. Since it is of dimension ≤ 1, it follows that this image has no isolated points, so is purely one-dimensional. Thus the image of C under the homotopy is also purely one-dimensional.

Now $\Upsilon_{bc} = \widehat{\Upsilon_{curves}} \cap \widehat{u}^{-1}(\Upsilon_{base})$; and by the inductive assumption (7) of Theorem 11.1.1, one may assume that Υ_{base} has pure dimension $n-1$. Since the morphism $V \to U$ restricts to a composition $V_0 \to U_0 \times \mathbb{P}^1 \to U_0$, where $V_0 \to U_0 \times \mathbb{P}^1$ is finite surjective, it follows from Corollary 9.7.4 that the map $\widehat{V_0} \to \widehat{U_0}$ is open. In particular the maps $\widehat{V_i \cup V_0} \to \widehat{U_0}$ are open. It follows easily that $\Upsilon_{bc} \cap \widehat{V_i}$ is of pure dimension n. □

11.5 THE TROPICAL HOMOTOPY

In this rather technical section we construct a homotopy in Γ_∞^w that we shall use in 11.6 in order to insure that the homotopy we build fixes pointwise its final image at every time.

By Theorem 6.2.8, there exists an A-definable, continuous, injective map $\alpha : \Upsilon_{bc} \to \Gamma_\infty^w$, with image $W \subset [0,\infty]^w$, where w is a finite A-definable set. We may assume for some coordinate x_i (resp. x_j), $x_i \circ \alpha$ (resp. $x_j \circ \alpha$) is the restriction of x_h (resp. x_v). Indeed, we may add two points h, v to w which we view as A-definable, i.e. fixed by the action of the Galois group and replace α by $x \mapsto (\alpha(x), x_h(x), x_v(x))$. We shall denote by $\underline{v}$ and $\underline{h}$ the projections $\Gamma_\infty^w \to \Gamma_\infty$ on the v and h coordinate, respectively.

We write $[x_i = x_j]$ for $\{a \in [0,\infty]^w : x_i(a) = x_j(a)\}$, and similarly $[x_i = 0]$, etc.

Since Υ_{bc} is σ-compact via (x_h, x_v), W is σ-compact via $(\underline{h}, \underline{v})$. In particular, $W \smallsetminus [\underline{v} = \infty]$ is σ-compact via $\underline{v}$, and hence closed in $\Gamma_\infty^w \smallsetminus [\underline{v} = \infty]$; so $W \cap \Gamma^w$ is closed in Γ^w.

We let G act on W, so that $\alpha : \Upsilon_{bc} \to \Gamma_\infty^w$ is equivariant. By re-embedding W in $\Gamma_\infty^{w \times G}$, via $w \mapsto (\sigma(w))_{\sigma \in G}$, we may assume G acts on the coordinate set w, and the induced action of G on Γ_∞^w extends the action of

G on W. We still denote by ξ_i the functions on W that are the composition of the restriction of ξ_i to Υ_{bc} with α^{-1}.

In Lemma 11.5.1, we shall show the existence, entirely within Γ^w_∞, of a definable deformation retraction from $(W \cap \Gamma^w) \cup [\underline{h} = \infty]$ to a definably compact subset W_0. Furthermore we shall show that when W has pure dimension n, one can insure $W_0 \cap W$ has also pure dimension n. Then, in Lemma 11.5.2, we shall extend this result to $(W^o \smallsetminus [\underline{v} = \infty]) \cup [\underline{h} = \infty]$, for some z-dense and z-open definable subset W^o of W. This will be used in an essential way in the final part of the proof given in 11.6.

LEMMA 11.5.1. *Let*

$$W' = (W \cap \Gamma^w) \cup [\underline{h} = \infty].$$

There exists an A-definable deformation retraction

$$H_\Gamma : [0, \infty] \times W' \to W'$$

whose image is a definably compact subset W_0 of W' and such that H_Γ leaves the ξ_i invariant, fixes $[\underline{h} = \infty]$, and is G-equivariant. Moreover, one may require the following to hold:

(1) *There exists an A-definable open subset W_o of W containing $W_0 \smallsetminus [\underline{h} = \infty]$ and $m \in \mathbb{N}$, $c \in \Gamma(A)$, such that $x_i \leq (m+1)x_h + c$ on W_o, for every $i \in w$.*

(2) *If W has pure dimension n, then $W_0 \cap W$ has also pure dimension n.*

In this lemma, we take 0 to be the initial point, ∞ the final point. On Γ_∞, we view ∞ as the unique simple point. In this sense the flow is still away from the simple points, as for the other homotopies. Moreover, starting at any given point, the flow will terminate at a finite time. The homotopy we obtain will in fact be a semigroup action, i.e. $H_\Gamma(s, H_\Gamma(t,x)) = H_\Gamma(s+t, x)$, in particular it will satisfy $(*)$ (in the form: $H_\Gamma(\infty, H_\Gamma(t,x)) = H_\Gamma(\infty, x)$).

Proof. For the convenience of the reader we shall divide the proof into 3 steps.

Step 1. *Preliminaries.* We start by choosing an A-definable cell decomposition $\mathcal{D}$ of Γ^w, compatible with $W \cap \Gamma^w$ and with $[x_a = 0]$ and $[x_a = x_b]$ where $a, b \in w$, and such that each ξ_i is linear on each cell of $\mathcal{D}$. We also assume $\mathcal{D}$ is invariant under both the Galois action of $\mathrm{Aut}(\mathrm{acl}(A)/A)$ and the G-action on w. This can be achieved as follows. Begin with a finite set of pairs $(\alpha_j, c_j) \in \mathbb{Q}^w \times \Gamma^w$, such that each of the subsets of Γ^w referred to above is defined by inequalities of the form $\alpha_j v - c_j \odot_j 0$, where $\odot_j$ is $<$ or $>$ or $=$. Take the closure of this set under the Galois action and the G-action. A cell of $\mathcal{D}$ is any nonempty set defined by conditions $\alpha_j v - c_j \odot_j 0$, where $\odot_j$ is any function from the set of indices to $\{<, >, =\}$. Such a cell is an open convex subset of its affine span.

Any bijection $b : w \to \{1, \dots, |w|\}$ yields a bijection $b_* : \Gamma^w \to \Gamma^{|w|}$; the image of c_j under these various bijections depends on the choice of b only up to reordering. Thus $b_*(c_j)$ gives a well-defined subset of Γ, which belongs to $\Gamma(A)$. Let $\mathbf{A}$ be the convex subgroup of $\Gamma = \Gamma(\mathbb{U})$ generated by $\Gamma(A)$, and let $B = \Gamma(\mathbb{U})/\mathbf{A}$. For each cell C of $\mathfrak{D}$, let βC be the image of C in B^w. Note that βC may have smaller dimension than C; notably, $\beta C = (0)$ iff C is bounded. At all events βC is a cell defined by homogeneous linear equalities and inequalities. When $\Gamma(A) \neq (0)$, βC is always a closed cell, i.e. defined by weak inequalities.

For any $C \in \mathfrak{D}$, let C_∞ be the closure of C in Γ_∞^w. Let $\mathfrak{D}_0$ be the set of cells $C \in \mathfrak{D}$ such that $C_\infty \setminus \Gamma^w \subset [\underline{h} = \infty]$. Equivalently, $C \in \mathfrak{D}_0$ if and only if for each $i \in w$, an inequality of the form $x_i \leq m\underline{h} + c$ holds on C, for some $m \in \mathbb{N}$ and $c \in \Gamma(A)$. Other equivalent conditions are that $x_i \leq m\underline{h}$ on βC for some i, or that there exists no $e \in \beta C$ with $\underline{h}(e) = 0$ but $x_i(e) \neq 0$. Let

$$W_0 = (W' \cap (\cup_{C \in \mathfrak{D}_0} C)) \cup [\underline{h} = \infty].$$

It is clear that W_0 is a definably compact subset of Γ_∞^w, contained in $W' = (W \cap \Gamma^w) \cup [\underline{h} = \infty]$.

More generally, define a quasi-ordering $\leq_C$ on w by: $i \leq_C j$ if for some $m \in \mathbb{N}$, $x_i(c) \leq m x_j(c)$ for all $c \in \beta C$. Since the decomposition respects the hyperplanes $x_i = x_j$, we have $i \leq_C j$ or $j \leq_C i$ or both. Thus $\leq_C$ is a linear quasi-order. Let $\beta' C = \beta C \cap [\underline{h} = 0]$. We have $\beta' C = 0$ iff $\underline{h}$ is $\leq_C$-maximal iff $C \in \mathfrak{D}_0$. If $C \in \mathfrak{D}_0$, let $e_C = 0$. Otherwise, $\beta' C$ is a nonzero rational linear cone, in the positive quadrant. Let e_C be the barycenter of $\beta' C \cap [\sum x_i = 1]$ (here we view $\beta' C$ as a cone in $\mathbb{Q}_+^w$). Thus e_C belongs to $\mathbb{Q}_+^w$ and is a nonzero element of $\beta' C$ which is G and Galois invariant.

For $t \in \Gamma$, we have $te_C := e_C t \in \Gamma^w$. If $e_C \neq 0$ then Γe_C is unbounded in Γ^w, so for any $x \in C$ there exists $t \in \Gamma$ such that $x - te_C \notin C$. Let $\tau(x)$ be the unique smallest such t. Note that $\tau(x) > 0$.

Step 2. *Construction of H_Γ and continuity.* We will now define $H_\Gamma : [0, \infty] \times C \to \Gamma^w$ separately on each cell $C \in \mathfrak{D}$ by induction on the dimension of C, as follows. If $C \in \mathfrak{D}_0$, $H_\Gamma(t, x) = x$. Assume $C \in \mathfrak{D} \setminus \mathfrak{D}_0$. If $x \in C$ and $t \leq \tau(x)$, let $H_\Gamma(t, x) = x - te_C$. So $H_\Gamma(\tau(x), x)$ lies in a lower-dimensional cell C'. For $t \geq \tau(x)$ let $H_\Gamma(t, x) = H_\Gamma(t - \tau(x), \tau(x))$. For fixed a, $H_\Gamma(t, a)$ thus traverses finitely many cells as $t \to \infty$, with strictly decreasing dimensions, thus ultimately reaching W_0.

We claim that H_Γ is continuous on $[0, \infty] \times \Gamma^w$. To see this fix $a \in C \in \mathfrak{D}$ and let $(t', a') \to (t, a)$. We need to show that $H_\Gamma(t', a') \to H_\Gamma(t, a)$. By curve selection it suffices to consider (t', a') varying along some line λ approaching (t, a). For some cell C' we have $a' \in C'$ eventually along this line.

If $a' \in W_0$ then $a \in W_0$ since W_0 is closed. In this case we have $H_\Gamma(a', t') = a'$, $H_\Gamma(a, t) = a$, and $a' \to a$ tautologically. Assume therefore that $a' \notin W_0$, so $C' \notin \mathcal{D}_0$ and $e' \neq 0$, where $e' = e_{C'}$.

Consider first the case: $t' \leq \tau(a')$ (cofinally along λ). Then by definition we have $H_\Gamma(t', a') = a' - t'e'$. Now C must be a boundary face of C', cut out from the closure of C' by certain hyperplanes $\alpha_j v - c_j = 0$ ($j \in J(C, C')$). We have $\alpha_j v = c_j$ for $v \in C$, and (we may assume) $\alpha_j v \geq c_j$ for $v \in C'$.

If $\gamma_j = \alpha_j e' > 0$ for some j, fix such a j. As $t' \leq \tau(a')$, we have $\alpha_j(a' - t'e') = \alpha_j a' - \gamma_j t' \geq c_j$, so $t' \leq \gamma_j^{-1}(\alpha_j a' - c_j)$. Now $a' \to a$ so $\alpha_j a' - c_j \to 0$. Thus $t' \to 0$, i.e. $t = 0$. So $H_\Gamma(t, a) = a$, and $H_\Gamma(t, a) - H_\Gamma(t', a') = a - (a' - t'e') = (a - a') + t'e \to 0$ (as $(t', a') \to (t, a)$ along λ).

The remaining possibility is that $\alpha_j e' = 0$ for each $j \in J(C, C')$. So $\alpha_j v = 0$ for each $v \in \beta' C'$. Hence $\beta' C' \subset \beta C$. Since $\beta' C \subset \beta' C'$, it follows that $\beta' C = \beta' C'$ and so $e_C = e_{C'}$. Now $(t, x) \mapsto x - te'$ is continuous on all of $\Gamma \times \Gamma^w$ so on $C \cup C'$, and hence again $H_\Gamma(t', a') \to H_\Gamma(t, a)$.

This finishes the case $t' \leq \tau(a')$. In particular, $\tau(a') \to t^*$ for some t^*, and letting $a'' = H_\Gamma(\tau(a'), a')$, $a'' \to H_\Gamma(t^*, a)$. Now by induction on the dimension of the cell C', we have $H_\Gamma(t' - \tau(a'), a'') \to H_\Gamma(t - t^*, a)$; it follows that $H_\Gamma(t', a') \to H_\Gamma(t, a)$. This shows continuity on $[0, \infty] \times \Gamma^w$.

Note that if $C \in \mathcal{D} \setminus \mathcal{D}_0$, then ξ_i depends only on coordinates x_i with $i \leq_C \underline{h}$. This follows from the fact that ξ_i is bounded on any part of C where $\underline{h}$ is bounded (by assumption $\xi_i^{-1}(\infty) \subset D_0$); so $\xi_i \leq m\underline{h}$ for some m, up to an additive constant. Since $x_i(e_C) = 0$ for $i \leq_C \underline{h}$, it follows that ξ_i is left unchanged by the homotopy on C. So along a path in the homotopy, ξ_i takes only finitely many values (one on each cell); being continuous, it must be constant. In other words the ξ_i are preserved. The closures of the cells are also preserved, hence, as $W \cap \Gamma^w$ is closed, $W \cap \Gamma^w$ is preserved by the homotopy.

Extend H_Γ to W' by letting $H_\Gamma(t, x) = x$ for $x \in W' \setminus \Gamma^w$. Thus, W_0 will be the image of the homotopy and by construction H_Γ fixes $[\underline{h} = \infty]$. We still have to prove that H_Γ is continuous at (t, a) for $a \in W' \setminus \Gamma^w$, i.e. $\underline{h}(a) = \infty$. We have to show that for a' close to a, for all t, $H_\Gamma(t, a')$ is also close to a. If $a' \notin \Gamma^w$ we have $H_\Gamma(t, a') = a'$. Assume $a' \in \Gamma^w$; so $a' \in C$ for some $C \in \mathcal{D}$. If $C \in \mathcal{D}_0$, again we have $H_\Gamma(t, a') = a'$. Otherwise, set $a'' = H_\Gamma(\tau(a'), a')$. Thus, $a'' \notin C$ and belongs to a cell of smaller dimension. We will show that $H_\Gamma(t, a')$ remains close to a for $t \leq \tau(a')$. In particular, a'' is close to a; so (inductively) $H_\Gamma(t, a'') = H_\Gamma(\tau(a') + t, a)$ is close to a. Thus it suffices to show for each coordinate $i \in w$ that $x_i(a')$ remains close to $x_i(a)$. If $i \leq_{C'} \underline{h}$ then the homotopy does not change $x_i(a')$ so (as a is fixed) we have $x_i(H_\Gamma(t, a')) = x_i(a') \to x_i(a) = x_i(H_\Gamma(t, a))$. So assume $h <_C i$. Since $\underline{h}(a) = \infty$ we have $\underline{h}(a') \to \infty$ and hence $x_i(a') \to \infty$. So $x_i(a) = \infty = x_i(H_\Gamma(t, a))$. For any $c = H_\Gamma(t, a')$, $t \leq \tau(a')$, we have $x_i(c) \geq \underline{h}(c)/m = \underline{h}(a')/m$ up to an additive constant. Since $a' \to a$, $\underline{h}(a')$ is large, so $x_i(c)$ is large, i.e. close to $x_i(a)$. This proves the continuity of

H_Γ on W'. This ends the proof of Lemma 11.5.1 except for the additional items.

Step 3. *End of the proof.* For (1), note that by construction, for each $i \in w$ there exists some $m_i \in \mathbb{N}$ and $c_i \in \Gamma(A)$ such that $x_i \leq m_i x_h + c_i$ on $W_0 \cap \Gamma^w$. Set $m = \max_i m_i$ and $c = \max_i c_i$. Now the open subset of $W \cap \Gamma^w$ defined by $W_o = \{x \in W \cap \Gamma^w ; x_i < (m+1)x_h + c, \forall i \in w\}$ does the job. Now let us prove that one can require (2). Set $M = |w|(m+1)$, $K = |w|c$ and let L be the hyperplane $\sum_i x_i = Mx_h + K$. Note that L is both G and Galois invariant. We now consider the cell decomposition $\mathcal{D}'$ generated by L and $\mathcal{D}$ and we denote by $\mathcal{D}'_0$ the corresponding set of "bounded" cells. We claim that replacing $\mathcal{D}$ by $\mathcal{D}'$ does the job. Let C be a cell in $\mathcal{D}'_0$ which is contained in $W \cap \Gamma^w$. Thus C lies in the closure of a cell C' in $\mathcal{D}$ of dimension n and contained in $W \cap \Gamma^w$. Let U be the half space defined by $\sum_i x_i < Mx_h + K$. Thus $C'' = U \cap C'$ is a cell in $\mathcal{D}'_0$ of dimension n contained in W' and C lies in the closure of C''. This shows that after replacing $\mathcal{D}$ by $\mathcal{D}'$, $W_0 \smallsetminus [\underline{h} = \infty]$ is of dimension n at every point. We still have to take care of $W \cap [\underline{h} = \infty]$. Let x be a point in $W \cap [\underline{h} = \infty]$. If some neighborhood of x in W is contained in $[\underline{h} = \infty]$, there is nothing to prove. Otherwise, x is in the closure of W', hence also in the closure of image of W' under the retraction attached to $\mathcal{D}'$, x being invariant under the retraction. Since that image has dimension n at all points, we are done. Finally note that it is possible to achieve (1) and (2) simultaneously. □

While the construction of the Γ-homotopy is essentially carried out in Lemma 11.5.1, we need to extend it to a more general situation in which, e.g. $W \cap \Gamma^w = \varnothing$, i.e. W lies entirely within the ∞-boundary of Γ^w_∞.

LEMMA 11.5.2. *There exists a z-dense and z-open A-definable subset W^o of W such that with*

$$W' = (W^o \smallsetminus [\underline{v} = \infty]) \cup [\underline{h} = \infty],$$

there exists an A-definable deformation retraction

$$H_\Gamma : [0, \infty] \times W' \to W'$$

whose image is a definably compact set W_0 of W' and such that H_Γ leaves the ξ_i invariant, fixes $[\underline{h} = \infty]$, and is G-equivariant. Moreover, one may require the following to hold:

1. *There exists an A-definable open subset W_o of W containing $W_0 \smallsetminus [\underline{h} = \infty]$, and $m \in \mathbb{N}$, $c \in \Gamma(A)$, for $i \in w$, such that $x_i \leq (m+1)x_h + c$ on W_o, for every $i \in w$.*

2. *Let $W = \cup_\nu W_\nu$ be the decomposition of W into z-components. For each ν such that W_ν has pure dimension n_ν, $W_0 \cap W_\nu$ has also pure dimension n_ν.*

Proof. First assume W is z-irreducible. Let w^o be the set of all $i \in w$ such that the i-th projection $\pi_i : W \to \Gamma_\infty$ does not take the constant value ∞ on W; the set w^o is Galois invariant. Clearly $\pi^o = \Pi_{i \in w^o} \pi_i$ is a homeomorphism between W and its image. Note that $\pi^o(W) \cap \Gamma^{w^o}$ is z-open and z-dense in $\pi^o(W)$, and disjoint from $[\underline{v} = \infty]$. Set $W^o = \pi^{o-1}(\pi^o(W) \cap \Gamma^{w^o})$. Thus, either $W^o \cap [\underline{v} = \infty] = \varnothing$ or W is contained in $[\underline{v} = \infty]$ (hence in $[\underline{h} = \infty]$). Set $W' = (W^o \smallsetminus [\underline{v} = \infty]) \cup [\underline{h} = \infty]$. In the first case, applying Lemma 11.5.1 to $\pi^o(W) \cap \Gamma^{w^o}$ and pulling back by π^o we obtain the required homotopy $H_\Gamma = H_{\Gamma, W} : [0, \infty] \times W' \to W'$. Furthermore one may require there exists an A-definable open subset W_o of W containing $W_0 \smallsetminus [\underline{h} = \infty]$, and $m \in \mathbb{N}$, $c \in \Gamma(A)$, for $i \in w_o$, such that $x_i \leq (m+1)x_h + c$ on W_o, for every $i \in w_o$. When $i \notin w_o$, $x_i \leq (m+1)x_h + c$ on W_o. Also one can require (2) holds. The second case is obvious (the homotopy is then the identity at all times).

In general let $W = \cup_\nu W_\nu$ be the decomposition of W into z-components. Define W_ν^o as above that and note that $W_\nu^o \cap W_{\nu'}^o = \varnothing$ if $\nu \neq \nu'$. Set $W^o = \cup_\nu W_\nu^o$. It is a z-dense, z-open subset of W. For each ν, let $H_{\Gamma, W_\nu} : [0, \infty] \times W'_\nu \to W'_\nu$ as above, with $W'_\nu = (W_\nu^o \smallsetminus [\underline{v} = \infty]) \cup [\underline{h} = \infty]$. The subsets W'_ν form a finite cover of W' by closed subsets. Hence the mappings H_{Γ, W_ν} glue to a continuous mapping $H_{\Gamma, W} : [0, \infty] \times W' \to W'$, because they all agree with the trivial retraction on $[\underline{h} = \infty]$ which is the intersection of the sets W'_ν. The process in Lemma 11.5.1 and in the first paragraph of the present lemma being entirely canonical, once an A-definable and G-invariant cell decomposition is chosen, the retraction $H_{\Gamma, W}$ obtained this way is A-definable and G-invariant. By construction the final image W_0 is definably compact. For the additional items, for each ν one has open subsets $W_{o,\nu}$ with corresponding m_ν and c_ν. One sets $W_o = \cup_{i \in w} W_{o,\nu}$, $m = \max m_\nu$, and $c = \max c_\nu$, which gives (1). By the construction in Lemma 11.5.1 it is clear one can require (2) at the same time. □

LEMMA 11.5.3. *Let Υ be an iso-definable Γ-internal subset of $\widehat{V}$. Let $\beta_0 : \widehat{V} \to [0, \infty]^{w_0}$ be a continuous A-pro-definable map, injective on Υ as provided by* Theorem 6.2.8. *Assume Υ is Zariski dense in $\widehat{V}$ in the sense of* 3.11. *Then we may enlarge w_0 to a finite A-definable set w such that β_0 factors through a continuous A-pro-definable map $\beta : \widehat{V} \to [0, \infty]^w$ (injective on Υ) such that:*

(1) *If O is a z-open z-dense subset of $\beta(\Upsilon)$, then $\beta^{-1}(O) \cap \Upsilon$ is a Zariski open dense subset of Υ.*

(2) *For any irreducible component V_i of V, $\beta(\Upsilon \cap \widehat{V_i})$ is a z-component of $\beta(\Upsilon)$.*

Proof. Let $V_1, \ldots, V_r$ be the irreducible components of V. For each V_j, let $x_j : V \to [0, \infty]$ be a schematic distance function to V_j. Set $\beta(x) = (\beta_0(x), x_1, \ldots, x_r)$. It follows from Lemma 10.4.3, that if W is a z-closed subset of $[0,\infty]^w$, then $\beta^{-1}(W)$ is Zariski closed. Thus, if $Z \subset Y$ is z-closed (resp. z-open) in $Y = \beta(\Upsilon)$, $\beta^{-1}(Z) \cap \Upsilon$ is Zariski closed (resp. open) in Υ. Let us prove (1). If $Z \subset Y$ is z-closed in Y and contains no z-component of Y, suppose $\beta^{-1}(Z)$ contains some $\widehat{V_{j_0}} \cap \Upsilon$. Then $\beta^{-1}(Z) \cup \cup_{j \neq j_0} \widehat{V_j}$ contains Υ, so $Z \cup \cup_{j \neq j_0}[x_j = \infty]$ contains Y. It follows that $\cup_{j \neq j_0}[x_j = \infty]$ contains Y already. But then as $\widehat{V_j} = \beta^{-1}([x_j = \infty])$ we have $\Upsilon \subset \cup_{j \neq j_0} \widehat{V_j}$, contradicting the hypothesis on Υ. For (2), let C_j, $j \in J$, denote the z-components of Y. We have $\Upsilon \cap \widehat{V_i} \subset \cup_{j \in J}(\Upsilon \cap \beta^{-1}(C_j))$. Since $\Upsilon \cap \widehat{V_i}$ is Zariski dense in $\widehat{V_i}$ and V_i is irreducible, it follows that, for some j_i, $\Upsilon \cap \widehat{V_i}$ is contained in the Zariski closed set $\Upsilon \cap \beta^{-1}(C_{j_i})$. Hence, $\beta(\Upsilon \cap \widehat{V_i})$ is contained in C_{j_i}. Since each $\beta(\Upsilon \cap \widehat{V_i})$ is z-closed in Y and the sets $\beta(\Upsilon \cap \widehat{V_i})$ are mutually not included one in another, it follows that $\beta(\Upsilon \cap \widehat{V_i}) = C_{j_i}$. □

11.6 END OF THE PROOF

In 11.4, we have constructed a continuous A-pro-definable retraction β_{bc} from $\widehat{V_0 \cup D_0} \to \Upsilon_{bc}$, sending v to the final value of $t \mapsto H_{bc}(t,v)$. Furthermore, by Lemma 11.4.1, Υ_{bc} is Zariski dense in $\widehat{V}$, and we may assume that, for every irreducible component V_i of V, $\Upsilon_{bc} \cap \widehat{V_i}$ is of pure dimension $n = \dim(V)$. By Theorem 6.2.8, there exists a continuous A-pro-definable map $\beta : \widehat{V} \to [0,\infty]^w$ for some finite A-definable set w, injective on Υ_{bc}. One denotes by α its restriction to Υ_{bc}. After enlarging w, we may assume we are in the setting of 11.5, in particular that with the notation therein, $\underline{v} = x_v, \underline{h} = x_h$ for some $h, v \in w$. Also, after adding schematic distance functions to the irreducible components of V, we may assume that the conclusions of Lemma 11.5.3 hold for β and Υ_{bc}. We set $W = \alpha(\Upsilon_{bc})$ and we define W^o, W', H_Γ, W_0, W_o, m and c as in Lemma 11.5.2.

Note that $V \setminus V_0$ contains no irreducible component of V (recall V_0 is the preimage of U_0 in V). Indeed, if V_i is an irreducible component of V, $D_0 \cap V_i$ is nonempty of dimension $n-1$ and u restricts to a finite morphism $D_0 \cap V_i \to U$, thus $u(D_0 \cap V_i)$ contains U_0. By Lemma 10.3.2 there exists an A-definable homotopy $H_{inf} : [0,\infty] \times \widehat{V} \to \widehat{V}$ respecting the functions ξ_i and the group action G and fixing pointwise $\widehat{D_0}$ with image contained in $\widehat{V_0 \cup D_0}$. (In fact, by Lemma 10.3.1 the image is contained in $\widehat{Z}$ with Z a v+g-closed bounded definable subset of V with $Z \cap (V \setminus V_0) \subset D_0$.) For each $i \in w$, set $\phi_i = \min(x_i, (m+1)x_h + c)$. Note that, outside D_0, the functions $\phi_i \circ \beta$ are v+g-continuous with values in Γ. Furthermore, the functions ϕ_i are definable over a finite Galois extension of A and permuted by the Galois group. Thus, by Lemma 10.3.2, we may also require that the functions $\phi_i \circ \beta$ are preserved by H_{inf} away from $\widehat{D_0}$, hence, since H_{inf} fixes pointwise $\widehat{D_0}$, that the functions $\phi_i \circ \beta$ are preserved by H_{inf} everywhere. Recall W_o is an open subset of W containing $W_0 \setminus [\underline{h} = \infty]$, so $\alpha^{-1}(W_o)$ is open in Υ_{bc}. Thus, $\alpha^{-1}(W_o)$ has pure dimension $n = \dim(V)$. Since the restriction of ϕ_i to W_o is just the i-th coordinate function, it follows from Proposition 8.3.1 (2) that $\alpha^{-1}(W_o)$ is fixed pointwise by H_{inf}. Hence so is $\alpha^{-1}(W_0 \smallsetminus [\underline{h} = \infty])$, and thus also $\alpha^{-1}(W_0)$. By construction H_{inf} satisfies $(*)$.

We will define h as the composition (or concatenation) of homotopies

$$h = H_\Gamma^\alpha \circ ((H_{\widetilde{base}} \circ H_{curves}) \circ H_{inf}) : I'' \times \widehat{V} \longrightarrow \widehat{V}$$

where H_Γ^α is to be constructed, and I'' denotes the generalized interval obtained by gluing $[\infty, 0]$, I' and $[0, \infty]$. Being the composition of homotopies satisfying $(*)$, h satisfies $(*)$.

Since the image of H_{inf} is contained in the domain of H_{bc}, the first composition makes sense.

The set W^o is a z-dense, z-open subset of W. Hence, by Lemma 11.5.3 (1), $\alpha^{-1}(W^o)$ is a Zariski open dense subset of Υ_{bc}. Let O be a Zariski dense open subset of V such that $\widehat{O} \cap \Upsilon_{bc} = \alpha^{-1}(W^o)$. By construction of H_{inf}, the image I_{inf} of H_{inf} is contained in $\widehat{O \cup D_0}$. Thus $\beta_{bc}(I_{inf})$ is a definably compact subset of $\beta^{-1}(W') \cap \Upsilon_{bc}$. Note that β restricts to a homeomorphism α_1 between this set and a definably compact subset W_1 of W. One sets $H_\Gamma^\alpha(t, x) = \alpha_1^{-1} H_\Gamma(t, \alpha_1(x))$: in short, H_Γ^α is H_Γ conjugated by α, restricted to an appropriate definably compact set. So h is well-defined by the above quadruple composition.

Since H_{inf} fixes $\alpha^{-1}(W_0)$, and W_0 is the image of H_Γ, H_{inf} fixes the image of h. On the other hand H_{bc} fixes Υ_{bc} and hence the subset $\alpha^{-1}(W_0) \subset \Upsilon_{bc}$. Thus h fixes its own image $\Upsilon = \alpha^{-1}(W_0)$. It follows from Proposition 4.2.9 that Υ is definably compact and α is a homeomorphism from Υ to the definably compact subset W_0 of Γ_∞^w.

We have thus constructed a homotopy $h : I'' \times \widehat{V} \to \widehat{V}$ satisfying the statement of the theorem together with conditions (1), (2) and (4). We shall now check that (3), (5), (6) and (7) also hold.

Let us start by checking (3), that is, h is Zariski generalizing, i.e. for any Zariski open subset U of V, $\widehat{U}$ is invariant under h. This property clearly holds for the first three homotopies in the concatenation; let us check it for H_Γ^α. By Corollary 10.4.6 it is enough to prove H_Γ is Zariski generalizing. Consider a definable continuous function $\eta : W' \to \Gamma_\infty$ such that $W' \smallsetminus \eta^{-1}(\infty) \neq \varnothing$. Pick a point x in W' with $\eta(x)$ finite. By construction of H_Γ, for some finite t_0, $H_\Gamma(t_0, x)$ lies in W_0. Thus, the function $t \mapsto \eta(H_\Gamma(t, x))$ can only take finite values for finite t, since a definable continuous function $[0, t_0] \to \Gamma_\infty$ which is nonconstant can take only finite values.

Let us now check (6), that is, Υ is the image of the set of simple points. Set $e = e_{I''}$. Let p be a point in $\widehat{V}$. Since Υ is iso-definable Γ-internal, by orthogonality to Γ there exists a definable subset D of V containing p such that $h(e, x) = h(e, p)$ for every (simple) point x of D.

We now prove (5). By Lemma 8.1.4 (5), integrating a function into $V^{\#}$ on an element of $V^{\#}$ gives an element of $V^{\#}$. We will use this repeatedly below. In particular by (6), it suffices for (5) to show that the image of the simple points lies in $V^{\#}$. Now (5) is clear for the inflation homotopy, as this homotopy is a finite cover of the standard affine homotopy $I \times \mathbb{A}^n \to \widehat{\mathbb{A}^n}$ (the image of a simple point being a tensor power of the image of a point on $\mathbb{A}^1$). By the remark on integration, precomposing with the inflation homotopy will not spoil (5). Composing with a homotopy taking place purely on the skeleton obviously does not add to the image of $h(e, V)$, as it adds no new points to this image. It remains to consider the inductive step. Inductively, we may assume (5) holds for the skeleton of the base homotopy. In relative dimension one, any element of $\widehat{V/U}$ is in fact in $V^{\#}$. Hence again by transitivity every element of V moves through $V^{\#}$ throughout the homotopy.

It remains to prove (7), i.e. that given a finite family of closed irreducible subvarieties W_i of V, one can assume $\Upsilon \cap \widehat{W_i}$ has pure dimension $\dim(W_i)$. We already proved one can achieve each $\Upsilon_{bc} \cap \widehat{V_i}$ is of pure dimension n. It follows that each $\alpha(\Upsilon_{bc} \cap \widehat{V_i})$ is of pure dimension n. By the conclusion of Lemma 11.5.3 (2) which holds for β and Υ_{bc}, the sets $\alpha(\Upsilon_{bc} \cap \widehat{V_i})$ are the z-components of W. It follows from Lemma 11.5.2 (2) that one can achieve that $\alpha(\Upsilon_{bc} \cap \widehat{V_i}) \cap W_0$ is of pure dimension n. Since α restricts to a homeomorphism between W_0 and Υ, it follows that each $\Upsilon \cap \widehat{V_i}$ is of pure dimension n. With these choices, for any W_i of dimension n, $\Upsilon \cap \widehat{W_i}$ has pure dimension $\dim(W_i)$. Let us now deal with the case where some W_i are of dimension $m_i < n$. We may require all such W_i are contained in the hypersurface D_0 considered in 11.2. All reductions go through and when at the end of 11.2 we replace V by V_1, it is enough to replace W_i by its strict transform. The restriction u_{W_i} of u to W_i is a finite morphism. Set $W_i' = u(W_i)$. By construction, the homotopies H_Γ^α, H_{curves} and H_{inf} fix pointwise the intersection of $\widehat{W_i}$ with their domains. Now note that the pseudo-Galois morphism $U' \to U$ considered in Theorem 6.4.4 may be chosen so to factor through any given finite surjective morphism $U'' \to U$. Thus, we may assume $D_0 \times_U U' \to U'$ is a generically trivial covering. Let W_i' be an irreducible component of $W_i \times_U U'$ and denote by C_i its image under the projection to U'. By the induction hypothesis, we may require the base homotopy h' at the beginning of 11.4 satisfies (7) for all C_i associated to some W_i of dimension $< n$. Let Υ_i' be the final image of C_i under the retraction h'. By hypothesis it has pure dimension m_i. Since $W_i' \to C_i$ is generically an isomorphism and Υ_i' is Zariski dense in C_i by induction, the same holds for the preimage Υ_i'' of Υ_i' in $\widehat{W_i'}$. The morphism $\widehat{W_i'} \to \widehat{W_i}$ being continuous with finite fibers, it follows that the image Υ_i of Υ_i'' in $\widehat{W_i}$ also has pure dimension m_i. By construction the final image of $\widehat{W_i}$ under $H_{\widetilde{base}}$ is equal to Υ_i, which proves (7).

This ends the proof of Theorem 11.1.1. □

REMARK 11.6.1. In the proof of Theorem 11.1.1 one uses the induction hypothesis for the base U, lifted to a certain o-minimal cover (using the same generalized interval). The homotopy on U is (in a certain order) lifted and composed with three additional homotopies: inflation, the relative curve homotopy, and the homotopy internal to Γ. Each of these use the standard interval from ∞ to 0 (in reverse order, in the case of the homotopy internal to Γ). The number $h(n)$ of basic intervals needed for an n-dimensional variety thus satisfies $h(1) = 1$, $h(n+1) \leq h(n) + 3$, so $h(n) \leq 3n - 2$.

For a homotopy whose interval cannot be contracted to a standard one consider $\mathbb{P}^1 \times \mathbb{P}^1$. With the natural choice of fibering in curves, the proof of Theorem 11.1.1 will work even without the inflation homotopy. It will lead to an iterated homotopy to a point: first collapse to $\{\text{point}\} \times \mathbb{P}^1$, then to $\{\text{point}\} \times \{\text{point}\}$.

11.7 VARIATION IN FAMILIES

Consider a commutative diagram

$$\begin{array}{ccc} X & \xrightarrow{\;h\;} & Y \\ & \searrow \quad \swarrow & \\ & T & \end{array}$$

of pro-definable maps, with T a definable set. We shall refer to the family of maps $h_\tau : X_\tau \to Y_\tau$ obtained by restriction to fibers above $\tau \in T$ as *uniformly pro-definable.*

Consider a situation where $(V, X) = (V_\tau, X_\tau)$ are given uniformly in a parameter τ, varying in a definable set T. For each τ, Theorem 11.1.1 guarantees the existence of a strong deformation retraction $h_\tau : I \times \widehat{X_\tau} \to \widehat{X_\tau}$, and a definable homeomorphism $j_\tau : W_\tau \to h_\tau(e_I, \widehat{X_\tau})$, with W_τ a definable subset of $\Gamma_\infty^{w(\tau)}$. Such statements are often automatically uniform in the parameter τ. For instance if X_τ, Y_τ are uniformly definable families of definable sets, and for each a there exists an a-definable bijection $X_a \to Y_a$, then automatically there must exist a uniformly definable bijection $h_\tau : X_\tau \to Y_\tau$. Indeed if H is the collection of all $\varnothing$-definable subsets of $X \times_T Y$, then for any $a \in T$, for some $h \in H$, $h_a : X_a \to Y_a$ is a bijection. By compactness the family of all formulas asserting that h_τ is not a bijection $X_\tau \to Y_\tau$ is inconsistent. Hence a finite subset is inconsistent; i.e. there exists a finite set $h^1, \ldots, h^r \in H$ such that for any $a \in T$, for some $i \leq r$, h_a^i is a bijection $X_a \to Y_a$. For any τ, let i_0 be the smallest $i \leq r$ such that h_τ^i is a bijection, and let $h_\tau : X_\tau \to Y_\tau$ be equal to $h_\tau^{i_0}$. More generally, if each h_a has some property P which is ind-definable (i.e. the family of all definable maps for which it holds is an ind-definable family), then one can find h such that each h_τ has this property. (See a fuller explanation in [26], introductory section on compactness and gluing.)

Here the pro-definable map h_τ is given by an infinite collection of definable maps, so compactness does not directly apply. Nevertheless the theorem is uniform in the parameter τ. The reason is that h_τ is determined by its restriction to the simple points, and on these, the homotopy moves along $V^\#$, which is endowed with a canonical strict ind-definable structure by 8.2. We state this as a separate proposition.

PROPOSITION 11.7.1. *Let V_τ be a quasi-projective variety, X_τ a definable subset of $V_\tau \times \Gamma_\infty^\ell$, definable uniformly in $\tau \in T$ over some base set A. Then there exists a uniformly pro-definable family $h_\tau : I \times \widehat{X_\tau} \to \widehat{X_\tau}$, a finite set $w(\tau)$, a definable set $W_\tau \subset \Gamma_\infty^{w(\tau)}$, and $j_\tau : W_\tau \to h_\tau(0, \widehat{X_\tau})$, pro-definable uniformly in τ, such that for each $\tau \in T$, h_τ is a deformation retraction, and $j_\tau : W_\tau \to h_\tau(0, \widehat{X_\tau})$ is a pro-definable homeomorphism. Moreover, we may require* (1) *and* (2) *of* Theorem 11.1.1 *to hold if the ξ_i and the group action are given uniformly, as can* (4), (5), (6) *and* (7).

Proof. For any $a \in T$, we have h_a, j_a with the stated properties, by Theorem 11.1.1. By Theorem 11.1.1 (5), h_a restricts to $h_a^\# : V_a \times I \to {V_a}^\#$. Note that in principle $I = I_a$ depends on a. However, as $\dim(V_a)$ is bounded by some m, it follows from Remark 11.6.1 that I_a is a union of at most $3m - 2$ copies of $[0, \infty]$; extending the homotopy trivially to be constant to the left, we may assume it is a gluing of exactly $3m - 2$ copies of this interval, so it does not depend on a. We have:

(1) Given finitely many A-definable functions $\xi_i : \coprod_{\tau \in T} V_\tau \to \Gamma_\infty$, one can choose h_a to respect the ξ_i, i.e. $\xi_i(h_a^\#(t, x)) = \xi_i(x)$ for all τ.

(2) Assume given, in addition, a finite algebraic group action on V_a given uniformly in a. Then the homotopy retraction can be chosen to be equivariant.

(4) Let $x \in X$ and let $c = h_a^\#(e_I, x)$ be the final image of x. Also let $t \in I$, and $p = h(t, x)$. Then for generic $y \models p$, $h_a^\#(e_I, y) = c$; i.e. $\models (d p_y) h_a^\#(e_I, y) = c$.

(7) Each irreducible component V' is left invariant by $h_a^\#$; and if $X \cap V'$ contains an open subset of V', then $h_a^\#(0, V')$ has pure dimension equal to $\dim(V')$.

(5′) $h_a^\#$ extends to a homotopy $H_a : \widehat{X_a} \to \widehat{X_a}$.

Now the validity of (5′) for $h_a^\#$ is an ind-definable property of a, by Proposition 9.9.1, and (1), (2) and (7) are obviously ind-definable (using the classical fact that the irreducible components of V_a are ACF-definable uniformly in a). Property (4) is also stated in an ind-definable way.

Hence by the compactness and gluing argument mentioned above, one can find a uniformly definable family h_τ with the same properties. Now let $H_\tau(p) = \int_{x \models p} h_\tau(x)$. By (5′), this is a homotopy $H_\tau : \widehat{X_\tau} \to \widehat{X_\tau}$. Property (5) of Theorem 11.1.1 holds by definition. Property (6) is proved in the same way as in Theorem 11.1.1. □

Remark 11.7.2. We proved above that irreducible components are preserved, but not the full Zariski generalization property Theorem 11.1.1 (3), as it is not an ind-definable property on the face of it. It can still be achieved uniformly; this can be seen in one of two ways:

- either by following the proof of (3), carrying the parameter τ along;
- or else by proving that a stronger ind-definable property holds; namely that there is a uniformly definable family of varieties, such that the Zariski closure of $h(x,t)$ is an element of this family, and is increasing with t along I. In the case of a definably compact set X contained in the smooth locus of V, the proof of Theorem 12.1.1 gives this in a very simple form: the Zariski closure of $h(x,t)$ is $\{x\}$ if $t = \infty$, and equals V otherwise.

Chapter Twelve

The smooth case

Summary. In this chapter we examine the simplifications occuring in the proof of the main theorem in the smooth case. We also note the birational character of the definable homotopy type in Remark 12.2.4.

12.1 STATEMENT

For definable sets avoiding the singular locus it is possible to prove the following variant of Theorem 11.1.1. The proof uses the same ingredients but is considerably simpler in that only birational versions of most parts of the construction are required. For clauses (1), (2) and (4), the homotopy internal to Γ is not required; and a global inflation homotopy is applied only once, rather than iterated at each dimension. For clause (3), a single final use of the Γ-homotopy is added.

Given an algebraic variety V over a field, one denotes by V_{sing} its singular locus, i.e. its nonsmooth locus.

THEOREM 12.1.1. *Let V be a quasi-projective variety over a valued field F and let X be a v-clopen F-definable subset of $V \smallsetminus V_{\text{sing}}$. Then there exists an F-definable homotopy $h : I \times \widehat{X} \to \widehat{X}$ between the identity and a continuous map to an F-iso-definable subset definably homeomorphic to a definable subset of $w' \times \Gamma^w$, for some finite F-definable sets w and w'.*

Moreover, one can require the following additional properties for h:

(1) *Given finitely many v-continuous F-definable functions $\xi_i : X \to \Gamma$, one can choose h to respect the ξ_i, i.e. $\xi_i(h(t,x)) = \xi_i(x)$ for all t.*

(2) *Assume given, in addition, a finite algebraic group action on V. Then the homotopy can be chosen to be equivariant.*

(3) *If $\widehat{X}$ is definably compact, h can be taken to be a deformation retraction.*

(4) *Clauses (3) to (6) of* Theorem 11.1.1 *hold. Also, if V has dimension d at each point $x \in X$, then each point q of the image of h is strongly stably dominated with* $\dim(q) = d$.

In particular this holds for $X = V$ when V is smooth.

By Remark 12.2.3, when $\widehat{X}$ is definably compact the conclusion is stronger than Theorem 11.1.1 in that the interval is the standard interval $[0, \infty]$. If $\widehat{X}$ is not definably compact, the conclusion is also weaker in that we do not assert that the final image is fixed by the homotopy.

The finite set w' can be dispensed with if $\Gamma(F) \neq (0)$, or if $\widehat{X}$ is connected, but not otherwise, as can be seen by considering the case when X is finite. Indeed, when $\Gamma(F) = (0)$ the only nonempty finite F-definable subset of Γ^n is $\{0\}$, but one can have arbitrarily large finite F-definable subsets in Γ^n_∞ for n large enough.

12.2 PROOF AND REMARKS

The proof depends on two lemmas. The first recaps the proof of Theorem 11.1.1, but on a Zariski dense open set V_0 only. The second uses smoothness to enable a stronger form of inflation, serving to move into V_0.

While the theorem requires the functions ξ_i to be v-continuous, this need not be assumed in Lemma 12.2.1 since any definable function is v-continuous on some Zariski dense open set. But then X need not be explicitly mentioned, since one can add the valuation of the characteristic function of X to the list of ξ_i. The proof of this lemma uses only an iteration of the curves homotopy, without inflation or the Γ-homotopy.

LEMMA 12.2.1. *Let V be a quasi-projective variety defined over F. Then there exists a Zariski open dense subset V_0 of V, and an F-definable deformation retraction $H : I \times \widehat{V_0} \to \widehat{V_0}$ whose image is an F-iso-definable subset S_0, definably homeomorphic to an F-definable subset of $w' \times \Gamma^w$, for some finite F-definable sets w' and w.*

Moreover:

(1) *Given finitely many F-definable functions $\xi_i : V \to \Gamma$, one can choose h to respect the ξ_i, i.e. $\xi_i(h(t, x)) = \xi_i(x)$ for all t.*

(2) *Assume given, in addition, a finite algebraic group action on V. Then V_0 and the deformation retraction can be chosen to be equivariant.*

Proof. Find a Zariski open V_1 with $\dim(V \smallsetminus V_1) < \dim(V)$, and a morphism $\pi : V_1 \to U$, with U normal, whose fibers are curves. Let h_{curves} be the homotopy described in 11.3. It is v+g-continuous outside some subvariety W of U with $\dim(W) < \dim(U)$; replace V_1 by $V_1 = V \smallsetminus \pi^{-1}(W)$. So h_{curves} is v+g-continuous on V_1 and its image S_1 is relatively Γ-internal over U. By (a greatly simplified version of) the results of 6.4, there exists a finite pseudo-Galois covering $U' \to U$, such that, setting $V_1' = U' \times_U V_1$ and $S_1' = U' \times_U S_1$, there exists a definable v+g-continuous morphism $g : V_1' \to U' \times \Gamma^n_\infty$ such that g induces a homeomorphism between $\widehat{S_1'}$ and its image in $\widehat{U'} \times \Gamma^n_\infty$.

Claim. *After replacing V_1 by a Zariski dense open subset, one may assume there exists a definable isomorphism between S_1' and an iso-definable subset of $\widehat{U'} \times \{1, \dots, N\} \times \Gamma^n$ relatively Γ-internal over U', for some positive integers N, n, inducing a homeomorphism between $\widehat{S_1'}$ and its image in $\widehat{U'} \times \{1, \dots, N\} \times \Gamma^n$.*

Proof of the claim. After removing a nowhere dense subvariety, we may assume V_1 is a disjoint union of irreducible components, and work within each component separately. So we may assume V_1 is irreducible. We may also assume V_1' is irreducible. The set of points of S_1' which are mapped by g to $U' \times \Gamma^n$ is Zariski open in S_1; thus, if it is nonempty it must be dense in V_1', and after shrinking V_1 again we may assume S_1' maps to $U' \times \Gamma^n$. Otherwise S_1' maps to $U' \times (\Gamma_\infty^n \setminus \Gamma^n)$. In this case we can remove a proper subvariety and decompose the rest into finitely many algebraic pieces, each mapping into one hyperplane at ∞ of Γ_∞^n. Then one concludes the proof by induction on n. □

We may thus assume there exists a definable isomorphism between S_1' and an iso-definable subset of $\widehat{U'} \times \{1, \dots, N\} \times \Gamma^n$ relatively Γ-internal over U', for some positive integers N, n, inducing a homeomorphism between $\widehat{S_1'}$ and its image in $\widehat{U'} \times \{1, \dots, N\} \times \Gamma^n$. By induction, there exists a Zariski dense open U_0' of U_0 and an F-definable deformation retraction $h' : I \times \widehat{U_0'} \to \widehat{U_0'}$ satisfying the conclusions of the lemma. Furthermore we may assume the pseudo-Galois covering $U' \to U$ restricts to a pseudo-Galois covering $U_0' \to U_0$ for some dense open subset U_0 of U and that h' is the lifting of an F-definable deformation retraction $h : I \times \widehat{U_0} \to \widehat{U_0}$ satisfying the conclusions of the lemma. Set $V_0 = \pi^{-1}(U_0)$. Using Theorem 6.4.4 as in 11.4, we may arrange that h lifts to a homotopy $H_{\widetilde{base}} : I \times (\widehat{S_1} \cap \widehat{V_0}) \to \widehat{S_1} \cap \widehat{V_0}$. The homotopies can be taken to meet conditions (1) and (2). Composing, we obtain a deformation retraction of $\widehat{V_0}$ to a subset S, and a homeomorphism $\alpha : S \to Z \subset \{1, \dots, M\} \times \Gamma^m$, defined over $\operatorname{acl}(A)$. We may assume $M > 1$. As in Theorem 6.2.8 we can obtain an A-definable homeomorphism into $(\{1, \dots, M\} \times \Gamma^m)^w$. □

Lemma 12.2.2. *Let V be a subvariety of $\mathbb{P}^n$, and let $a \in V$ be a smooth point. Then the standard metric on $\mathbb{P}^n$ restricts to a good metric on some v-open neighborhood of a in V.*

Proof. For sufficiently large α, the set of points of distance $\geq \alpha$ from a may be represented as the $\mathcal{O}$-points of a scheme over $\mathcal{O}$ with good reduction, whose special fiber is irreducible, in fact a linear variety.

This can be done as follows. We may assume $V \subset \mathbb{A}^n$, and $a = (0)$. As a is smooth, V is a complete intersection near 0, and we may localize further and assume it is cut out by polynomials $f, \dots, h$ in affine coordinates $x_1, \dots, x_n$, whose number ℓ is the codimension of V.

We can write $f = f_1 + f_2$, where f_1 is linear and f_2 consists of higher degree terms and similarly for $g, \ldots, h$. The vectors $f_1, \ldots, h_1$ generate an ℓ-dimensional subspace of the space with basis $x_1, \ldots, x_n$.

By performing row operations, we may assume $f_1, \ldots, h_1$ have coefficients in $\mathcal{O}$, and further that their coefficient vectors generate a lattice of rank ℓ in $\mathcal{O}^n$. (In fact, permuting the variables if necessary, and performing row operations, we can arrange that modulo $\mathcal{O}x_{\ell+1} + \ldots + \mathcal{O}x_n$ we have $f_1 = x_1, \ldots, h_1 = x_\ell$.)

Of course, the nonlinear coefficients of $f, \ldots, h$ have coefficients in the field K, some having valuation as negative as $-\mathrm{val}(c)$ say, where $c \in \mathcal{O}$. Let $F(x) = c^{-1}f(cx), \ldots, H(x) = c^{-1}h(cx)$. The intersection of V with $c\mathcal{O}^n$ is isomorphic to the intersection of $(F, \ldots, H)$ with $\mathcal{O}^n$. But it is clear that $F, \ldots, H$ have coefficients in $\mathcal{O}$, and that they cut out a smooth scheme S_c over $\mathcal{O}$.

For this c or for any c' with $\mathrm{val}(c') \geq \mathrm{val}(c)$, $S_c(K)$ clearly admits a unique generic type, dominated by the generic type of the linear variety $S_c(k)$, via the residue map. □

Proof of Theorem 12.1.1. Let V_0, H and S_0 be as provided by Lemma 12.2.1. As in the first few lines of the proof of Theorem 11.1.1, we may choose a projective embedding equivariant with respect to the finite group action of (2). By Lemma 12.2.2, for any $x \in X$, the standard metric d on $\mathbb{P}^n$ restricts to a good metric on some v-open neighborhood of x. Thus, there exists a definable function $\rho : X \to [0; \infty)$ which is locally bounded and such that for any $x \in X$ and any $t \geq \rho(x)$, $B(x; d, t)$ is affine and has a unique generic type which we shall denote by $p(x, t)$. Since X is v-open and the functions ξ_i are v-continuous, we may assume that for $t \geq \rho(x)$, $p(x, t)$ lies in $\widehat{X}$ and $t \mapsto \xi_i(p(x, t))$ is constant. Since X is v-closed, by Lemma 10.1.8 there exists a v+g-continuous function $g : X \to [0; \infty)$ such that for every $x \in X$, $\rho(x) \leq g(x)$. For $t \in [0, \infty]$ and $x \in X$, set $H_{inf}(x, t) = p(x, \max(t, g(x)))$. It is a v+g-continuous definable function $[0, \infty] \times X \to \widehat{X}$ which extends to a homotopy $H_{inf} : [0, \infty] \times \widehat{X} \to \widehat{X}$. Note that the image of H_{inf} is contained in $\widehat{V_0}$. Define h as the composition of H and H_{inf}.

For clause (3), to ensure that the composition is also a deformation retraction, we compose with an additional homotopy internal to Γ as in Theorem 11.1.1.

The verification that the image of closed points is strongly stably dominated is as in Theorem 11.1.1; moreover the homotopies of Lemma 12.2.2 are Zariski generalizing, while the inflation homotopy Lemma 12.2.1 has final image consisting of points of maximal dimension; this proves (4). □

REMARK 12.2.3. Under the hypotheses of Theorem 12.1.1, if $\widehat{X}$ is definably compact, the interval I can in fact be taken to be $[0, \infty]$. We sketch the argument. The proof above yields a composition of homotopies $H_\Gamma \circ H_m \circ$

$\cdots H_1 \circ H_{inf}$, where the H_i for $i = 1, \ldots, m$ are relative curve homotopies using intervals $[0, \infty]$ oriented from ∞ to 0, H_{inf} uses a similar interval $[0, \infty]$, and H_Γ, the homotopy internal to Γ, uses the same interval oriented in the opposite direction.

For $k = 0, \ldots, m$, set $H^k = H_k \circ \cdots \circ H_{inf}$, with $H^0 = H_{inf}$, and denote by S_k the final image of H^k. We wish to show by induction on k that the interval of H^k can be contracted to a standard interval $[0, \infty]$. It suffices to replace H_k by a homotopy whose time interval is a closed interval in Γ, by showing that for some α_k, for all $t > \alpha_k$ and all $x \in S_k$, $H_k(t, x) = x$.

If we write X as a finite union of definably compact subsets X_ν of affine open subsets of V, and show that the statement holds for each X_ν, then it holds for X. In this way we can reduce to an affine situation.

Each $a \in S_k$ is a strongly stably dominated point. It is possible to find an étale neighborhood V' of X in V and morphisms $f : V' \to W$ and $g : W \to U$ such that $W \subset U \times \mathbb{A}^1$ and g is the projection, $(g \circ f)_*(a) = a'$ is a generically stable type on U, $a = \int_{a'} h$ where h is a definable map $U \to \widehat{\mathbb{A}^1}$, and H_k is compatible with the standard homotopy on $\mathbb{A}^1$, relative to U. The decomposition $f : V' \to W$ and $g : W \to U$ is part of the construction of the homotopy, while the integral decomposition of a over a' follows from the strong stable domination of a (cf. Proposition 8.2.5). Moreover, as a lies in the final image of the inflation homotopy, the Zariski closure has dimension $\dim(V)$, and it follows that one can take $h(u)$ to be the generic type of a closed ball which is not reduced to a point. Moreover the radius of this ball is a continuous definable function on U. By definable compactness, it is bounded above on S_k, say $\leq \alpha_k$. It follows that $H_k(t, x) = x$ for $t > \alpha_k, x \in S_k$. This allows us to collapse the interval of H^m to a standard interval $[0, \infty]$.

Recall now the homotopy within Γ. The composed curve homotopies $H_m \circ \cdots H_1$ act on a certain affine $\tilde{V}$, with final image $\tilde{S} \cong \Omega$; Ω is a definable subset of Γ_∞^w. The homotopy H_Γ takes Ω to a definably compact set S_Γ. At this point, H_{inf} is chosen so as to fix S_Γ. The final image of the composition $H_m \circ \cdots H_1 \circ H_{inf}$ is the definably compact set S_m. Now H_Γ is applied, with time interval $[0, \infty]$. But H_Γ moves each point of S_m into S_Γ in finite time. Since S_m is definably compact, there is some time t_Γ such that by time t_Γ, each point of S_m is moved by H_Γ into S_Γ (and then frozen). Thus if H'_Γ is the restriction of H_Γ to time interval $[0, t_\Gamma]$, then the composition $H_\Gamma \circ H_m \circ \cdots H_1 \circ H_{inf}$ also has final image fixed by H_{inf} and by each H_i and H_Γ. This gives a homotopy whose time interval is the concatenation of $[0, \infty]$ with $[0, t_\Gamma]$; this is again isomorphic to $[0, \infty]$.

REMARK 12.2.4 (A birational invariant). It follows from the proof of Theorem 12.1.1 that the definable homotopy type of $\widehat{V \smallsetminus V_{\text{sing}}}$ (or more generally of $\widehat{X \smallsetminus V_{\text{sing}}}$ when X is a v-clopen definable subset of V) is a birational invariant of V (of the pair (V, X)). This rather curiously complements a theorem of Thuillier [39].

As a referee pointed out, this remark requires only the inflation homotopy. Let us spell this out without X, to simplify notation. It suffices to show that if U is a smooth variety and W a Zariski dense open subset, then $\widehat{U}$ and $\widehat{W}$ are pro-definably homotopy equivalent. Indeed, let $H = H_{inf}$ be the inflation homotopy on U as considered in the proof of Theorem 12.1.1 and denote by Z its image. Note that if $z \in Z$, then $H(t, z) = z$ for all t. By density of simple points we may assume $z = H(t_0, z_0)$ with z_0 a simple point and t_0 the stopping time provided by the cut-off function. It is enough to prove that if M is a base, $z = H(t_0, z_0)$, $x \models z|M$, and y realizes the generic type over $M(x)$, of the ball of valuative radius t around x, then $y \models z|M$. Indeed y still falls in the ball of valuative radius $t_0 = \mathrm{val}(c_0)$ around z_0, and has the same image as x under the dominating function $\mathrm{res}(c_0^{-1}y)$. Thus, H provides a deformation retraction of $\widehat{U}$ to Z. Since $Z \subset \widehat{W}$, the restriction of H to $\widehat{W}$ provides a deformation retraction of $\widehat{W}$ to Z. Thus, $\widehat{U}$ and $\widehat{W}$ are both definably homotopy equivalent to Z.

Chapter Thirteen

An equivalence of categories

Summary. In this chapter we deduce from Theorem 11.1.1 an equivalence of categories between the homotopy category of definable subsets of quasi-projective varieties over a given valued field and the homotopy category of definable subsets of some Γ_∞^w.

13.1 STATEMENT OF THE EQUIVALENCE OF CATEGORIES

Let F be a valued field. We fix a base set $A \subset \mathrm{VF} \cup \Gamma$ with $F = \mathrm{VF}(A)$. Let V be an algebraic variety over F; by a *semi-algebraic* subset of $\widehat{V}$ we mean a subset of the form $\widehat{X}$, where X is a definable subset of V. If X is A-definable, we say $\widehat{X}$ is *A-semi-algebraic.*

Let C_{VF} be the category of semi-algebraic subsets of $\widehat{V}$, with V a quasi-projective variety over F; the morphisms are pro-definable continuous maps. We could also say that the objects are definable subsets of V, but the morphisms $U \to U'$ are still pro-definable continuous maps $\widehat{U} \to \widehat{U'}$.

Let C_Γ be the category of definable subsets X of Γ_∞^w (for various definable finite sets w), with definable continuous maps. Any such map is piecewise given by an element of $\mathrm{GL}_w(\mathbb{Q})$ composed with a translation, and with coordinate projections and inclusions $x \mapsto (x, \infty)$ and $x \mapsto (x, 0)$. Let C_Γ^i be the category of topologically Γ-internal subsets X of $\widehat{V}$, for various varieties V, with continuous definable maps.

These categories can be viewed as ind-pro definable: more precisely Ob_C is an ind-definable set, and for $X, Y \in \mathrm{Ob}_C$, $\mathrm{Mor}(X, Y)$ is a pro-ind-definable set. The three categories admit natural functors to the category TOP of topological spaces with continuous maps. But usually we will be interested only in the subcategory consisting of A-definable objects and morphisms. It can be defined in the same way as in the first place, only replacing definability by A-definability. We shall denote these categories by C_{VF}^A, C_Γ^A and $C_\Gamma^{A,i}$.

There is a natural functor $\iota : C_\Gamma^A \to C_\Gamma^{A,i}$, commuting with the natural functors to TOP; namely, given an A-definable subset $X \subset \Gamma_\infty^n$, let $\iota(X) = \{p_\gamma : \gamma \in X\}$, where p_γ is as defined above Lemma 3.5.2. By Lemma 3.5.2 and Lemma 3.5.3, the map $\gamma \mapsto p_\gamma$ induces a homeomorphism $X \to \iota(X)$.

LEMMA 13.1.1. *The functor $\iota : C_\Gamma^A \to C_\Gamma^{A,i}$ is an equivalence of categories.*

Proof. It is clear that the functor is fully faithful. Essential surjectivity follows from Theorem 6.2.9 and Remark 6.2.10. □

We now consider the corresponding homotopy categories HC^A_{VF}, HC^A_Γ and $HC^{A,i}_\Gamma$. These categories have the same objects as the original ones, but the morphisms are factored out by (strong) homotopy equivalence. Namely two morphisms f and g from X to Y are identified if there exists an A-definable generalized interval $I = [i_I, e_I]$ and a continuous A-pro-definable map $h : X \times I \to Y$ with $h_{i_I} = f$ and $h_{e_I} = g$. One may verify that composition preserves equivalence; the image of Id_X is the identity morphism in the category.

The equivalence ι above induces an equivalence $HC^A_\Gamma \to HC^{A,i}_\Gamma$. As a reader pointed out, the same retraction was considered by Berkovich in the setting of Berkovich spaces.

LEMMA 13.1.2. *Let w be an A-definable finite set. For an A-definable subset $X \subset \Gamma^w_\infty$, let $C(X) = \{x \in \mathbb{A}^w : \mathrm{val}(x) \in X\}$. Then the inclusion $\iota(X) \subset \widehat{C(X)}$ is a homotopy equivalence.*

Proof. For $t \in [0, \infty]$ one sets $H_0 = G_m(\mathcal{O})$, $H_\infty = \{1\}$, and for $t > 0$, with $t = \mathrm{val}(a)$, H_t denotes the subgroup $1 + a\mathcal{O}$ of $G_m(\mathcal{O})$. For x in $C(X)$ one denotes by $p(H_t x)$ the unique H_t-translation invariant stably dominated type on $H_t x$. In this way one defines an A-definable homotopy $[0,\infty] \times C(X) \to \widehat{C(X)}$ by sending (x, t) to $p(H_t x)$, whose canonical extension $[0,\infty] \times \widehat{C(X)} \to \widehat{C(X)}$ is a deformation retraction with image $\iota(X)$. □

We shall prove the following statement in the next section:

THEOREM 13.1.3. *The categories HC^A_Γ and HC^A_{VF} are equivalent by an equivalence respecting the subcategories of definably compact objects.*

13.2 PROOF OF THE EQUIVALENCE OF CATEGORIES

To prove Theorem 13.1.3, we introduce a category C^A_2 defined as follows. Objects of C^A_2 are pairs (X, π), with X an object of C^A_{VF} and $\pi : X \to X$ a continuous A-definable retraction with topologically Γ-internal image, which is homotopic to the identity $\mathrm{Id} : X \to X$ via an A-definable homotopy $h : I \times X \to X$ with $h_{i_I} = \mathrm{Id}$, $h_{e_I} = \pi$, and $\pi \circ h_t = h_t \circ \pi = \pi$ for every t in I. A morphism $f : (X, \pi) \to (X', \pi')$ in C^A_2 is a continuous A-definable map $f : X \to X'$ such that $f \circ \pi = \pi' \circ f$. We define a homotopy equivalence relation $\sim_2$ on $\mathrm{Mor}_{C^A_2}((X, \pi), (X', \pi'))$ by $f \sim_2 g$ if there exists a continuous A-definable $h : I \times X \to X'$, with $h_{i_I} = f$ and $h_{e_I} = g$, such that $h_t \circ \pi = \pi' \circ h_t$ for all t. Note that $f \sim_2 f \circ \pi$ and $f \sim_2 \pi' \circ f$. In particular, $f \sim_2 \pi' \circ f \circ \pi$. Again one checks that this is a congruence and that one can define a quotient category denoted by HC^A_2.

There is an obvious functor $C_2^A \to C_{\mathrm{VF}}^A$ forgetting π, and also a functor $C_2^A \to C_\Gamma^{A,i}$, mapping (X, π) to $\pi(X)$. One checks that the natural maps on morphisms are well-defined and that they induce functors $HC_2^A \to HC_{\mathrm{VF}}^A$ and $HC_2^A \to HC_\Gamma^{A,i}$. To prove the theorem, it suffices therefore to prove, keeping in mind Lemma 13.1.1, that each of these two functors is essentially surjective and fully faithful, and to observe that they restrict to functors on the definably compact objects, essentially surjective on definably compact objects.

(If the categories are viewed as ind-pro-definable, these functors are morphisms of ind-pro-definable objects, but we do not claim that a direct definable equivalence exists.)

LEMMA 13.2.1. *The functor $HC_2^A \to HC_{\mathrm{VF}}^A$ is surjective on objects, and fully faithful.*

Proof. Surjectivity on objects is given by Theorem 11.1.1. Consider (X, π) and (X', π') in $\mathrm{Ob}HC_2^A = \mathrm{Ob}C_2^A$. Let $f : X \to X'$ be a morphism of C_{VF}^A. Then the composition $\pi' \circ f \circ \pi$ is homotopy equivalent to f, since $\pi \sim \mathrm{Id}_X$ and $\pi' \sim \mathrm{Id}_{X'}$, and is a morphism of C_2^A. This proves surjectivity of $\mathrm{Mor}_{HC_2^A}((X, \pi), (X', \pi')) \to \mathrm{Mor}_{HC_{\mathrm{VF}}^A}(X, X')$. For injectivity, let $f, g : (X, \pi) \to (X', \pi')$ with $f \sim g$ in C_{VF}^A. Thus, $\pi' \circ f \circ \pi$ and $\pi' \circ g \circ \pi$ are homotopic in C_2^A. Since $f \sim_2 \pi' \circ f \circ \pi$ and $g \sim_2 \pi' \circ g \circ \pi$, it follows that $f \sim_2 g$. □

LEMMA 13.2.2. *The functor $HC_2^A \to HC_\Gamma^{A,i}$ is essentially surjective and fully faithful.*

Proof. To prove essential surjectivity it suffices to consider objects of the form $\iota(X)$, with $X \in \mathrm{Ob}C_\Gamma^A$. For these, Lemma 13.1.2 does the job.

Let $(X, \pi), (X', \pi') \in \mathrm{Ob}HC_2^A = \mathrm{Ob}C_2^A$. Let $g : \pi(X) \to \pi'(X')$ be a morphism of $C_\Gamma^{A,i}$. Then $g \circ \pi : X \to X'$ is a morphism of C_2^A. So even $\mathrm{Mor}_{C_2^A}((X, \pi), (X', \pi')) \to \mathrm{Mor}_{C_\Gamma^{A,i}}(X, X')$ is surjective.

To prove injectivity, suppose g_1 and $g_2 : X \to X'$ are C_2^A-morphisms, and $h : I \times \pi(X) \to \pi'(X')$ is a homotopy between $g_1|\pi(X)$ and $g_2|\pi(X)$. We wish to show that g_1 and g_2 are C_2^A-homotopic. Now for $i = 1, 2$, g_i and $\pi' \circ g_i \circ \pi$ have the same image in $\mathrm{Mor}(\pi(X), \pi'(X'))$, and there is a homotopy between g_i and $\pi' \circ g_i \circ \pi$, $i = 1, 2$, as remarked before. So we may assume $g_i = \pi' \circ g_i \circ \pi$ for $i = 1, 2$. Define $H : I \times X \to X'$ by $H(t, x) = \pi' \circ h(t, \pi(x))$. This is a C_2^A-homotopy between g_1 and g_2 showing that g_1 and g_2 have the same class as morphisms in HC_2^A. □

REMARK 13.2.3. Note that in the definition of the category HC_Γ^A one cannot replace Γ_∞ by Γ. Indeed, consider the triangle T in Γ_∞^2 consisting in those (x, y) with $0 \leq x, y \leq \infty$ belonging to one of the lines $y = 0$, $x = y$, and $x = \infty$. There does not exist a homotopy equivalence $g : T \to T'$ with T'

a definable subset of some Γ^n (or some Γ^w with finite definable w). Indeed, assume such a g exists and consider a homotopy inverse $f : T' \to T$. Note that any definable subset $X \neq T$ of T which is definably connected retracts to a point. It follows that any homotopy equivalence $T \to T$ is surjective, so $f \circ g$ should be surjective. In particular, f should be surjective. On the other hand, T' should be definably path connected, hence definably connected. But a continuous surjective definable $f : T' \to T$ with T' a definably connected subset of some Γ^n cannot exist, since $(y \circ f)^{-1}(\infty)$ would be a nontrivial clopen.

13.3 REMARKS ON HOMOTOPIES OVER IMAGINARY BASE SETS

Note that Theorem 11.1.1 is valid over an arbitrary base set A, including imaginaries, when X is a constructible subset of V. More generally, if X and the ξ_i are defined over $A \cap (\mathrm{VF} \cup \Gamma)$, the theorem follows, simply by applying it over $A' = A \cap (\mathrm{VF} \cup \Gamma)$.

Is Theorem 11.1.1 true in full generality over an arbitrary base? Here is an indication that the answer may be positive, at least over a finite extension. Assume (V, X) are given as in Theorem 11.1.1, but over a base A including imaginary elements. A homotopy h_c is definable over additional field parameters c, satisfying the conclusion of Theorem 11.1.1 over $A(c)$. By the uniformity results of 11.7, there exists an A-definable set Q such that any parameter $c \in Q$ will do. One can find a definable type q on Q, over a finite extension A' of A (i.e. $A' = A(a'), a' \in \mathrm{acl}(A)$). We know that $q = \int_r f$, with r an A-definable type on Γ^n, and f an A-definable r-germ of a function into $\widehat{Q}$. Define $h(t, v) = \lim_{u \in r} \int_{c \models f(u)} h_c(t, v)$. Then $h(t, v)$ is an A'-definable homotopy. The final image of h is clearly Γ-parameterized, and has property (5) of Theorem 11.1.1; isotriviality, as well as the condition of being topologically Γ-internal, should follow.

Chapter Fourteen

Applications to the topology of Berkovich spaces

Summary. In this final chapter we deduce from our main results general tameness statements about the topology of Berkovich spaces. In Theorem 14.2.1 we prove the existence of strong retractions to skeleta for analytifications of definable subsets of quasi-projective varieties. Theorem 14.2.3 is about functoriality and birationality statements for these retractions. In Theorem 14.2.4, we show that, in the compact case, these analytifications are homeomorphic to the projective limit of embedded finite simplicial complexes, under a compactness assumption. In Theorem 14.3.1 we prove finiteness of homotopy types in families in a strong sense. We prove local contractibility in Theorem 14.4.1 and a result on homotopy equivalence of upper level sets of definable functions in Theorem 14.4.4. All these results are based on a certain surjection from the stable completion of a variety over a maximal immediate extension of the algebraic closure of a field F, to the Berkovich space of that variety over F. In the final section, we describe an injection in the opposite direction (over an algebraically closed field) which in general provides an identification between points of Berkovich analytifications and Galois orbits of stably dominated points.

14.1 BERKOVICH SPACES

Set $\mathbb{R}_\infty = \mathbb{R} \cup \{\infty\}$. Let F be a valued field with $\mathrm{val}(F) \subset \mathbb{R}_\infty$, and let $\mathbf{F} = (F, \mathbb{R})$ be viewed as a substructure of a model of ACVF (in the VF and Γ-sorts). Here $\mathbb{R} = (\mathbb{R}, +)$ is viewed as an ordered abelian group.

Let V be an algebraic variety over F, and let X be an $\mathbf{F}$-definable subset of the variety V; or more generally, of $V \times \Gamma_\infty^n$. We define the Berkovich space $B_{\mathbf{F}}(X)$ to be the space of types over $\mathbf{F}$, in X, that are almost orthogonal to Γ. Thus for any $\mathbf{F}$-definable function $f : X \to \Gamma_\infty$ and any $a \models p$, we have $f(a) \in \Gamma_\infty(\mathbf{F}) = \mathbb{R}_\infty$. So $f(a)$ does not depend on a, and we denote it by $f(p)$. We endow $B_{\mathbf{F}}(X)$ with a topology by defining a pre-basic open set to have the form $\{p \in X \cap U : \mathrm{val}(f)(p) \in W\}$, where U is an affine open subset of V, f is regular on U, and W is an open subset of $\mathbb{R}_\infty$. A basic open set is a finite intersection of pre-basic ones. This construction is functorial, thus, if $f : X \to X'$ is an $\mathbf{F}$-definable morphism between $\mathbf{F}$-definable subsets of algebraic varieties over F, one denotes by $B_{\mathbf{F}}(f) : B_{\mathbf{F}}(X) \to B_{\mathbf{F}}(X')$ the induced morphism. When we wish to consider $q \in B_{\mathbf{F}}(X)$ as a type, rather than a point, we will write it as $q|\mathbf{F}$.

When V is an algebraic variety over F, $B_{\mathbf{F}}(V)$ can be identified with the underlying topological space of the Berkovich analytification V^{an} of V.

Recall that the underlying set of V^{an} may be described as the set of pairs (x, u_x) with x a point (in the schematic sense) of V and $u_x : F(x) \to \mathbb{R}_\infty$ a valuation extending val on the residual field $F(x)$, cf. [13]. Such a pair (x, u_x) determines a rational point $c_x \in V(F(x))$ whose type p_x belongs to $B_{\mathbf{F}}(V)$. This correspondence is clearly bijective and a homeomorphism. It follows from Theorems 3.4.8 (i) and 3.5.1 (i) of [3] that V^{an} is Hausdorff, since under our conventions an algebraic variety is always assumed to be separated. When X is an $\mathbf{F}$-definable subset of V, $B_{\mathbf{F}}(X)$ is a semi-algebraic subset of $B_{\mathbf{F}}(V)$ in the sense of [12]; conversely any semi-algebraic subset has this form.

An element of $B_{\mathbf{F}}(X)$ has the form $\mathrm{tp}(a/\mathbf{F})$, where $\mathbf{F}(a)$ is an extension whose value group remains $\mathbb{R}$. To see the relation to stably dominated types, note that if there exists an $\mathbf{F}$-definable stably dominated type p with $p|\mathbf{F} = \mathrm{tp}(a/\mathbf{F})$, then p is unique; in this case the Berkovich point can be directly identified with this element of $\widehat{X}$. If there exists a stably dominated type p defined over a finite Galois extension F' of F, $\mathbf{F}' = (F', \mathbb{R})$, with $p|\mathbf{F} = \mathrm{tp}(a/\mathbf{F})$, then the Galois orbit of p is unique; in this case the relation between Berkovich points and points of $\widehat{X}$ is similar to the relation between closed points of $\mathrm{Spec}(V)$ and points of $V(F^{\mathrm{alg}})$. In general the Berkovich point of view relates to ours in rather the same way that Grothendieck's schematic points relate to Weil's points of the universal domain. We proceed to make this more explicit.

Let K be a maximally complete algebraically closed field, containing F, with value group $\mathbb{R}$, and residue field equal to the algebraic closure of the residue field of F. Such a K is unique up to isomorphism over $\mathbf{F}$ by Kaplansky's theorem, and it will be convenient to pick a copy of this field K and denote it F^{max}.

We have a restriction map from types over F^{max} to types over $\mathbf{F}$. On the other hand we have an injective restriction map from stably dominated types defined over F^{max}, to types defined over F^{max}. Composing these maps, we obtain a map from the set of stably dominated types in X defined over F^{max} to the set of types over $\mathbf{F}$ on X whose image is contained in $B_{\mathbf{F}}(X)$. Indeed, if q lies in the image of this map, then $q = \mathrm{tp}(c/\mathbf{F})$ for some c with $\mathrm{tp}(c/F^{max})$ orthogonal to Γ, and it follows that $\Gamma(\mathbf{F}(c)) \subset \Gamma(F^{max}(c)) = \Gamma(F^{max}) = \Gamma(\mathbf{F})$. This defines a continuous map

$$\pi_X : \widehat{X}(F^{max}) \to B_{\mathbf{F}}(X).$$

We shall sometimes omit the subscript when there is no ambiguity.

LEMMA 14.1.1. *Let X be an $\mathbf{F}$-definable subset of an algebraic variety over F. The mapping $\pi : \widehat{X}(F^{max}) \to B_{\mathbf{F}}(X)$ is surjective. In case $F = F^{max}$, π is a homeomorphism.*

Proof. Suppose $q = \mathrm{tp}(c/\mathbf{F})$ is almost orthogonal to Γ. Let $L = F(c)^{max}$. Then $\Gamma(\mathbf{F}) = \Gamma(\mathbf{F}(c)) = \Gamma(L)$. The field F^{max} embeds into L over $\mathbf{F}$; taking it so embedded, let $p = \mathrm{tp}(c/F^{max})$. Then p is almost orthogonal to Γ, and $q = p|\mathbf{F}$. Since F^{max} is maximally complete, p is orthogonal to Γ, cf. Theorem 2.9.2.

In case $F = F^{max}$, π is also injective since $p|F$ determines p, for a stably dominated type based on F. Thus π is a continuous bijection; since in this case the definitions of the topologies coincide on both sides, it is a homeomorphism. □

Recall 3.3, and the remarks on definable topologies there.

PROPOSITION 14.1.2. *Let X be an $\mathbf{F}$-definable subset of an algebraic variety V over F. Let $\pi : \widehat{V}(F^{max}) \to B_{\mathbf{F}}(V)$ be the natural map. Then $\pi^{-1}(B_{\mathbf{F}}(X)) = \widehat{X}(F^{max})$, and $\pi : \widehat{X}(F^{max}) \to B_{\mathbf{F}}(X)$ is a closed map. Moreover, the following conditions are equivalent:*

(1) *$\widehat{X}$ is definably compact;*

(2) *X is bounded and v+g-closed;*

(3) *$\widehat{X}(F^{max})$ is compact;*

(4) *$B_{\mathbf{F}}(X)$ is compact;*

(5) *$B_{\mathbf{F}}(X)$ is closed in $B_{\mathbf{F}}(V')$, where V' is any complete F-variety containing V.*

The natural map $B_{\mathbf{F}'}(X) \to B_{\mathbf{F}}(X)$ is also closed, if $F \leq F'$ and $\Gamma(F') \leq \mathbb{R}$. In particular, $B_{\mathbf{F}}(X)$ is closed in $B_{\mathbf{F}}(V)$ iff $B_{\mathbf{F}'}(X)$ is closed in $B_{\mathbf{F}'}(V)$.

Proof. The equality $\pi^{-1}(B_{\mathbf{F}}(X)) = \widehat{X}(F^{max})$ is clear from the definitions. Let us consider the five conditions.

The equivalence of (1) and (2) is already known by Theorem 4.2.20.

Assume (2). We wish to prove (3) over F^{max}. As X is bounded, there exists a finite affine cover $V = \cup V_i$, closed immersions $g_i : V_i \to \mathbb{A}^n$, and balls $B_i = \{x \in \mathbb{A}^n : v(x_j) \geq b_i\}$, such that $X \subset \cup_i g_i^{-1}(B_i)$. It suffices to prove (3) for $X \cap g_i^{-1}(B_i)$. Thus we may assume $X \subset B = \{x \in \mathbb{A}^n : v(x_i) \geq b\}$.

By Lemma 14.1.1, the natural map $\widehat{B}(F^{max}) \to B_{F^{max}}(B)$ is a homeomorphism. Let us first prove that this space is compact. Consider the polynomial ring $A = F^{max}[X_1, \dots, X_n]$. Each element $p \in B_{F^{max}}(B)$ determines a map $v_p : A \to \mathbb{R}_\infty$. This provides an embedding $\Phi : B_{F^{max}}(B) \to \mathrm{Fn}(A, \mathbb{R}_\infty)$, with $\mathrm{Fn}(A, \mathbb{R}_\infty)$ the space of functions from A to $\mathbb{R}_\infty$. If one endows $\mathrm{Fn}(A, \mathbb{R}_\infty)$ with the Tychonoff topology, Φ induces a homeomorphism between $B_{F^{max}}(B)$ and its image $\Phi(B_{F^{max}}(B))$. For f in A, denote by d_f the degree of f, by a_f the smallest valuation of a coefficient of f, and set $b_f = bd_f + a_f$. Since $v_p(f) \geq b_f$ for any $p \in B_{F^{max}}(B)$, $\Phi(B_{F^{max}}(B))$ is contained in $\prod_{f \in A}[b_f, \infty]$, which is compact by Tychonoff's theorem. On the other hand, $\Phi(B_{F^{max}}(B))$ is clearly closed, being the set of functions

$u : A \to \mathbb{R}_\infty$ such that $u(fg) = u(f) + u(g)$, $u(f+g) \geq \min(u(f), u(g))$, u restricts to val on F^{max}, and $u(X_i) \geq b$ for every i. It follows that $B_{F^{max}}(B)$ is compact. The definable set X, being v+g-closed in B, is a positive Boolean combination of algebraic equalities $f_i = 0$ and weak inequalities $\mathrm{val}(g_i) \leq \mathrm{val}(h_i)$ by Proposition 3.7.3. Thus $\Phi(B_{F^{max}}(X))$ is the subset of $\Phi(B_{F^{max}}(B))$ similarly defined by the conditions $u(f_i) = \infty$ and $u(g_i) \leq u(h_i)$, hence is closed. It follows that $\widehat{X}(F^{max}) = B_{F^{max}}(X)$ is compact. This gives (3).

If (3) holds, then (4) also, since $\pi(\widehat{X}(F^{max})) = B_{\mathbf{F}}(X)$. If V' is any complete F-variety containing V, the inclusion $B_{\mathbf{F}}(X) \to B_{\mathbf{F}}(V')$ is continuous, and $B_{\mathbf{F}}(V')$ is Hausdorff, so (4) implies (5).

On the other hand if (1) fails, let V' be some complete variety containing V. There exists an F^{max}-definable type on $\widehat{X}$ with limit point q in $\widehat{V'} \smallsetminus \widehat{X}$. So $\pi(q)$ is in $B_{\mathbf{F}}(V')$ and in the closure of $B_{\mathbf{F}}(X)$, but not in $B_{\mathbf{F}}(X)$. This finishes the proof of the equivalence of (1-5).

Now the restriction of a closed map π to a set of the form $\pi^{-1}(W)$ is always closed, as a map onto W. So to prove the closedness property of π, we may take $X = V$, and moreover by embedding V in a complete variety we may assume V is complete. In this case $X = V$ is v+g-closed and bounded, so $\widehat{X}(F^{max})$ is compact by condition (3). As $B_{\mathbf{F}}(X)$ is Hausdorff, π is closed. The proof that $B_{\mathbf{F}'}(X) \to B_{\mathbf{F}}(X)$ is also closed is identical, and taking $X = V$ we obtain the statement on the base invariance of the closedness of X. We could alternatively use the proof of Lemma 3.5.4. □

PROPOSITION 14.1.3. *Assume X and W are $\mathbf{F}$-definable subsets of some algebraic variety over F.*

(1) *Let $h_0 : X \to \widehat{W}$ be an $\mathbf{F}$-definable function. Then h_0 induces functorially a function $\widetilde{h} : B_{\mathbf{F}}(X) \to B_{\mathbf{F}}(W)$ such that $\pi_W \circ h_0 = \widetilde{h} \circ \pi_X \circ i$, with $i : X \to \widehat{X}$ the canonical inclusion.*

(2) *Any continuous $\mathbf{F}$-definable function $h : \widehat{X} \to \widehat{W}$ induces a continuous function $\widetilde{h} : B_{\mathbf{F}}(X) \to B_{\mathbf{F}}(W)$ such that $\pi_W \circ h = \widetilde{h} \circ \pi_X$.*

(3) *The same applies if either X or W is a definable subset of Γ_∞^n and we read $B_{\mathbf{F}}(X) = X(\mathbf{F})$, respectively $B_{\mathbf{F}}(W) = W(\mathbf{F})$.*

Proof. Define $\widetilde{h} : B_{\mathbf{F}}(X) \to B_{\mathbf{F}}(W)$ similarly as in 3.8. Namely, let $p \in B_{\mathbf{F}}(X)$. We view p as a type over $\mathbf{F}$, almost orthogonal to Γ. Say $p|\mathbf{F} = \mathrm{tp}(c/\mathbf{F})$. Let $d \models h_0(c)|\mathbf{F}(c)$. Since $h_0(c)$ is stably dominated, $\mathrm{tp}(d/\mathbf{F}(c))$ is almost orthogonal to Γ, hence so is $\mathrm{tp}(cd/\mathbf{F})$, and thus also $\mathrm{tp}(d/\mathbf{F})$. Let $\widetilde{h}(c) = \mathrm{tp}(d/\mathbf{F}) \in B_{\mathbf{F}}(W)$. Then $\widetilde{h}(c)$ depends only on $\mathrm{tp}(c/\mathbf{F})$, so we can let $\widetilde{h}(p) = \widetilde{h}(c)$.

For the second part, let $h_0 = h|X$ be the restriction of h to the simple points. It is v+g-continuous and by Lemma 3.8.5, h is the unique continuous extension of h_0. Define $\widetilde{h}$ as in (1). Let $\pi_X : \widehat{X}(F^{max}) \to B_{\mathbf{F}}(X)$ and

$\pi_W : \widehat{W}(F^{max}) \to B_{\mathbf{F}}(W)$ be the restriction maps as above. It is clear from the definition that $\widetilde{h}(\pi_X(p)) = \pi_W(h(p))$. (In case F^{max} is nontrivially valued, this is also clear from the density of simple points, since $\widetilde{h} \circ \pi_X$ and $\pi_W \circ h$ agree on the simple points of $\widehat{X}(F^{max})$.)

It remains to prove continuity. By the discussion above, π_X is a surjective and closed map. Let Z be a closed subset of $B_{\mathbf{F}}(X)$. By continuity of $\pi_W \circ h$, $\pi_X^{-1}(\widetilde{h}^{-1}(Z)) = h^{-1}(\pi_W^{-1}(Z))$ is closed, hence $\pi_X(\pi_X^{-1}(\widetilde{h}^{-1}(Z))) = \widetilde{h}^{-1}(Z)$ is closed.

(3) The proof goes through in both cases. □

If $f : X \to Y$ is an $\mathbf{F}$-definable map and b is a point in Y, we denote by X_b the fiber $f^{-1}(b)$ over b. Similarly, if q is a point of $B_{\mathbf{F}}(Y)$, $B_{\mathbf{F}}(X)_q$ denotes the fiber over q of the induced mapping $B_{\mathbf{F}}(X) \to B_{\mathbf{F}}(Y)$.

LEMMA 14.1.4. *Let X be an $\mathbf{F}$-definable subset of $V \times \Gamma_\infty^n$ with V a variety over F.*

(1) *Let $f : X \to Y$ be an $\mathbf{F}$-definable map, with Y an $\mathbf{F}$-definable subset of some variety over F. Let $q \in B_{\mathbf{F}}(Y)$, and assume U is an $\mathbf{F}$-definable subset of X, and $\widehat{U_b}$ is closed in $\widehat{X_b}$ for any $b \models q|\mathbf{F}$. Then $B_{\mathbf{F}}(U)_q$ is closed in $B_{\mathbf{F}}(X)_q$.*

(2) *Similarly if $g : X \to \mathbb{R}_\infty$ is an $\mathbf{F}$-definable function, and $\widehat{g}|\widehat{X_b}$ is continuous for any $b \models q|\mathbf{F}$, then $B_{\mathbf{F}}(g)$ induces a continuous map on $B_{\mathbf{F}}(X)_q \to \mathbb{R}_\infty$.*

(3) *More generally, if $g : X \to V'$ is an $\mathbf{F}$-definable map into some variety V', and $g|X_b$ is v+g-continuous for any $b \models q|\mathbf{F}$, then $B_{\mathbf{F}}(g)$ induces a continuous map $B_{\mathbf{F}}(X)_q : B_{\mathbf{F}}(X)_q \to B_{\mathbf{F}}(Z)$.*

Proof. Indeed if $r \in B_{\mathbf{F}}(X)_q \smallsetminus B_{\mathbf{F}}(U)_q$, let $c \models r|\mathbf{F}$, $b = f(c)$. We have $c \in X_b \smallsetminus U_b$, so there exists a definable function $\alpha_b : X_b \to \Gamma_\infty$ and an open neighborhood E_c of $\alpha_b(c)$ such that $\alpha_b^{-1}(E_b) \subset X_b \smallsetminus U_b$. By Lemma 3.5.4, α_b can be taken to be $\mathbf{F}(b)$-definable, and in fact to be a continuous function of the valuations of some F-definable regular functions, and elements of $\Gamma(\mathbf{F})$. There exists an $\mathbf{F}$-definable function α on X with $\alpha_b = \alpha|X_b$. Now α separates r from $B_{\mathbf{F}}(U)_q$ on $B_{\mathbf{F}}(X)_q$, showing that U is closed in $B_{\mathbf{F}}(X)_q$.

The statement on continuity (2) follows immediately: if Z is a closed subset of Γ_∞, then $g^{-1}(Z) \cap \widehat{X_b}$ is closed in each $\widehat{X_b}$, hence $g^{-1}(Z) \cap B_{\mathbf{F}}(U)_q$ is closed.

The more general statement (3) follows since to show that a map into $B_{\mathbf{F}}(Z)$ is continuous, it suffices to show that the composition with $B_{\mathbf{F}}(s)$ is continuous for any definable, continuous $s : Z' \to \Gamma_\infty$, with Z' Zariski open in Z. □

The following lemma will be applied when W is also over Y and $h : X \to \widehat{W/Y}$; but a referee pointed out that the more general statement is also valid, and simpler.

LEMMA 14.1.5. *Let X, Y and W be $\mathbf{F}$-definable subsets of some algebraic variety over F. Let $f : X \to Y$ be a v+g-continuous, $\mathbf{F}$-definable map, and $h : X \to \widehat{W}$ an $\mathbf{F}$-definable map inducing $H : \widehat{X/Y} \to \widehat{W}$. Assume $H|\widehat{X_b}$ is continuous for every $b \in Y$. Then for any $q \in B_{\mathbf{F}}(Y)$, h induces a continuous function $\widetilde{h}_q : B_{\mathbf{F}}(X)_q \to B_{\mathbf{F}}(W)$.*

Proof. The topology on $B_{\mathbf{F}}(W)$ is the coarsest one such that $B_{\mathbf{F}}(g)$ is continuous for any v+g-continuous definable $g : W \to \Gamma_\infty$. Composing with $B_{\mathbf{F}}(g)$, we see that we may assume $W = \Gamma_\infty$. We have $h : X \to \Gamma_\infty$, inducing $H : \widehat{X/Y} \to \Gamma_\infty$, and we assume $H|\widehat{X_b}$ is continuous for $b \in Y$. We have to show that a continuous $\widetilde{h}_q : B_{\mathbf{F}}(X)_q \to \Gamma_\infty$ is induced.

In case the map $\widehat{X} \to \Gamma_\infty$ induced from h is continuous, by Proposition 14.1.3 $\widetilde{h}$ is continuous, and hence the restriction to each fiber $B_{\mathbf{F}}(X)_q$ is continuous.

In general, let X' be the graph of h. The projection $X' \to \Gamma_\infty$ being v+g-continuous, a natural, continuous function $B_{\mathbf{F}}(X')_q \to \mathbb{R}_\infty$ is induced, by the above special case. It remains to prove that the projection map $B_{\mathbf{F}}(X')_q \to B_{\mathbf{F}}(X)_q$ is a homeomorphism with inverse induced by $x \mapsto (x, h(x))$. When $q = b \in Y$ is a simple point, this follows from the continuity of $H|\widehat{X_b}$. Hence by Lemma 14.1.4, it is true in general. □

In the Berkovich category, as in 3.9 and throughout the paper, by deformation retraction we mean a strong deformation retraction. We continue to write $\pi : \widehat{V}(F^{max}) \to B_{\mathbf{F}}(V)$ for the natural map, defined above Lemma 14.1.1.

COROLLARY 14.1.6.

(1) *Let X be an $\mathbf{F}$-definable subset of some algebraic variety over F. Let $h : I \times \widehat{X} \to \widehat{X}$ be an $\mathbf{F}$-definable deformation retraction, with image $h(e_I, \widehat{X}) = Z$. Let $\mathbf{I} = I(\mathbb{R}_\infty)$ and $\mathbf{Z} = \pi(Z(F^{max}))$. Then h induces a deformation retraction $\widetilde{h} : \mathbf{I} \times B_{\mathbf{F}}(X) \to B_{\mathbf{F}}(X)$ with image $\mathbf{Z}$.*

(2) *Let $X \to Y$ be an $\mathbf{F}$-definable morphism between $\mathbf{F}$-definable subsets of some algebraic variety over F. Let $h : I \times \widehat{X/Y} \to \widehat{X/Y}$ be an $\mathbf{F}$-definable deformation retraction satisfying $(*)$, with fibers h_y having image Z_y. Let $q \in B_{\mathbf{F}}(Y)$. Then h induces a deformation retraction $\widetilde{h}_q : \mathbf{I} \times B_{\mathbf{F}}(X)_q \to B_{\mathbf{F}}(X)_q$, with image $\mathbf{Z}_q$.*

(3) *Assume in addition there exists a definable $\Upsilon \subset \Gamma_\infty^n$ and definable homeomorphisms $\alpha_y : Z_y \to \Upsilon$, given uniformly in y. Then $\mathbf{Z}_q \cong \Upsilon$. More generally if $\Upsilon \subset \Gamma_\infty^w$ with w a finite, Galois invariant subset of*

a finite field extension F' of F, $\alpha_y : Z_y \to \Upsilon$, then $\mathbf{Z}_q \cong \Upsilon/G$, where $G = \mathrm{Gal}(F'/F)$ is acting naturally on w.

Proof. (1) follows from Proposition 14.1.3; the statement on the image is easy to verify. (2) follows similarly from Lemma 14.1.5. For (3), define $\beta : X \to \Upsilon$ by $\beta(x) = \alpha_y(h(e_I, x))$ for $x \in X_y$, e_I being the final point of I. Then $\alpha_y^{-1} \circ \beta(x) = h(e_I, x)$, $\beta(h(t,x)) = \beta(x)$, $\beta(\alpha_y^{-1}(x)) = x$. Applying $B_{\mathbf{F}}$ and restricting to the fiber over q we obtain continuous maps β, α_y^{-1} by Lemma 14.1.4; the identities survive, and give the result. □

14.2 RETRACTIONS TO SKELETA

Let V be an algebraic variety over a valued field F with $\mathrm{val}(F) \subset \mathbb{R}_\infty$ and let S be an $\mathbf{F}$-iso-definable Γ-internal subset of $\widehat{V}$. According to Theorem 6.2.9, there exists an $\mathbf{F}$-definable embedding $S \to \Gamma_\infty^w$, where w is a finite set. Let F' be a finite Galois extension of F, such that $\mathrm{Aut}(F^{\mathrm{alg}}/F')$ fixes each point of w. We shall say S *splits* over F'. Then there exists an $\mathbf{F}'$-definable embedding $S \to \Gamma_\infty^n$, $n = |w|$. It follows that $S(\mathbf{F}'') = S(\mathbf{F}')$ whenever $F'' \geq F'$ is a valued field extension with $\Gamma(F'') \subset \mathbb{R}$. The image $S_{\mathbf{F}}$ of S in $B_{\mathbf{F}}(X)$ is thus homeomorphic to $S(\mathbf{F}')/\mathrm{Gal}(F'/F)$. The image $S_{\mathbf{F}''}$ of S in $B_{\mathbf{F}''}(X)$ is homeomorphic to $S(\mathbf{F}')$. Note that the canonical map $\widehat{V}(F^{max}) \to B_{\mathbf{F}'}(V)$ restricts to an injective map on S, since $S(F^{max}) \subset S(\mathbf{F}')$.

For our purposes, a *$\mathbb{Q}$-tropical structure* on a topological space X is a homeomorphism of X with a subspace $\mathbf{S}$ of $[0,\infty]^n$ defined as a finite Boolean combination of equalities or inequalities between terms $\sum \alpha_i x_i + c$ with $\alpha_i \in \mathbb{Q}, \alpha_i \geq 0, c \in \mathbb{R}$. Since $\mathbf{S}$ is definable in $(\mathbb{R}, +, \cdot)$, X is homeomorphic to a finite simplicial complex. Recall that a valued field extension L of a valued field F is called an *Abhyankar extension* if the transcendence degree of L/F is equal to the sum of the transcendence degree of the residue field extension and the $\mathbb{Q}$-rank of $\Gamma(L)/\Gamma(F)$.

From Theorem 11.1.1 and Corollary 14.1.6 we obtain:

THEOREM 14.2.1. *Let X be an $\mathbf{F}$-definable subset of a quasi-projective algebraic variety V over a valued field F with $\mathrm{val}(F) \subset \mathbb{R}_\infty$. There exists a (strong) deformation retraction $H : \mathbf{I} \times B_{\mathbf{F}}(X) \to B_{\mathbf{F}}(X)$, whose image $\mathbf{Z}$ is of the form $S_{\mathbf{F}}$ with S an $\mathbf{F}$ iso-definable Γ-internal subset of $\widehat{V}$. Thus, $\mathbf{Z}$ has a $\mathbb{Q}$-tropical structure, in particular it is homeomorphic to a finite simplicial complex. Furthermore each point q of $\mathbf{Z}$, as a type over $\mathbf{F}$, extends to a unique stably dominated type p and this type is strongly stably dominated. Restricted to F, q determines an Abhyankar extension of the valued field F.*

Proof. Let S be the final image provided by Theorem 11.1.1 assuming (5) holds. Thus S consists of strongly stably dominated types and we have an $\mathbf{F}$-definable homeomorphism $h : W \to S$, where W is a subset of Γ_∞^w, with w a finite $\mathbf{F}$-definable set. So for $a \in W(\mathbf{F}) = W(\mathbb{R})$, $p = h(a)$ is strongly stably dominated over $\mathbf{F}$, and extends the restriction to $\mathbf{F}$, which is the image in $\mathbf{Z}$ of $h(a)$. For the last point, p is defined over $F \cup A$ where A is a finitely generated $\mathbb{Q}$-subspace of $\mathbb{R}$. Let F' be an Abhyankar extension of F, with value group equal to $\mathrm{val}(F) + A$. Then $F'(p)$ is Abhyankar over F', and hence over F. □

EXAMPLE 14.2.2. Let us revisit the elliptic curve example of Example 7.5.2 in the Berkovich setting. Assume for instance $F = \mathbb{Q}_3$ and set $\lambda = 3$. So C_3 is the projective model of the curve $y^2 = x(x-1)(x-3)$. We have seen in Example 7.5.2 that its skeleton K' in $\widehat{C_3}$ is a combinatorial circle. This circle admits a $\mathbb{Q}_3$-definable embedding in $\Gamma^{\{i,-i\}}$, it splits over $\mathbb{Q}_3(i)$ and conjugation acts on it by exchanging the points in the fibers of $K' \to K$. Thus, for $F = \mathbb{Q}_3(i)$, $B_{\mathbf{F}}(C_3)$ has the homotopy type of a circle, while for $F' = \mathbb{Q}_3$, $B_{\mathbf{F}'}(C_3)$ retracts to a segment, thus is contractible.

We now state some functorial properties of the deformation retraction above. Like Theorem 14.2.1, these were proved by Berkovich assuming the base field F is nontrivially valued, and that U and V can be embedded in proper varieties which admit a pluri-stable model over the ring of integers of F. We thank Vladimir Berkovich for suggesting these statements to us.

Whenever we write $B_{\mathbf{F}}(V)$, we assume the valuation on F is real valued, allowing the case that the valuation is trivial. If $\mathbf{F}'$ is an extension of $\mathbf{F}$, we write $B_{\mathbf{F}'}(U)$ for $B_{\mathbf{F}'}(U \otimes F')$.

THEOREM 14.2.3. *Let U and V be quasi-projective algebraic varieties over a valued field F with value group contained in $\mathbb{R}$. Let X and Y be $\mathbf{F}$-definable subsets of U and V, respectively.*

(1) *There exists a finite separable extension F' of F such that, for any non-archimedean field F'' over F', the canonical map $B_{\mathbf{F}''}(X) \to B_{\mathbf{F}'}(X)$ is a homotopy equivalence. In fact, there exists a deformation retraction of $B_{\mathbf{F}'}(X)$ to $\mathbf{Z}'$ as in* Theorem 14.2.1 *that lifts to a deformation retraction of $B_{\mathbf{F}''}(X)$ to $\mathbf{Z}''$, for which the canonical map $\mathbf{Z}'' \to \mathbf{Z}'$ is a homeomorphism.*

(2) *There exists a finite separable extension F' of F such that, for any non-archimedean field extension F'' of F', the canonical map $B_{\mathbf{F}''}(X \times Y) \to B_{\mathbf{F}''}(X) \times B_{\mathbf{F}''}(Y)$ is a homotopy equivalence.*

(3) *Let U be smooth and U' be a dense open subset of U. Then the canonical embedding $B_{\mathbf{F}}(U') \to B_{\mathbf{F}}(U)$ is a homotopy equivalence.*

Proof. Let us prove (1). The homotopy of Theorem 11.1.1 is F-definable, and so functorial on F''-points for any $F'' \geq F$. Denote by S its final image. Choose a finite Galois extension F' of F that splits S. For any $F \leq F' \leq F''$, the homotopy of $B_{\mathbf{F}''}(X)$ is compatible with the homotopy of $B_{\mathbf{F}'}(X)$ via the natural map $B_{\mathbf{F}''}(X) \to B_{\mathbf{F}'}(X)$ (restriction of types). The final image of the homotopies is respectively $S_{F''}$ and $S_{F'}$; we noted that these are homeomorphic images of S as soon as F' splits S and hence homeomorphic via the natural map.

(2) follows similarly from Corollary 9.8.6 (which was devised precisely with the present motivation) and its proof. Indeed, as in the proof of Corollary 9.8.6, let us consider definable deformations retractions for X and Y with final images S and T. Recall the homotopy equivalence $\widehat{X \times Y} \to \widehat{X} \times \widehat{Y}$ in Corollary 9.8.6 was part of a commutative diagram

$$\begin{array}{ccc} \widehat{X \times Y} & \longrightarrow & S \otimes T \\ {\scriptstyle \pi_X \times \pi_Y}\downarrow & & \downarrow{\scriptstyle \pi_S \times \pi_T} \\ \widehat{X} \times \widehat{Y} & \longrightarrow & S \times T, \end{array}$$

whose horizontal morphisms are definable retractions and that $\pi_S \times \pi_T$ was proven to be a homeomorphism. Choose a finite Galois extension F' which splits both S and T (in fact it would be enough to require F' to split one of S and T). It is then clear that for any $F'' \geq F'$, the homotopy equivalence $\widehat{X \times Y} \to \widehat{X} \times \widehat{Y}$ induces a homotopy equivalence $B_{\mathbf{F}''}(X \times Y) \to B_{\mathbf{F}''}(X) \times B_{\mathbf{F}''}(Y)$.

(3) follows directly from Remark 12.2.4. □

The following result was previously known when X is a smooth projective curve [3].

THEOREM 14.2.4. *Let X be an $\mathbf{F}$-definable subset of a quasi-projective algebraic variety V over a valued field F with $\mathrm{val}(F) \subset \mathbb{R}_\infty$ and assume $B_{\mathbf{F}}(X)$ is compact. Then there exists a family $(X_i)_{i \in I}$ of finite simplicial complexes of dimension $\leq \dim V$, embedded in $B_{\mathbf{F}}(X)$, where I is a directed partially ordered set, such that X_i is a subcomplex of X_j for $i < j$, with deformation retractions $\pi_{i,j} : X_j \to X_i$ for $i < j$, and deformation retractions $\pi_i : B_{\mathbf{F}}(X) \to X_i$ for $i \in I$, satisfying $\pi_{i,j} \circ \pi_j = \pi_i$ for $i < j$, such that the canonical map from $B_{\mathbf{F}}(X)$ to the projective limit of the spaces X_i is a homeomorphism.*

Proof. Let the index set I consist of all $\mathbf{F}$-definable continuous maps $j : \widehat{X} \to \widehat{X}$, such that there exists an $\mathbf{F}$-definable deformation retraction $H_j : I \times \widehat{V} \to \widehat{V}$ as in Theorem 11.1.1, restricting to a deformation retraction $H_j^X : I \times \widehat{X} \to \widehat{X}$ such that $j(x) = H_j^X(e_I, x)$. Here we insist that H_j satisfies clause (7) of Theorem 11.1.1 for the irreducible components of V. Let us denote by T_j the final image of H_j and by S_j that of H_j^X. Thus $S_j = j(\widehat{X}) = \widehat{X} \cap T_j$. Let X_j denote the image of $S_j(F^{max})$ in $B_{\mathbf{F}}(X)$. Thus X_j is homeomorphic to $S_j(\mathrm{acl}(\mathbf{F}))/\mathrm{Gal}(F^{\mathrm{alg}}/F)$. Say that $j_1 \leq j_2$

if $S_{j_1} \subset S_{j_2}$. In this case, $j_1|S_{j_2} : S_{j_2} \to S_{j_1}$ is a deformation retraction through the homotopy $j_2 \circ H_{j_1}(t, \cdot)$.

Let π_{j_1,j_2} be the induced map $X_{j_2} \to X_{j_1}$. It is a deformation retraction. Let us prove the system is directed, i.e. given j_1 and j_2 there exists j_3 with $j_1, j_2 \leq j_3$. To see this, for $j = j_1, j_2$, let $\alpha_j : T_j \to \Gamma_\infty^{w_j}$ be a definable injective map, with w_j a finite F-definable set, and let j_3 belong to a homotopy H_{j_3} respecting the functions $x \mapsto \alpha_{j_1}(H_{j_1}(e_I, x))$, $x \mapsto \alpha_{j_2}(H_{j_2}(e_I, x))$ and preserving the irreducible components of V. Then by Proposition 8.3.1 (2), since H_{j_3} satisfies clause (7) of Theorem 11.1.1 for the irreducible components of V, H_{j_3} fixes T_{j_1} and T_{j_2} pointwise, thus $H_{j_3}^X$ fixes S_{j_1} and S_{j_2} pointwise and the image of j_3 includes them both.

We have a natural surjective map $\pi_j : B_{\mathbf{F}}(X) \to X_j$ for each j, induced by the mapping j; it satisfies $\pi_{i,j} \circ \pi_j = \pi_i$ for $i < j$ and it is a deformation retraction.

This yields a continuous map from $\theta : B_{\mathbf{F}}(X) \to \varprojlim_j X_j$. The image is dense since each π_j is surjective; as $B_{\mathbf{F}}(X)$ is compact the image is closed, so θ is surjective. We now show that θ is injective. Let $p \neq q \in B_{\mathbf{F}}(X)$; view them as types almost orthogonal to Γ. For any open affine U and regular f on U, for some α, either $x \notin U$ is in p or $\mathrm{val}(f) = \alpha$ is in p; this is because p is almost orthogonal to Γ. Thus as $p \neq q$, for some open affine U and some regular f on U, either $p \in U$ and $q \notin U$, or vice versa, or $p, q \in U$ and for some regular f on U, $f(x) = \alpha \in p$, $f(x) = \beta \in q$, with $\alpha \neq \beta$. Let H be as in Theorem 11.1.1 respecting U and $\mathrm{val}(f)$, and let j be a corresponding retraction. Then clearly $\pi_j(p) \neq \pi_j(q)$. Thus, θ is a continuous bijection and by compactness it is a homeomorphism. □

REMARK 14.2.5. Let Σ be (image of) the direct limit of the X_i in $B_{\mathbf{F}}(X)$. Note that Σ contains all rigid points of $B_{\mathbf{F}}(X)$ (that is, images of simple points under the mapping π in Lemma 14.1.1): this follows from Theorem 11.1.1, by finding a homotopy to a skeleton S_x fixing a given simple point x of $\widehat{X}$. We are not certain whether Σ can be taken to be the whole of $B_{\mathbf{F}}(X)$. But given a stably dominated type p on X, letting $S_p = S_x$ for $x \models p$ and averaging the homotopies with image S_x over $x \models p$, we obtain a definable homotopy whose final image is a continuous, definable image of S_p. In this way we can express $B_{\mathbf{F}}(X)$ as a direct limit of a system of finite simplicial complexes, with continuous transition maps.

14.3 FINITELY MANY HOMOTOPY TYPES

We will now prove that a uniform family of Berkovich spaces runs through only finitely many homotopy types.

In the definable setting, for stable completions, the situation is different. Consider a family of triangles in Γ^2; they may be the skeleta of elliptic curves, and so homotopy equivalent to them. Two triangles are definably homotopy equivalent iff they are definable homeomorphic. But there are many definable homeomorphism types of triangles, or even of segments; indeed $[0, \alpha]$ and $[0, \beta]$ are definably homeomorphic iff β is a rational multiple of α.

On the other hand, if we expand Γ to be a model of the theory RCF of real closed fields, then it is known that only finitely many homeomorphism types appear in a given definable family. Using the uniform version of Theorem 11.1.1, this extends to uniformly definable families of stable completions.

For applications to Berkovich spaces in terms of the usual topological homotopy type, or even homeomorphism type of skeleta, the expansion to RCF is harmless. In the setting of stable completions, we explain in Remark 14.3.4 how it can be avoided.

Part (1) of the following theorem is a special case of part (2); we single it out as we will prove it first. We consider a uniformly definable family of definable subset of $\mathbb{P}^m$.

THEOREM 14.3.1. *Let V be a variety defined over a valued field F. Let Y be an $\mathbf{F}$-definable subset of $V \times \Gamma^r$, for some r, and let X be an $\mathbf{F}$-definable subset of $Y \times \mathbb{P}^m$ for some m. Denote by $f : X \to Y$ the projection on the first factor.*

(1) *For $b \in Y$, let $X_b = f^{-1}(b)$. Then there are finitely many possibilities for the homotopy type of $B_{\mathbf{F}(b)}(X_b)$, as b runs through Y. More generally, let $U_1 \subset \ldots \subset U_\ell = X$ be a chain of $\mathbf{F}$-definable sets. Then as b runs through Y there are finitely many possibilities for the homotopy type of the tuple $(B_{\mathbf{F}(b)}(X_b \cap U_i))$.*

(2) *For any valued field extension $F \leq F'$ with $\Gamma(F') \leq \mathbb{R}$ and $q \in B_{\mathbf{F}'}(Y)$, let $B_{\mathbf{F}'}(X)_q$ denote the fiber over q of the canonical map $B_{\mathbf{F}'}(X) \to B_{\mathbf{F}'}(Y)$. Then there are only finitely many possibilities for the homotopy type of $B_{\mathbf{F}'}(X)_q$ as q runs over $B_{\mathbf{F}'}(Y)$ and F' over extensions of F. More generally, let $U_1 \subset \ldots \subset U_\ell = X$ be a chain of $\mathbf{F}$-definable sets. Then as q runs over $B_{\mathbf{F}'}(Y)$ and F' over extensions of F there are finitely many possibilities for the homotopy type of the tuple $(B_{\mathbf{F}'(b)}(X_b \cap U_i))$.*

Proof. Let us start by proving the first statement in (1) under the assumption that for any $b \in Y$, X_b is Zariski closed in $\mathbb{P}^m$.

According to the uniform version of Theorem 11.1.1, Proposition 11.7.1, there exists an $\mathbf{F}$-definable map $W \to Y$ with finite fibers $W(b)$ over $b \in Y$, and uniformly in $b \in Y$ an $\mathbf{F}(b)$-definable homotopy retraction h_b on X_b preserving the given data, with final image Z_b, and an $\mathbf{F}(b)$-definable homeomorphism $\phi_b : Z_b \to S_b \subset \Gamma_\infty^{W(b)}$.

CLAIM. *We may find, definably uniformly in b, an $\mathbf{F}(b)$-definable subset $T_b \subset \Gamma_\infty^n$, a finite $\mathbf{F}(b)$-definable set $W_!(b)$, and for $w \in W_!(b)$, a definable* *homeomorphism* $\psi_w : Z_b \to T_b$, *such that* $H_b = \{\psi_{w'}^{-1} \circ \psi_w : w, w' \in W_!(b)\}$ *is a group of homeomorphisms of* Z_b, *and* $H'_b = \{\psi_{w'} \circ \psi_w^{-1} : w, w' \in W_!(b)\}$ *is a group of homeomorphisms of* T_b.

Proof of the claim. In fact for a fixed b, one can pick some $W(b)$-definable homeomorphism ψ_b of Z_b onto a definable subspace of Γ_∞^n; let $\Xi_b = \{\psi_w : w \in W_!(b)\}$ be the set of automorphic conjugates of ψ_b over $\mathbf{F}(b)$; and verify that H_b is a finite group, Ξ_b is a principal torsor for H_b, and so H'_b is also a finite group (isomorphic to H_b). Thus, for a fixed b, one can do the construction as stated, obtaining the stated properties. Now the fact that the ψ_w are conjugates of ψ_b is not an ind-definable property of b. But the consequences mentioned in the claim—that ψ_w is a definable homeomorphism, and the compositional properties—are clearly ind-definable, and in fact definable, properties of b. Hence by the compactness and gluing argument we may find $W_!(b)$ and Ξ_b uniformly in b, with the required properties. In particular, there exists an $\mathbf{F}$-definable map $W_! \to Y$ with fibers $W_!(b)$ over $b \in Y$. □

By stable embeddedness of Γ, and elimination of imaginaries in Γ, we may write $T_b = T_{\rho(b)}$ where $\rho : Y \to \Gamma^m$ is a definable function. Let Γ^* be an expansion of Γ to RCF. Then by Remark 14.3.2, T_b runs through finitely many Γ^*-definable homeomorphism types as b runs through Y. Similarly, the pair (T_b, H'_b) runs through finitely many Γ^*-definable equivariant homeomorphism types (e.g. we may find an H'_b-invariant cell decomposition of T_b and describe the action combinatorially). In particular, for $b \in Y$, $(T_b(\mathbb{R}), H'_b)$ runs through finitely many homeomorphism types (i.e. isomorphism types of pairs (U, H) with U a topological space, H a finite group acting on U by auto-homeomorphisms).

By Corollary 14.1.6 we have, for $b \in Y$, a deformation retraction of $B_{\mathbf{F}(b)}(X_b)$ to $B_{\mathbf{F}(b)}(Z_b)$. Pick $w \in W_!(b)$, and let $W^*(b)$ be the set of realizations of $\mathrm{tp}(w/\mathbf{F}(b))$. If $w, w' \in W^*(b)$ then $w' = \sigma(w)$ for some automorphism σ fixing $\mathbf{F}(b)$; we may take it to fix Γ too. It follows that $\psi_w^{-1} \circ \psi_{w'} = \sigma | Z_b$. Conversely, if σ is any automorphism of $W_!(b)$, it may be extended by the identity on Γ, and it follows that $\psi_{\sigma(w)} = \psi_w \circ \sigma$; so $W^*(b)$ is a torsor of $H^*(b) = \{\psi_w^{-1} \circ \psi_{w'} : w, w' \in W^*(b)\}$, which is a group. Let $H_*(b) = \{\psi_w \circ \psi_{w'}^{-1} : w, w' \in W^*(b)\}$. It follows that $H_*(b)$ is a group, and for any $w \in W^*(b)$, ψ_w induces a bijection $Z_b/H^*(b) \to T_b/H_*(b)$; moreover it is the same bijection, i.e. it does not depend on the choice of $w \in W^*(b)$.

We are interested in the case $\Gamma(\mathbf{F}(b)) = \Gamma(\mathbf{F}) = \mathbb{R}$. In this case, since $H^*(b)$ acts by automorphisms over $\mathbf{F}(b)$, two $H^*(b)$-conjugate elements of Z_b have the same image in $B_{\mathbf{F}(b)}(X_b)$. On the other hand two non-conjugate elements have distinct images in $T_b/H_*(b)$, and so cannot have the same image in $B_{\mathbf{F}(b)}(X_b)$. It follows that $B_{\mathbf{F}(b)}(Z_b)$, $Z_b(\mathbf{F}(b))/H^*(b)$ and $T_b(\mathbb{R})/H_*(b)$ are canonically isomorphic. By compactness and definable compactness considerations one deduces that these isomorphisms between

$B_{\mathbf{F}(b)}(Z_b)$, $Z_b(\mathbf{F}(b))/H^*(b)$ and $T_b(\mathbb{R})/H_*(b)$ are in fact homeomorphisms. It is only for this reason that we required X_b to be Zariski closed in the beginning of the proof.

The number of possibilities for $H_*(b)$ is finite and bounded, since H'_b is a group of finite size, bounded independently of b, and $H_*(b)$ is a subgroup of H'_b. Since the number of equivariant homeomorphism types of $(T_b(\mathbb{R}), H'_b)$ is bounded, we are done with the first statement in (1).

With the help of Corollary 14.1.6, this proof goes through for non-simple Berkovich points too. Let $q \in B_{\mathbf{F}}(Y)$, and view it as a type over $\mathbf{F}$. By Corollary 14.1.6 (2), $B_{\mathbf{F}}(X)_q$ has the homotopy type of $\mathbf{Z}_q$. Let $b \models q$, pick $w \in W_!(b)$ and let notation be as above. Let $b' = (b, w)$ and let $q' = \mathrm{tp}(b, w/\mathbf{F})$. Let $X' = X \times_Y W_!$. By Corollary 14.1.6 (2) applied to the pullback of the retraction $I \times \widehat{X/Y} \to \widehat{X/Y}$ to $\widehat{X'/W_!}$, $B_{\mathbf{F}}(X')_{q'}$ retracts to a space $\mathbf{Z}_{q'}$ which is homeomorphic to $T_b(\mathbb{R})$. By the same reasoning as above, it follows that $\mathbf{Z}_q$ is homeomorphic to $\mathbf{Z}_{q'}$ *modulo* a certain subgroup $H^*(b)$ of H_b, and also homeomorphic to T_b modulo $H_*(b)$ for a certain subgroup of H'_b, so again the number of possibilities is bounded. This holds uniformly when F is replaced by any valued field extension, and the first statement in (2) follows.

The proof goes through directly to provide the generalization to chains. In particular we can now remove the hypothesis that X_b is Zariski closed in $\mathbb{P}^m$, after replacing $U_1 \subset \ldots \subset U_\ell = X$ by $U_1 \subset \ldots \subset U_m \subset U_{\ell+1} = Y \times \mathbb{P}^m$. □

REMARK 14.3.2. In the expansion of Γ to a real closed field, definable subsets of Γ_∞^n are locally contractible and definably compact subsets of Γ_∞^n admit a definable triangulation, compatible with any given definable partition into finitely many subsets. By taking the closure in case the sets are not compact, it follows that given a definable family of semi-algebraic subsets of $\mathbb{R}_\infty^n$, there exist a finite number of rational polytopes (with some faces missing), such that each member of the family is homeomorphic to at least one such polytope. In particular the number of definable homotopy types is finite. In fact it is known that the number of definable homeomorphism types is finite. See [9], [11].

REMARK 14.3.3. Eleftheriou has shown [15] that there exist abelian groups interpretable in $\mathrm{Th}(\mathbb{Q}, +, <)$ that cannot be definably and homeomorphically embedded in affine space within DOAG. By Theorem 6.2.8, the skeleta of abelian varieties can be so embedded. It would be good to bring out the additional structure they have that ensures this embedding.

REMARK 14.3.4. Let us explain how to avoid the use of the expansion to RCF in the setting of stable completions. It is shown in the thesis of Eleftheriou, see also [16] p. 1115, that a definable subset X of Γ^n may be partitioned into finitely many linear cells. This decomposition is defined by some formulas $\phi_j(x, \alpha_X)$ requiring some parameters α_X (the coefficients in the linear equations are from $\mathbb{Q}$; the parameters refer to the inhomogeneous part of the equations). One can easily determine a 0-definable set P such that $\alpha_X \in P$, and such that for $\alpha \in P$, the formulas $\phi_j(x, \alpha)$ determine a cell complex with the same adjacencies. Choosing $\beta \in P(\mathbb{R}, +, <)$, and letting X_β be defined by the formulas $\phi_j(x, \beta)$, we obtain a topological space whose homeomorphism type clearly does not depend on the choice of β. By refining one can see that it also does not depend on the choice of the formulas ϕ_j (though strictly speaking, that is not needed for our finiteness statements). We call this homeomorphism type the *combinatorial homeomorphism type* of X. For instance, all triangles have the same combinatorial homeomorphism type, though as explained above they have distinct definable homotopy type. Now any definable family of definable subsets of Γ^n runs through a finite number of combinatorial homeomorphism types. It follows that for any definable family of quasi-projective varieties, there exists a finite set Ω of combinatorial homeomorphism types such that the stable completion of any variety in the family admits a skeleton with combinatorial homeomorphism type in Ω. Similar considerations and finiteness statements apply to Γ^n_∞, and filtered definable spaces $(X, X_1, \ldots, X_n)$ where for $X \subset \Gamma^n_\infty$, we let X_k be the subset of points exactly k of whose coordinates are ∞.

14.4 MORE TAME TOPOLOGICAL PROPERTIES

THEOREM 14.4.1 (Local contractibility). *Let X be an $\mathbf{F}$-definable subset of an algebraic variety V over a valued field F with* $\mathrm{val}(F) \subset \mathbb{R}_\infty$. *The space $B_{\mathbf{F}}(X)$ is locally contractible.*

Proof. We may assume V is affine. Since the topology of $B_{\mathbf{F}}(X)$ is generated by open subsets of the form $B_{\mathbf{F}}(X')$ with X' definable in X, it is enough to prove that every point x of $B_{\mathbf{F}}(X)$ admits a contractible neighborhood. By Theorem 11.1.1 and Corollary 14.1.6, there exists a strong deformation retraction $H : I \times B_{\mathbf{F}}(X) \to B_{\mathbf{F}}(X)$ with image a subset Υ which is homeomorphic to a semi-algebraic subset of some $\mathbb{R}^n$. Denote by ϱ the retraction $B_{\mathbf{F}}(X) \to \Upsilon$. By (4) in Theorem 11.1.1 one may assume that $\varrho(H(t,x)) = \varrho(x)$ for every t and x. Recall that any semi-algebraic subset Z of $\mathbb{R}^n$ is locally contractible: one may assume Z is bounded, then its closure $\overline{Z}$ is compact and semi-algebraic and the statement follows from the existence of triangulations of $\overline{Z}$ compatible with the inclusion $Z \hookrightarrow \overline{Z}$ and having any given point of Z as vertex. It is thus possible to pick a contractible neighborhood U of $\varrho(x)$ in Υ. Since the set $\varrho^{-1}(U)$ is invariant by the homotopy H, it retracts to U, hence is contractible. □

REMARK 14.4.2. As noted in Remark 14.6.6, if x is an Abhyankar point in the sense of Definition 14.6.4, it follows from Theorem 8.4.2 and Proposition 14.6.5, together with the proof of Theorem 14.4.1, that x admits a basis of neighborhoods that strongly retracts to x.

REMARK 14.4.3. Berkovich proved in [5] and [6] local contractibility of smooth non-archimedean analytic spaces, and raised the question of the singular case. His proof uses de Jong's results on alterations.

Let us give another application of our results, in the spirit of results of Abbes and Saito [1] 5.1 and Poineau [35] Théorème 2.

THEOREM 14.4.4. *Let X be an $\mathbf{F}$-definable subset of a quasi-projective algebraic variety over a valued field F with $\mathrm{val}(F) \subset \mathbb{R}_\infty$ and let $G : X \to \Gamma_\infty$ be an $\mathbf{F}$-definable map. Consider the corresponding map $\mathbf{G} : B_{\mathbf{F}}(X) \to \mathbb{R}_\infty$. Then there is a finite partition of $\mathbb{R}_\infty$ into intervals such that the fibers of $\mathbf{G}$ over each interval have the same homotopy type. Also, if one sets $B_{\mathbf{F}}(X)_{\leq \varepsilon}$ to be the preimage of $(-\infty, \varepsilon]$, there exists a finite partition of $\mathbb{R}_\infty$ into intervals such that for each interval I the inclusion $B_{\mathbf{F}}(X)_{\leq \varepsilon} \to B_{\mathbf{F}}(X)_{\leq \varepsilon'}$, for $\varepsilon < \varepsilon'$ both in I, is a homotopy equivalence.*

Proof. Consider a strong deformation retraction of $\widehat{X}$ leaving the fibers of G invariant, as provided by Theorem 11.1.1. By Corollary 14.1.6 it induces a retraction ϱ of $B_{\mathbf{F}}(X)$ onto a subset Υ such that there exists a homeomorphism $h : \Upsilon \to S$ with S a semi-algebraic subset of some $\mathbb{R}^n$. By construction $\mathbf{G}$ factors as $\mathbf{G} = g \circ \varrho$ with g a function $S \to \mathbb{R}_\infty$. Furthermore, we may assume that $g' := h^{-1} \circ g$ is a semi-algebraic function S. Thus, it is enough to prove that there is a finite partition of $\mathbb{R}_\infty$ into intervals such that the fibers of g' over each interval have the same homotopy type and that if $S_{\leq \varepsilon}$ is the locus of $g' \leq \varepsilon$, there exists a finite partition of $\mathbb{R}_\infty$ into intervals such that for each interval I the inclusion $S_{\leq \varepsilon} \to S_{\leq \varepsilon'}$, for $\varepsilon < \varepsilon'$ both in I, is a homotopy equivalence. But such statements are well-known in o-minimal geometry, cf., e.g., [9] Theorem 5.22. □

14.5 THE LATTICE COMPLETION

The previous constructions depended on a canonical map from the stable completion to the Berkovich space. In this section we will introduce a different and more direct connection between the Berkovich space and the stable completion. Our construction involves a preliminary base change to a canonical *completion* of the given base, one involving imaginary elements from the sort S_n, as well as the field sort and $S_1 = \Gamma$.

Let F be a valued field. The usual completion of F as a valued field can be viewed as a subfield of a maximal immediate extension F^{max} of F^{alg}, consisting of "rigid" points, i.e. points invariant under $\mathrm{Aut}(F^{max}/F)$. The completion is well-defined up to a *unique* F-isomorphism; in particular there is no dependence on the choice of F^{max}. The completion is functorial in extensions that do not augment the value group at ∞.

We now take the sorts S_n into consideration. Recall the linear topology of 5.1. Define the *lattice completion* $\overline{F}$ of F to consist of the completion F^c of F in the field sort, and the closure of $S_n(F^c)$ in $S_n(F^{max})$ in the S_n-sorts. As the linear topology is Hausdorff, it is clear that points of $S_n(\overline{F})$ are fixed by $\mathrm{Aut}(F^{max}/F)$. Thus, up to a unique isomorphism over F, the lattice completion $\overline{F}$ is well-defined and independent of the choice of F^{max}. In fact it is functorial for extensions that do not augment the value group at 0 or at ∞.

Let L be a valued field. If Λ and Λ' are two lattices in L^n, there exists $M \in \mathrm{GL}_n(L)$ with $M\Lambda = \Lambda'$; $\mathrm{val}(\det(M))$ does not depend on the choice of M, we call it the *relative volume* and denote it by $\mathrm{vol}(\Lambda', \Lambda)$. Thus, if one sets $\mathrm{vol}(\Lambda') = \mathrm{vol}(\Lambda', \mathcal{O}^n)$, we may also write the relative volume as $\mathrm{vol}(\Lambda') - \mathrm{vol}(\Lambda)$. We say a family of lattices is *directed* (respectively, reverse directed) if any two lattices in the family is contained in (respectively, contain) a third.

LEMMA 14.5.1. *Let L be a valued field and let Λ be a rank n sublattice of L^n. Consider a directed family (Λ_i) of rank n sublattices of Λ. Assume* $\mathrm{vol}(\Lambda_i) - \mathrm{vol}(\Lambda) \to 0$ *in* $\Gamma(L)$. *Then* $\Lambda_i \to \Lambda$ *in* $S_n(L)$.

Proof. Let w_i be the seminorm corresponding to Λ_i, and w to Λ. We have to show that for any v we have $w_i(v) \to w(v)$; equivalently if $w(v) = 0$ we have to show that $w_i(v) \to 0$. Let $\alpha_i = \mathrm{vol}(\Lambda_i) - \mathrm{vol}(\Lambda)$; so $\alpha_i \to 0$. We claim that $0 \geq w_i(v) \geq -\alpha_i$ from which the statement follows. To see this, fix i and set $\Lambda_i = \Lambda'$, $w_i = w'$, $\alpha_i = \alpha$. Since Λ and Λ' can be simultaneously diagonalized in some basis, we may assume that $\Lambda = \mathcal{O}^n$ and $\Lambda' = \oplus \mathcal{O} d_k$. Since $\mathrm{val}(d_k) \geq 0$ and $\sum_k \mathrm{val}(d_k) = \alpha$, we have $0 \leq \mathrm{val}(d_k) \leq \alpha$. It follows that for any v with $w(v) = 0$, $0 \geq w'(v) \geq -\alpha$. □

LEMMA 14.5.2. *Assume L is maximally complete. Consider a family of rank n lattices Λ_i in L^n, directed under inclusion or reverse inclusion. Assume, for any subspace U of L^n, that* $\mathrm{vol}(\Lambda_i \cap U) \to \gamma_U \in \Gamma(L)$. *Then there exists a unique rank n lattice Λ in L^n with $\Lambda_i \to \Lambda$. Moreover,* $\mathrm{vol}(\Lambda_i) - \mathrm{vol}(\Lambda) \to 0$.

Proof. Uniqueness is clear since the linear topology is Hausdorff. To prove the remaining assertions, it suffices, by Lemma 14.5.1, to find a rank n lattice Λ in L^n such that $\mathrm{vol}(\Lambda_i, \Lambda) \to 0$ in $\Gamma(L)$, and Λ contains the Λ_i in the inclusion case (respectively, in the reverse inclusion case, is contained in the Λ_i).

Consider first the one-dimensional case. Then $\Lambda_k = \{x : \mathrm{val}(x) \geq \alpha_k\}$ with $\alpha_k \to \gamma$. We set $\Lambda = \{x : \mathrm{val}(x) \geq \gamma\}$, and the statement is clear. Note that $\Lambda = \cap_k \Lambda_k$ in case the Λ_k form a descending chain.

Recall that when V is an n-dimensional vector space, with dual V^*, the dual of a lattice Λ in V is $\Lambda^* := \{x \in V^* : (\forall y \in \Lambda) \operatorname{val}(x \cdot y) \geq 0\}$. Beginning with $\mathcal{O}^n$ as the standard lattice of K^n, we take the standard lattice of the dual space to be the dual lattice of the standard lattice of a given space, and the standard lattice of a subspace U to be the intersection with U of the standard lattice, and of a quotient V/U to be the image of the standard lattice. Duality reverses inclusion and volume, i.e. $\operatorname{vol}(\Lambda_1^*, \Lambda_2^*) = -\operatorname{vol}(\Lambda_1, \Lambda_2)$. Also, $\operatorname{vol}(\Lambda^* \cap U^{\perp}) = -\operatorname{vol}(\Lambda + U) = -\operatorname{vol}(\Lambda) + \operatorname{vol}(\Lambda \cap U)$; so the convergence assumption goes through to the dual. Thus by passing to duals if necessary, it suffices to prove the statement in the case that the Λ_i are reverse directed.

In this case, let Λ be the intersection of all Λ_i. We argue first that Λ spans L^n. Thus fix a subspace U of L^n of dimension $n-1$; we have to show it does not contain Λ. By induction, $\Lambda_i \cap U$ converges to a lattice Λ' of U. Modulo U, the lattices $\Lambda_i + U$ have volume $\operatorname{vol}(\Lambda_i) - \operatorname{vol}(\Lambda_i \cap U)$ which converges to $\gamma - \operatorname{vol}(\Lambda')$, and it follows that they contain some nonzero element $c + U$. Now viewing $c + U$ as a coset of U in L^n, maximal completeness implies that $(\cap_i \Lambda_i) \cap (c + U) \neq \varnothing$, so as $c + U$ is disjoint from U, we see that Λ is not contained in U. Thus indeed Λ spans L, i.e. $L\Lambda = L^n$. In particular, $\Lambda \cap W \neq (0)$ for any one-dimensional W (if $0 \neq w \in W$, then $w \in c\Lambda$ for some $c \in K$ so $c^{-1}w \in W \cap \Lambda$).

To see that Λ is a lattice, since L is maximally complete, it suffices to show that $\Lambda \cap W$ is a lattice for any one-dimensional W. This is clear by the one-dimensional case treated above. □

REMARK 14.5.3. The convergence assumption in Lemma 14.5.2 holds automatically when the value group is $\mathbb{R}$, provided $|\operatorname{vol}(\Lambda_i)|$ is bounded in $\Gamma(L)$. Indeed for any subspace U, $\operatorname{vol}(\Lambda_i \cap U)$ and $\operatorname{vol}(\Lambda_i/U)$ are both monotone in Λ_i (e.g. both increasing if the Λ_i are increasing) and their sum $\operatorname{vol}(\Lambda_i)$ is bounded (in absolute value), hence they both tend towards a real limit value.

14.6 BERKOVICH POINTS AS GALOIS ORBITS

In this section we fix a valued field F with $\operatorname{val}(F) \subset \mathbb{R}_\infty$. Let V be an algebraic variety over F. For any base set A containing F with $\Gamma(A) = \mathbb{R}$, we define $B_A(V)$ to be the space of types on V over A that are almost orthogonal to Γ. We shall be concerned with the case when $A = \overline{F}$.

LEMMA 14.6.1. *Let V be an algebraic variety over F. The restriction map $B_{\overline{F}}(V) \to B_{\mathbf{F}}(V)$ is bijective.*

Proof. Since the mapping $\widehat{V}(F^{max}) \to B_{\mathbf{F}}(V)$ factors through the map $\widehat{V}(F^{max}) \to B_{\overline{F}}(V) \to B_{\mathbf{F}}(V)$, surjectivity follows from Lemma 14.1.1.

For injectivity, consider q' and q'' in $B_{\overline{F}}(V)$ with the same restriction to $B_{\mathbf{F}}(V)$. Let $c' \models q'$ and $c'' \models q''$. We can find models M' and M'' containing respectively $\overline{F}(c')$ and $\overline{F}(c'')$, and with value group $\mathbb{R}$. Let K' and K'' be maximally complete algebraically closed valued fields, with value group $\mathbb{R}$, containing respectively $\overline{F}(c')$ and $\overline{F}(c'')$. By enlarging one of them, we may assume their residue fields have the same transcendence degree over the residue field of F, so that they are isomorphic over $\mathbf{F}$. Since $\mathrm{tp}(c/\mathbf{F}) = \mathrm{tp}(c'/\mathbf{F})$, there exists an $\mathbf{F}$-isomorphism $\beta : K' \to K''$ with $\beta(c) = c'$. But $\beta|\overline{F}$ must be the identity, since each point of $\overline{F}$ is the unique limit point of some sequence (or net) of elements of $\mathbf{F}$ or $S_n(\mathbf{F})$. Thus $q' = \mathrm{tp}(c'/\overline{F}) = \mathrm{tp}(c''/\overline{F}) = q''$. □

Lemma 14.6.2. *Let V be an algebraic variety over F. The natural map $\widehat{V}(\overline{F}) \to B_{\overline{F}}(V)$ sending q to $q|\overline{F}$ is surjective. Hence when F is algebraically closed, it is bijective.*

Proof. We may assume F is complete as a valued field. Let $p \in B_{\overline{F}}(V)$. We may assume that V is affine and that p is Zariski dense in V. We first give the argument assuming $\mathrm{val}(F)$ is dense in $\mathbb{R}$; the general case differs only notationally, and will be explained below.

Let H be the affine coordinate ring of V, $H = \cup_{d \geq 0} H_d$ where H_d is the space of polynomials of degree at most d, modulo those vanishing on V. Let $M_d = \{f \in H_d(F) : (\mathrm{val}(f) \geq 0) \in p\}$ and let $\mathfrak{F}_d$ be the family of all lattices of H_d, generated by a finite subset of M_d. We view it as ordered by inclusion.

We wish to show that $\mathfrak{F}_d$ admits a limit lattice containing it. For this purpose we may replace $\mathfrak{F}_d$ by the subfamily of elements of $\mathfrak{F}_d$ containing some fixed such lattice Λ_0; then this family has a lower bound. Let us prove the existence of an upper bound. The category of complete $\mathbb{R}$-valued fields admits algebraically independent amalgamation; the one-dimensional case is easy, and the general case follows inductively using [22] Lemma 6.18. Hence, using the Zariski density of p, one can find realizations $c_1, \ldots, c_N$ of p in some $\mathbb{R}$-valued field, where $N = \dim(H_d)$, such that no nonzero element of H_d vanishes on all the c_i. Then $(c_1, \ldots, c_N)$ generate a lattice in H_d^*, whose dual lattice contains M_d and hence all elements of $\mathfrak{F}_d$. In particular, $\mathrm{vol}(\Lambda)$ is bounded above and below, for $\Lambda \in \mathfrak{F}_d$. By Lemma 14.5.2 and Remark 14.5.3, $\mathfrak{F}_d$ has a unique limit lattice Λ_d in F^{max}. A code for Λ_d lies in $\overline{F}$ by definition of the latter.

Let w_d be the seminorm on H_d corresponding to Λ_d. We now prove that these seminorms satisfy the condition in Remark 5.1.5, namely that for any $f_i \in H_{d_1}, f_2 \in H_{d_2}$ we have $w_{d_1+d_2}(f_1 f_2) = w_{d_1}(f_1) + w_{d_2}(f_2)$, for every d_1 and d_2. Indeed, find a sequence (or net) of lattices $\Omega_j \in \mathfrak{F}_{d_1+d_2}$ such that $\Omega_j \to \Lambda_{d_1+d_2}$ while $\Omega_j \cap H_{d_i} \to \Lambda_{d_i}$. Let w'_j be the seminorm corresponding to Ω_j; then $w'_j(f_1 f_2) \to w_{d_1+d_2}(f_1 f_2)$ while $w'_j(f_i) \to w_{d_i}(f_i)$, and the condition follows by continuity. One deduces from Remark 5.1.5 that there exists a unique q in $\widehat{V}(\overline{F})$ such that $J_d(q) = \Lambda_d$, for every d, in the notation of 5.1.

Note that $\Lambda_d \cap H_d(F) = M_d$. Indeed by definition we have $M_d \subset \Lambda_d$. On the other hand if $f \in H_d(F)$ and $f \notin M_d$ then $p(x) \vdash \operatorname{val} f(x) \leq \alpha$ for some negative $\alpha \in \mathbb{Q}$, and it follows by continuity that $w_d(f) \leq \alpha$ so $f \notin \Lambda_d$. Thus $q|B_F = p|B_F$. In fact since $\operatorname{val}(F)$ is dense in $\mathbb{R}$ it follows that $q|B_{\mathbf{F}} = p$, and the statement follows.

When $\operatorname{val}(F)$ is not necessarily dense in $\mathbb{R}$, we must define $\mathcal{F}_d$ as the family of all lattices $\oplus_{i=1}^N \alpha_i \mathcal{O} f_i$, where $f_1, \ldots, f_N$ is a basis for $H_d(F)$, $p(x) \vdash \operatorname{val}(f_i(x)) = -\alpha_i$, and $\gamma\mathcal{O} := \{x : \operatorname{val}(x) \geq \gamma\}$. These lattices are still defined over $\mathbf{F}$ (though they may not have a basis in F), and the proof goes through as before. □

REMARK 14.6.3. At this point we recover the functorial base change of [36]. Namely when F is algebraically closed, it follows from Lemma 14.6.1 and Lemma 14.6.2 that we have a canonical bijection between $B_{\mathbf{F}}(V)$ and $\widehat{V}(\overline{F})$ and thus, by canonical extension of stably dominated types, we get a canonical map from $B_{\mathbf{F}}(V)$ to $\widehat{V}(\mathbb{U})$. In fact we discovered the approach of this section upon reflecting upon Poineau's theorem. Another proof was independently given in [2].

Let V be a variety over F. Let $G = \operatorname{Aut}(F^{\mathrm{alg}}/F^{\mathrm{h}})$ be the absolute Galois group of the Henselization F^{h} of F; so G acts also on $\overline{F^{\mathrm{alg}}}$. Recall that G is the group of valued field automorphisms of F^{alg} over F, and $B_{\mathbf{F}}(V) = B_{\mathbf{F}^{\mathrm{alg}}}(V)/G$, with $\mathbf{F}^{\mathrm{alg}}$ the structure $(F^{\mathrm{alg}}, \mathbb{R})$. Composing the maps of Lemma 14.6.1 and Lemma 14.6.2 one gets a G-equivariant bijection

$$\widehat{V}(\overline{F^{\mathrm{alg}}}) \longrightarrow B_{\mathbf{F}^{\mathrm{alg}}}(V)$$

which one checks to be a homeomorphism and whose inverse induces a natural homeomorphism

$$\varrho : B_{\mathbf{F}}(V) \longrightarrow \widehat{V}(\overline{F^{\mathrm{alg}}})/G.$$

DEFINITION 14.6.4. We call a point of $B_{\mathbf{F}}(V)$ which restricted to F determines an Abhyankar extension of the valued field F an *Abhyankar point* of $B_{\mathbf{F}}(V)$.

PROPOSITION 14.6.5. *The homeomorphism ϱ induces a bijection between Abhyankar points of $B_{\mathbf{F}}(V)$ and G-orbits of strongly stably dominated points of $\widehat{V}(\overline{F^{\mathrm{alg}}})$.*

Proof. Let p be an Abhyankar point in $B_{\mathbf{F}}(V)$. Let $c \models p|F$ and $d = \operatorname{trdeg}_F F(c)$. Then there exist F-definable functions $f_1, \ldots, f_k$ and $g_1, \ldots, g_\ell$, with $k + \ell = d$, such that $f_1(c), \ldots, f_k(c)$ are elements of the residue field, algebraically independent over the residue field of F and $g_1(c), \ldots, g_\ell(c)$ are elements of the value group, linearly independent over $\mathrm{val}(F)$. Now over $\mathbf{F}$, $f_1(c), \ldots, f_k(c), g_1(c), \ldots, g_\ell(c)$ are algebraically independent elements of $\mathrm{RES}_{\mathbf{F}}$, hence $\mathrm{tp}(c/\mathbf{F})$ is strongly stably dominated. For the other direction see the proof of Theorem 14.2.1. □

REMARK 14.6.6. The direct connection between Berkovich points and stably dominated points over the lattice completion provided by the homeomorphism ϱ immediately yields another proof of Theorem 14.2.1. Simply, given V over F and data over $\mathbf{F}$, the homotopy of Theorem 11.1.1 is defined over $\mathbf{F}$; and being $\mathbf{F}$-definable, in particular $\overline{F}$-definable, it takes points of $\widehat{V}(\overline{F})$ to points of $\widehat{V}(\overline{F})$, so it restricts to a homotopy on $\widehat{V}(\overline{F})$; the isomorphism ϱ translates this to a homotopy on $B_{\mathbf{F}}(V)$.

Moreover, by Theorem 8.4.2 and Proposition 14.6.5, one may ask that the homotopy fix any given Abhyankar point p of $B_{\mathbf{F}}(V)$. Thus the proof of Theorem 14.4.1, along with the analogous fact in the o-minimal case, shows that p admits a basis of neighborhoods that strongly retract to p.

Bibliography

[1] A. Abbes, T. Saito, *Ramification of local fields with imperfect residue fields*, Amer. J. Math. **124** (2002), 879–920.

[2] I. Ben Yaacov, *Tensor products of valued fields*, Bull. Lond. Math. Soc. **47** (2015), 42–46.

[3] V.G. Berkovich, *Spectral theory and analytic geometry over non-archimedean fields*, Mathematical Surveys and Monographs, 33. American Mathematical Society, Providence, RI, 1990.

[4] V.G. Berkovich, *Étale cohomology for non-archimedean analytic spaces*, Inst. Hautes Études Sci. Publ. Math. **78** (1993), 5–161.

[5] V.G. Berkovich, *Smooth p-adic analytic spaces are locally contractible*, Invent. Math. **137** (1999), 1–84.

[6] V.G. Berkovich, *Smooth p-adic analytic spaces are locally contractible. II*, in Geometric aspects of Dwork theory, Vol. I, II (Walter de Gruyter, Berlin, 2004), 293–370.

[7] C.C. Chang, H.J. Keisler, *Model theory*, Third edition. Studies in Logic and the Foundations of Mathematics, 73. North-Holland Publishing Co., Amsterdam, 1990.

[8] Z. Chatzidakis, E. Hrushovski, *Model theory of difference fields*, Trans. Amer. Math. Soc. **351** (1999), 2997–3051.

[9] M. Coste, *An introduction to o-minimal geometry*, Dip. Mat. Univ. Pisa, Dottorato di Ricerca in Matematica, Istituti Editoriali e Poligrafici Internazionali, Pisa (2000).

[10] L. van den Dries, *Dimension of definable sets, algebraic boundedness and Henselian fields*, Ann. Pure Appl. Logic **45** (1989), 189–209.

[11] L. van den Dries, *Tame topology and o-minimal structures*, Cambridge Univ. Press, New York, 1998.

[12] A. Ducros, *Parties semi-algébriques d'une variété algébrique p-adique*, Manuscripta Math. **111** (2003), 513–528.

[13] A. Ducros, *Espaces analytiques p-adiques au sens de Berkovich*, Séminaire Bourbaki. Vol. 2005/2006. Astérisque **311** (2007), 137–176.

[14] A. Ducros, *Les espaces de Berkovich sont modérés, d'après E. Hrushovski et F. Loeser*, Séminaire Bourbaki. Vol. 2011/2012. Astérisque **352** (2013), 459–507.

[15] P. Eleftheriou, *A semi-linear group which is not affine*, Ann. Pure Appl. Logic **156** (2008), 287–289.

[16] P. Eleftheriou, S. Starchenko, *Groups definable in ordered vector spaces over ordered division rings*, J. Symbolic Logic **72** (2007), 1108–1140.

[17] M. Fried, M. Jarden, *Field arithmetic*, Ergebnisse der Mathematik und ihrer Grenzgebiete (3) 11. Springer-Verlag, Berlin, 1986.

[18] A. Grothendieck, *Éléments de géométrie algébrique. II. Étude globale élémentaire de quelques classes de morphismes*, Inst. Hautes Études Sci. Publ. Math. **8** (1961), 5–222.

[19] D. Haskell, E. Hrushovski, D. Macpherson, *Definable sets in algebraically closed valued fields: elimination of imaginaries*, J. Reine Angew. Math. **597** (2006), 175–236.

[20] D. Haskell, E. Hrushovski, D. Macpherson, *Stable domination and independence in algebraically closed valued fields*, Lecture Notes in Logic, 30. Association for Symbolic Logic, Chicago, IL; Cambridge University Press, Cambridge, 2008.

[21] E. Hrushovski, *Computing the Galois group of a linear differential equation*, in Differential Galois theory (Będlewo, 2001), 97–138, Banach Center Publ., 58, Polish Acad. Sci., Warsaw, 2002.

[22] E. Hrushovski, *The first order theory of the Frobenius automorphisms*, preprint, available at http://www.ma.huji.ac.il/~ehud/FROB.pdf.

[23] E. Hrushovski, *Valued fields, metastable groups*, preprint, available at http://www.math.jussieu.fr/~loeser/mst.pdf.

[24] E. Hrushovski, *On finite imaginaries*, in Logic Colloquium 2006, 195–212, Lect. Notes Log., 32, Assoc. Symbol. Logic, Chicago, IL, 2009.

[25] E. Hrushovski, *Imaginaries and definable types in algebraically closed valued fields*, in Valuation theory in interaction, EMS Series of Congress reports, 297–319 (2014).

[26] E. Hrushovski, D. Kazhdan, *Integration in valued fields*, in Algebraic geometry and number theory, Progress in Mathematics 253, 261–405 (2006), Birkhäuser.

[27] E. Hrushovski, A. Pillay, *On* NIP *and invariant measures*, J. Eur. Math. Soc. **13** (2011), 1005–1061.

[28] R. Huber, M. Knebusch, *On valuation spectra*, Recent advances in real algebraic geometry and quadratic forms (Berkeley, CA, 1990/1991; San Francisco, CA, 1991), 167–206, Contemp. Math., 155, Amer. Math. Soc., Providence, RI, 1994.

[29] M. Kamensky, *Ind- and pro- definable sets*, Ann. Pure Appl. Logic **147** (2007), 180–186.

[30] M. Kontsevich, Y. Soibelman, *Affine structures and non-archimedean analytic spaces*, in The unity of mathematics, Progress in Mathematics 244, 321–385 (2006), Birkhäuser.

[31] D. Mumford, *Abelian varieties*, Tata Institute of Fundamental Research Studies in Mathematics vol. 5, Oxford University Press, 1970.

[32] Y. Peterzil, C. Steinhorn, *Definable compactness and definable subgroups of o-minimal groups*, J. London Math. Soc. **59** (1999), 769–786.

[33] A. Pillay, *An introduction to stability theory*, Dover Books on Mathematics, 2008, reprinted from the original published in Oxford Logic Guides, 8, Oxford University Press, 1983.

[34] A. Pillay, *Model theory and stability theory, with applications in differential algebra and algebraic geometry*, in Model theory and applications to algebra and analysis, volume 1, LMS Lecture Notes Series 349, 2008 (edited by Chatzidakis, Macpherson, Pillay, Wilkie), 1–23.

[35] J. Poineau, *Un résultat de connexité pour les variétés analytiques p-adiques: privilège et noethérianité*, Compos. Math. **144** (2008), 107–133.

[36] J. Poineau, *Les espaces de Berkovich sont angéliques*, Bull. Soc. Math. France **141** (2013), 267–297.

[37] B. Poizat, *Groupes stables*, Nur Al-Mantiq Wal-Ma'rifah, Villeurbanne, France, 1987. English translation: *Stable groups*, Mathematical Surveys and Monographs, 87. American Mathematical Society, Providence, RI, 2001.

[38] J.-P. Serre, *Lectures on the Mordell-Weil theorem*, Aspects of Mathematics, Vieweg, Braunschweig, 1997.

[39] A. Thuillier, *Géométrie toroïdale et géométrie analytique non archimédienne. Application au type d'homotopie de certains schémas formels*, Manuscripta Math. **123** (2007), 381–451.

[40] V. Voevodsky, A. Suslin, E. Friedlander, *Cycles, transfers and motivic homology theories*, Annals of Math Studies vol. 143, Princeton University Press, Princeton, 2000.

[41] P. Winkler, *Model-completeness and Skolem expansions*, Model theory and algebra (memorial tribute to Abraham Robinson), p. 408–463. Lecture Notes in Math., Vol. 498, Springer, Berlin, 1975.

[42] M. Ziegler, *A language for topological structures which satisfies a Lindström-theorem*, Bull. Amer. Math. Soc. **82** (1976), 568–570.

Index

List of notations